AF389320

Using Remote Sensing Techniques to Improve Hydrological Predictions in a Rapidly Changing World

Using Remote Sensing Techniques to Improve Hydrological Predictions in a Rapidly Changing World

Editors

Yongqiang Zhang
Dongryeol Ryu
Donghai Zheng

MDPI • Basel • Beijing • Wuhan • Barcelona • Belgrade • Manchester • Tokyo • Cluj • Tianjin

Editors
Yongqiang Zhang
Chinese Academy of Sciences
China

Dongryeol Ryu
The University of Melbourne
Austrilia

Donghai Zheng
Chinese Academy of Sciences
China

Editorial Office
MDPI
St. Alban-Anlage 66
4052 Basel, Switzerland

This is a reprint of articles from the Special Issue published online in the open access journal *Remote Sensing* (ISSN 2072-4292) (available at: https://www.mdpi.com/journal/remotesensing/special_issues/rs_hydrological_predictions).

For citation purposes, cite each article independently as indicated on the article page online and as indicated below:

LastName, A.A.; LastName, B.B.; LastName, C.C. Article Title. *Journal Name* **Year**, *Volume Number*, Page Range.

ISBN 978-3-0365-2331-6 (Hbk)
ISBN 978-3-0365-2332-3 (PDF)

Contents

About the Editors

Yongqiang Zhang has 24 years of experience in hydrological modelling and remote sensing hydrology. He is currently a Distinguished Professor in Hydrology and Water Resources in the Institute of Geographic Sciences and Natural Resources Research, Chinese Academy of Sciences. He worked in CSIRO as the Research Scientist, Senior Research Scientist and Principal Research Scientist in 2006–2018. He is the prestigious Alexander von Humboldt Fellow. He has won 12 professional awards, including 2012 GN Alexander Medal. Professor Zhang is a global leader in hydrology and water resource studies that incorporate remote sensing techniques into hydrological modelling. Dr Zhang has 152 ISI publications, including journal papers published in Nature stables (*Nature Climate Change* (IF = 24.624), *Nature Communications* (IF = 14.526)) and top field research journals; *Water Resources Research* (IF = 5.240), *Journal of Hydrology* (IF = 5.722), *Journal of Geophysical Research: Atmospheres* (IF = 4.261), *Remote Sensing of Environment* (IF = 10.164), and *Global Change Biology* (IF = 10.863). He has more than 80 journal papers published in top journals, of which 54 journal papers have been published since 2016. He has Google Scholar citations of >10,000, ISI Web of Science citations of >6140 and a H-index of 41. His major research contributions include developing novel hydrological modelling approaches by using remote sensing techniques to significantly improve runoff (and streamflow and water availability) simulations and predictions in gauged and ungauged catchments, and the development of the PML evapotranspiration model, a representative diagnostic model to evaluate IPCC global climate model simulations. Professor Zhang is serving for six peer-reviewed journals: including as Associate Editor of the *Journal of Hydrology*, the *Journal of Geophysical Research: Atmospheres* and *Science of Remote Sensing*, and as an Editorial Board Member of *Remote Sensing of Environment*.

Dongryeol Ryu is an Associate Professor in the Department of Infrastructure Engineering, The University of Melbourne, Australia. Previously, he worked as a Research Physical Scientist in the Hydrology and Remote Sensing Lab., US Department of Agriculture (USDA), Agricultural Research Services (ARS). He is a NASA Earth Science Graduate Fellow and a recipient of the University of California, Irvine Medal fellowship. He currently leads the Environmental Sensing and Modelling Lab. Specialized in remote sensing of land surface variables such as soil moisture and vegetation characteristics, their spatial and temporal variability, and hydrological applications. He also investigates the roles of land surface changes in hydroclimate processes. Dongryeol received his Ph.D. degree in Earth System Science from the University of California, Irvine in 2006, and the B.S. and M.S. degrees in Geological Sciences from Seoul National University, Korea.

Donghai Zheng has 10 years of experience in land surface modelling and microwave remote sensing of soil moisture and freeze–thaw dynamics. He is currently a Professor in the Institute of Tibetan Plateau Research, Chinese Academy of Sciences. He worked in ITC of University of Twente as the Researcher in 2015–2018. Professor Zheng is a recognized international expert in the use of microwave remote sensing and land surface model to quantify soil freeze–thaw process and water cycle over the Tibetan Plateau (TP). He has 40+ ISI papers published in research field top journals (e.g., *Remote Sensing of Environment* (IF = 10.164), *IEEE Transactions on Geoscience and Remote Sensing* (IF = 5.6), *Journal of Geophysical Research: Atmospheres* (IF = 4.261)). The main achievements of his work are (i) highlighting the importance of coupled heat and water exchange and soil freeze–thaw processes for quantifying water cycle over the TP, (ii) revealing microwave emission characteristics of frozen layered soil via conducting an ground-based L-band radiometry (ELBARA-III) field campaign on the TP since 2016 and developing an integrated land surface and microwave radiative transfer model, and (iii) quantifying the sampling depth of L-band radiometry and retrieving liquid water content for both frozen and thawed soil conditions. The ELBARA-III field site is selected as calibration/validation site for ESA's SMOS and NASA's SMAP satellite missions.

 *remote sensing*

Editorial

Using Remote Sensing Techniques to Improve Hydrological Predictions in a Rapidly Changing World

Yongqiang Zhang [1,*], Dongryeol Ryu [2] and Donghai Zheng [3]

[1] Key Laboratory of Water Cycle and Related Land Surface Processes, Institute of Geographic Sciences and Natural Resources Research, Chinese Academy of Sciences, Beijing 100101, China

[2] Department of Infrastructure Engineering, Faculty of Engineering and Information Technology, The University of Melbourne, Melbourne, VIC 3010, Australia; dryu@unimelb.edu.au

[3] National Tibetan Plateau Data Center, State Key Laboratory of Tibetan Plateau Earth System, Resource and Environment, Institute of Tibetan Plateau Research, Chinese Academy of Sciences, Beijing 100101, China; zhengd@itpcas.ac.cn

[*] Correspondence: yongqiang.zhang2014@gmail.com or zhangyq@igsnrr.ac.cn; Tel.: +86-10-64856515

Abstract: Remotely sensed geophysical datasets are being produced at increasingly fast rates to monitor various aspects of the Earth system in a rapidly changing world. The efficient and innovative use of these datasets to understand hydrological processes in various climatic and vegetation regimes under anthropogenic impacts has become an important challenge, but with a wide range of research opportunities. The ten contributions in this Special Issue have addressed the following four research topics: (1) Evapotranspiration estimation; (2) rainfall monitoring and prediction; (3) flood simulations and predictions; and (4) monitoring of ecohydrological processes using remote sensing techniques. Moreover, the authors have provided broader discussions, on how to make the most out of the state-of-the-art remote sensing techniques to improve hydrological model simulations and predictions, to enhance their skills in reproducing processes for the fast-changing world.

Keywords: remote sensing; model; hydrological prediction; climate change; land use change; evapotranspiration

Citation: Zhang, Y.; Ryu, D.; Zheng, D. Using Remote Sensing Techniques to Improve Hydrological Predictions in a Rapidly Changing World. *Remote Sens.* **2021**, *13*, 3865. https://doi.org/10.3390/rs13193865

Received: 22 September 2021
Accepted: 26 September 2021
Published: 27 September 2021

Publisher's Note: MDPI stays neutral with regard to jurisdictional claims in published maps and institutional affiliations.

1. Introduction

We are living in a world where geophysical datasets, particularly, remote sensing datasets, are created at fast increasing rates [1]. The efficient and innovative use of these datasets for understanding hydrological processes in various climatic and vegetation regimes under anthropogenic influence has become an important challenge, which offers a wide range of research opportunities at the same time [2,3]. This is particularly urgent for the hydrological research community at large who has relied on both spatially distributed and lumped hydrological models for hydrological simulations/predictions over the last several decades [4]. The demand for increasingly accurate water information spatially distributed at high resolution requires a deeper understanding of the underlying processes and more skillful predictions, at resolutions that do not commensurate with the traditional hydrological data. This is an important challenge to the conventional hydrological modelling.

To address these challenges, efforts need to be made to innovatively integrate various remote sensing techniques into hydrological simulations and predictions, which is more critical in a changing world. This Special Issue of Remote Sensing contributes towards this aim through investigations on how a better and smarter use of high-to-moderate-resolution remote sensing datasets can improve hydrological simulations and predictions. Ten peer-reviewed papers are published in this Issue, which are grouped into four categories using remote sensing techniques:

- Evapotranspiration estimation;
- Rainfall monitoring and prediction;
- Flood simulations and predictions; and
- Monitoring ecohydrological processes.

2. Contributed Papers

2.1. Estimating Evapotranspiration

In this category, two papers are published. Zhang and Song et al. [5] estimated the Evapotranspiration (ET) of Beijing city at a 10-m resolution, using vegetation information from Sentinel-2 satellite data. For the first time, monthly ET at a 10×10-m resolution for the Beijing Sponge City was estimated, using a water-carbon coupling model (PML-V2) driven by the 10-m resolution of the Sentinel-2-based Leaf Area Index (LAI), Sentinel-2-based land cover classification data, and surface meteorological data. Model validations show that the Sentinel-2-based LAI has a good correlation (R^2 = 0.74) with the observed values. The PML-V2-estimated ET compares well (R^2 = 0.64–0.90) with the flux measurements at three fields within the Beijing region. Model simulation results show that LAI, ET, and the Gross Primary Productivity (GPP) are 0.83 $m^2\ m^{-2}$, 1.6 mm d^{-1}, and 2.8 gC $m^{-2}\ d^{-1}$, respectively, during June 2018. Water bodies, including lakes and rivers, show the highest ET of >8 mm d^{-1}, followed by mixed forest and croplands that have ET varying in 4–6 mm d^{-1} and grasslands with ET varying in 2–4 mm d^{-1}. Compared to these land cover types, the impervious surface occupying ~60% of the sponge city areas shows the smallest ET of <2.0 mm d^{-1}. This study demonstrates that it is feasible to use the high-resolution satellite data to have detailed simulations of hydrological processes in urban ecosystems. Jepsen et al. [6] analyzed the suitability of remotely sensed ET for calibrating a hydrological model in the upper Kings River watershed (3999 km^2) of California's Sierra Nevada, a snow-influenced watershed in a Mediterranean climate. They compared a spatiotemporal pattern of ET from a remote-sensing product, MODIS MOD16A2, to that from a hydrological model (SWAT) calibrated against an observed streamflow. The ET estimates from both MOD16A2 and SWAT modelling were evaluated against observations from three flux towers at elevations of 1160–2700 m. It was found that the SWAT-modelled ET performs better than MOD16A2 ET, indicated by the Nash-Sutcliffe efficiency (+0.36 versus −0.43) and error in the elevational trend (+7.7% versus +81%). For this particular modelling experiment, the authors concluded that it is challenging for the remotely sensed ET product used for watershed-model parameter estimation. By analyzing ET-weather relationships, the authors found that the relatively large errors in MODIS ET may be related to weather-based corrections to water limitation not representative of the hydrology of this snow-influenced, Mediterranean-climate area. Therefore, attention should be paid when using global ET products (such as MOD16A2) at a watershed scale, particularly when the watershed involves snowmelt processes. Moreover, it is necessary to have bias corrections of the ET data before use for model parameterization [7].

2.2. Rainfall Monitoring and Prediction

In this category, two papers are published. Han et al. [8] investigated the potential to use next-generation millimeter-wave backhaul technologies for rainfall monitoring in a dense urban environment. Traditionally, microwave backhaul links are mainly used for communications between cellular base stations. In this study, the authors used the links for measuring the path-averaged rain rate. In particular, they investigated the rain attenuation characteristics in Gothenburgh, Sweden using the new microwave backhaul techniques at different mmWave frequencies and link lengths. They found that estimating the path-average rainfall using mmWave links is very effective. The mmWave link measurement-derived rain rate is very well correlated (R = 0.8–0.9) to the local measurement from rainfall gauges. Their study indicates that there is a great potential to use the mmWave links for monitoring rainfall in urban areas. Liu et al. [9] investigated how the assimilation frequency of radar reflectivity affects the rain storm prediction in the Daqinghe basin of northern China, using the Three-Dimensional Variational Data Assimilation (3DVAR) system of the Weather Research Forecast (WRF) model. Their results show that the WRF-3DVAR system noticeably improves its performance for predicting the location, tendency, and development of rain storm, using the assimilation of radar reflectivity and Global Telecommunication System (GTS) data collectively. Moreover, this study suggests that it is important to validate and correct the assimilated measurement data before performing data assimilation, which can benefit not only prediction accuracy but also assimilation efficiency.

2.3. Flood Simulations and Predictions

In this category, four papers are published. Ma et al. [10] conducted a flash flood warnings study in Yunnan Province, China using NASA's Integrated Multi-Satellite Retrievals for Global Precipitation Measurement (GPM-IMERG) precipitation products. They tested two GPM-IMERG products: The near-real-time IMERG Early run product (IMERG-E) and the post-real-time IMERG Final run product (IMERG-F) with a 6-h temporal resolution. The results show that MERG-F is better than IMERG-E over the study area, indicated by an hourly R of 0.46 and relative bias of 23%. Furthermore, the IMERG-F results are well corresponding to the gauge data when using the Rainfall Triggering Index (RTI) model for calibration, suggesting that MERG-F is suitable for flash flood warnings. Wang et al. [11] investigated if the prediction of flash flood can be improved for mountainous catchments of northern China using the WRF-3DVAR module through coupled atmospheric-hydrologic systems. Compared to the baseline (openloop) model run, the assimilation improved the accuracy of rainfall accumulation as well as provided more accurate flood forecasting. Based on the grid-based Hebei model, an atmospheric-hydrological coupling system was established and performed by predicting the flash flood as well as obtaining the best performance of Nash-Sutcliffe Efficiency (NSE) = 0.874 after assimilation, compared to NSE = 0.64 before assimilation. Moreover, the authors pointed out the need to carefully transfer hydrological parameters, since the locally derived hydrodynamic parameters may not be applicable to mesoscale areas. Zhu et al. [12] proposed a modelling framework for the urban flood analysis in ungauged catchments using short-term and high-resolution rainfall data. The framework includes three steps. First, generate extreme rainfall events using a rainfall generator named RainyDay together with a 9-year record of hourly, 0.1° remotely sensed rainfall data. Second, simulate runoff using an Urban Hydrological Model (SWMM) under different rainfall return periods and durations. Third, analyze urban flood using flood indicators, such as flood time, maximum rainfall rates, and total maximum rainfall volume. This framework was tested in Guagzhou city, China. The results show that a combination of RainyDay and short-term remotely sensed rainfall data can expand the rainfall records for urban hydrological simulations and predictions. Furthermore, the proposed framework shows a good performance for runoff process simulation, especially for high return periods or long durations (NSE > 0.90), demonstrating that the proposed framework has potential for urban flood analysis in ungauged catchments. Yasir et al. [13] analyzed and simulated runoff processes in a heavily regulated river basin, the Lhasa River where there are Zhikong and Pangduo hydropower dams changing hydrological regimes. The analysis indicates that the Lhasa River streamflow shows stronger variations during 2000–2016, compared to pre-2000. The Zhikong hydropower plant and the Pangduo power plant began operations in 2006 and 2013, respectively, which strongly influence the Lhasa River streamflow. The modelling results indicate that the SWAT hydrological model is capable of simulating the streamflow under reservoir influence, and can be used for predicting future streamflow. The predicted streamflow has a similar behavior to the observation while decreasing in the years from 2017 to 2025, indicating that the hydrological regimes in this region are simultaneously affected by climate change and anthropogenic impacts. Yang et al. [14] proposed a new method using an Unmanned Aerial Vehicle (UAV) combined with the incipient motion of stone to calculate the peak streamflow of ephemeral rivers in northwestern China. Critical initial velocities of moving stones were estimated using two methods: Logarithmic and exponential velocity distribution methods. Their results indicate that the exponential velocity distribution method outperforms the logarithmic method. Under high flood events (>20 m^3/s), the proposed method can achieve model errors less than 10%. Under a low streamflow condition (1 m^3/s), the accuracies are relatively low. Nevertheless, this study provides an alternative way to calculate streamflow in ungagged rivers, which are particularly useful for peak flood estimation. The proposed approach is easy to apply and has potential for large-scale application, considering the quick advance of the UAV technology.

2.4. Monitoring Ecohydrological Processes

In this category, one paper is published. Qiao et al. [15] investigated how water transfer in arid northwestern China influences the wetland ecosystem via surface and groundwater interactions. In arid and semiarid regions, water transfer is a useful way to prevent vegetation degradations and maintain healthy ecosystems. This study analyzed the spatiotemporal pattern of vegetation coverage before and after water transfer in Qingtu Lake and the surrounding area. The results show that water transfer from the upstream contributes to the expansion of water bodies and vegetation, particularly for the condition of fractional vegetation coverage of 30–50%. The groundwater and soil water content increase can remain at high levels for the following months after water transfer, suggesting that the transferred water can be stored as ground water or soil water due to the strong surface and subsurface interactions, which provide water use for vegetation in the following year.

3. Editorial Summary and Comments

There are various ways that remote sensing techniques are used to improve hydrological simulations and predictions, enhancing our collective efforts. The efforts can be summarized into, but not limited to, the following categories: (1) Detecting hydrological and other related changes using state-of-the-art remote sensing techniques; (2) mapping eco-hydrological and hydrological processes and their driving factors using large samples and high-resolution datasets; (3) understanding hydrological processes in a rapidly changing world using hydrological modelling together with high-to-moderate-resolution (several meters to hundred meters) remote sensing data; (4) improving hydrological prediction skills by modifying hydrological model structures to incorporate remote sensing data and using various model calibrations against remote sensing data; (5) developing hydrological modelling frameworks using advanced cloud cluster computation techniques and tools, such as the Google Earth Engine; (6) using remote sensing data together with data assimilation or/and machine learning techniques to improve predictions of various hydrological variables and hydrological signatures; and (7) using remote sensing techniques for water-related studies, such as on the water–food–energy security nexus. Although the ten papers published in this Special Issue only cover parts of the summarized categories, we believe that continued efforts in using remote sensing techniques in hydrology definitely promote the development of hydrology, particularly in the fields of fusion of hydrological model and remote sensing and ground observations that cover all of the categories 1–7 (Figure 1).

In fact, the first guest editor and his team (we) had put lots of efforts in these categories and shared some researches below. For instance, we used the Google Earth Engine platform to develop a carbon-water coupled model (PML-V2) for estimating the actual evapotranspiration and gross primary production products across the global land surface with 500 m and 8-day resolution for the period of 2002 to 2020 [16]. This is a particular example followed in the categories of 2, 3, and 5. Another example of category 4 is to develop state-of-the-art model-data fusion techniques for predicting runoff in ungauged catchments. We used remotely sensed ET data only to calibrate hydrological model parameters. Since it does not require observed streamflow data for model calibration, it has the great potential runoff prediction in poorly gauged or ungauged regions. We demonstrate that this approach is very useful in Australia [17] and China [7] after bias corrections. Last but not least, we modify the rainfall-runoff modelling structure for better incorporating remote sensing data. The reasoning is that traditional rainfall-runoff models do not have a structure to simulate the impact of Land Use and Land Cover Change (LUCC), and are not reliable to simulate hydrological processes with rapid LUCC. We modified traditional rainfall-runoff models by changing their submodule for describing soil moisture and actual evapotranspiration processes [18,19]. The modified rainfall-runoff models improve hydrological simulations noticeably in the catchments experiencing rapid land cover changes [18].

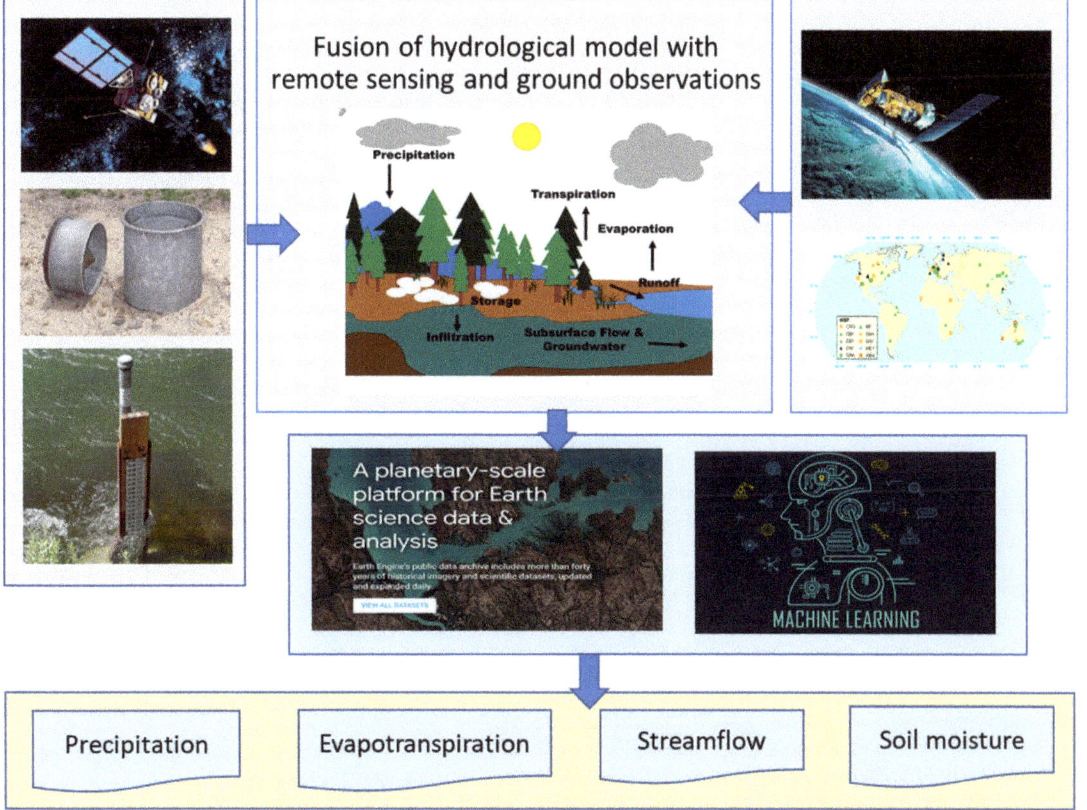

Figure 1. Conceptualized summary of model and data fusion techniques for improving hydrological predictions.

4. Conclusions

The world is quickly changing with land surface conditions changing dramatically due to anthropogenic impacts over the last two decades. Correspondingly, geophysical datasets, particularly, remote sensing datasets, are created at fast increasing rates. It is challenging to efficiently and innovatively use these datasets for understanding hydrological processes in various climatic and vegetation regimes under anthropogenic impacts, which also offer a wide range of research opportunities. To address these challenges, efforts need to be undertaken to use various remote sensing techniques to improve hydrological simulations and predictions in a changing world. Ten peer-reviewed papers were published in this Special Issue, and can be summarized into the following four categories:

- Estimating evapotranspiration;
- Rainfall monitoring and prediction;
- Flood simulations and predictions; and
- Monitoring ecohydrological processes.

The ten papers presented in this Special Issue reflect the efforts for improving hydrological simulations and predictions using various remote sensing techniques. The papers published in this issue advance the remote sensing of hydrology by applying a new measurement approach, such as UAV or model-data fusions. Though the published ten papers in this Special Issue only cover parts of the summarized categories, we believe that continuous efforts in using remote sensing techniques in hydrology definitely promote hydrology. Furthermore, the authors more broadly discuss how to smartly use the state-of-the-art remote sensing techniques to improve hydrological model simulations and predictions to tackle a quickly changing world.

Author Contributions: Y.Z. conceived and led the development of the Special Issue and this paper; D.R. and D.Z. each contributed to the writing of this paper. All authors have read and agreed to the published version of the manuscript.

Funding: This study is supported by the CAS Pioneer Talent Program and the National Natural Science Foundation of China (grant no. 41971032).

Institutional Review Board Statement: Not applicable.

Informed Consent Statement: Not applicable.

Data Availability Statement: Not applicable.

Acknowledgments: As guest editors of this Special Issue, we thank the journal editors and all the authors submitting manuscripts to this Special Issue. Our thanks are extended to the referees who put great efforts in reviewing the submissions, which is the cornerstone for the high-quality publications of Remote Sensing.

Conflicts of Interest: The authors declare no conflict of interest.

References

1. Pekel, J.-F.; Cottam, A.; Gorelick, N.; Belward, A.S. High-resolution mapping of global surface water and its long-term changes. *Nature* **2016**, *540*, 418–422. [CrossRef] [PubMed]
2. Rodell, M.; Famiglietti, J.S.; Wiese, D.N.; Reager, J.T.; Beaudoing, H.K.; Landerer, F.W.; Lo, M.H. Emerging trends in global freshwater availability. *Nature* **2018**, *557*, 650–659. [CrossRef] [PubMed]
3. Pulliainen, J.; Luojus, K.; Derksen, C.; Mudryk, L.; Lemmetyinen, J.; Salminen, M.; Ikonen, J.; Takala, M.; Cohen, J.; Smolander, T.; et al. Patterns and trends of Northern Hemisphere snow mass from 1980 to 2018. *Nature* **2020**, *581*, 294–298. [CrossRef] [PubMed]
4. Pascolini-Campbell, M.; Reager, J.T.; Chandanpurkar, H.A.; Rodell, M. A 10 per cent increase in global land evapotranspiration from 2003 to 2019. *Nature* **2021**, *593*, 543–547. [CrossRef] [PubMed]
5. Zhang, X.; Song, P. Estimating Urban Evapotranspiration at 10m Resolution Using Vegetation Information from Sentinel-2: A Case Study for the Beijing Sponge City. *Remote Sens.* **2021**, *13*, 2048. [CrossRef]
6. Jepsen, S.M.; Harmon, T.C.; Guan, B. Analyzing the Suitability of Remotely Sensed ET for Calibrating a Watershed Model of a Mediterranean Montane Forest. *Remote Sens.* **2021**, *13*, 1258. [CrossRef]
7. Huang, Q.; Qin, G.; Zhang, Y.; Tang, Q.; Liu, C.; Xia, J.; Chiew, F.H.S.; Post, D. Using Remote Sensing Data-Based Hydrological Model Calibrations for Predicting Runoff in Ungauged or Poorly Gauged Catchments. *Water Resour. Res.* **2020**, *56*, e2020WR028205. [CrossRef]
8. Han, C.; Huo, J.; Gao, Q.; Su, G.; Wang, H. Rainfall Monitoring Based on Next-Generation Millimeter-Wave Backhaul Technologies in a Dense Urban Environment. *Remote Sens.* **2020**, *12*, 1045. [CrossRef]
9. Liu, Y.; Liu, J.; Li, C.; Yu, F.; Wang, W. Effect of the Assimilation Frequency of Radar Reflectivity on Rain Storm Prediction by Using WRF-3DVAR. *Remote Sens.* **2021**, *13*, 2103. [CrossRef]
10. Ma, M.; Wang, H.; Jia, P.; Tang, G.; Wang, D.; Ma, Z.; Yan, H. Application of the GPM-IMERG Products in Flash Flood Warning: A Case Study in Yunnan, China. *Remote Sens.* **2020**, *12*, 1954. [CrossRef]
11. Wang, W.; Liu, J.; Li, C.; Liu, Y.; Yu, F. Data Assimilation for Rainfall-Runoff Prediction Based on Coupled Atmospheric-Hydrologic Systems with Variable Complexity. *Remote Sens.* **2021**, *13*, 595. [CrossRef]
12. Zhu, Z.; Yang, Y.; Cai, Y.; Yang, Z. Urban Flood Analysis in Ungauged Drainage Basin Using Short-Term and High-Resolution Remotely Sensed Rainfall Records. *Remote Sens.* **2021**, *13*, 2204. [CrossRef]
13. Yasir, M.; Hu, T.; Abdul Hakeem, S. Impending Hydrological Regime of Lhasa River as Subjected to Hydraulic Interventions—A SWAT Model Manifestation. *Remote Sens.* **2021**, *13*, 1382. [CrossRef]

14. Yang, S.; Li, C.; Lou, H.; Wang, P.; Wang, J.; Ren, X. Performance of an Unmanned Aerial Vehicle (UAV) in Calculating the Flood Peak Discharge of Ephemeral Rivers Combined with the Incipient Motion of Moving Stones in Arid Ungauged Regions. *Remote Sens.* **2020**, *12*, 1610. [CrossRef]
15. Qiao, S.; Ma, R.; Sun, Z.; Ge, M.; Bu, J.; Wang, J.; Wang, Z.; Nie, H. The Effect of Water Transfer during Non-growing Season on the Wetland Ecosystem via Surface and Groundwater Interactions in Arid Northwestern China. *Remote Sens.* **2020**, *12*, 2516. [CrossRef]
16. Zhang, Y.; Kong, D.; Gan, R.; Chiew, F.H.S.; McVicar, T.R.; Zhang, Q.; Yang, Y. Coupled estimation of 500 m and 8-day resolution global evapotranspiration and gross primary production in 2002–2017. *Remote. Sens. Environ.* **2019**, *222*, 165–182. [CrossRef]
17. Zhang, Y.; Chiew, F.H.S.; Liu, C.; Tang, Q.; Xia, J.; Tian, J.; Kong, D.; Li, C. Can Remotely Sensed Actual Evapotranspiration Facilitate Hydrological Prediction in Ungauged Regions Without Runoff Calibration? *Water Resour. Res.* **2020**, *56*. [CrossRef]
18. Luan, J.; Zhang, Y.; Tian, J.; Meresa, H.; Liu, D. Coal mining impacts on catchment runoff. *J. Hydrol.* **2020**, *589*. [CrossRef]
19. Zhang, Y.; Chiew, F.H.S.; Zhang, L.; Li, H. Use of Remotely Sensed Actual Evapotranspiration to Improve Rainfall-Runoff Modeling in Southeast Australia. *J. Hydrometeorol.* **2009**, *10*, 969–980. [CrossRef]

Article

Urban Flood Analysis in Ungauged Drainage Basin Using Short-Term and High-Resolution Remotely Sensed Rainfall Records

Zhihua Zhu [1,2], Yueying Yang [1,2], Yanpeng Cai [1,2,*] and Zhifeng Yang [1,2]

[1] Guangdong Provincial Key Laboratory of Water Quality Improvement and Ecological Restoration for Watersheds, Institute of Environmental and Ecological Engineering, Guangdong University of Technology, Guangzhou 510006, China; zhihuazhu@gdut.edu.cn (Z.Z.); 2111907046@mail2.gdut.edu.cn (Y.Y.); zfyang@gdut.edu.cn (Z.Y.)

[2] Southern Marine Science and Engineering Guangdong Laboratory (Guangzhou), Guangzhou 511458, China

[*] Correspondence: yanpeng.cai@gdut.edu.cn

Abstract: Analyzing flooding in urban areas is a great challenge due to the lack of long-term rainfall records. This study hereby seeks to propose a modeling framework for urban flood analysis in ungauged drainage basins. A platform called "RainyDay" combined with a nine-year record of hourly, 0.1° remotely sensed rainfall data are used to generate extreme rainfall events. These events are used as inputs to a hydrological model. The comprehensive characteristics of urban flooding are reflected through the projection pursuit method. We simulate runoff for different return periods for a typical urban drainage basin. The combination of RainyDay and short-record remotely sensed rainfall can reproduce recent observed rainfall frequencies, which are relatively close to the design rainfall calculated by the intensity-duration-frequency formula. More specifically, the design rainfall is closer at high (higher than 20-yr) return period or long duration (longer than 6 h). Contrasting with the flood-simulated results under different return periods, RainyDay-based estimates may underestimate the flood characteristics under low return period or short duration scenarios, but they can reflect the characteristics with increasing duration or return period. The proposed modeling framework provides an alternative way to estimate the ensemble spread of rainfall and flood estimates rather than a single estimate value.

Keywords: urban flood; design rainfall; ungauged drainage basin; RainyDay; IDF formula

 check for **updates**

Citation: Zhu, Z.; Yang, Y.; Cai, Y.; Yang, Z. Urban Flood Analysis in Ungauged Drainage Basin Using Short-Term and High-Resolution Remotely Sensed Rainfall Records. *Remote Sens.* **2021**, *13*, 2204. https://doi.org/10.3390/rs13112204

Academic Editors: Yongqiang Zhang, Donghai Zheng and Dongryeol Ryu

Received: 13 May 2021
Accepted: 2 June 2021
Published: 4 June 2021

Publisher's Note: MDPI stays neutral with regard to jurisdictional claims in published maps and institutional affiliations.

1. Introduction

Under the combined influences of global climate change and rapid urban development, the occurred frequency of record-breaking rainfall events has increased significantly [1,2]. Floods caused by extreme rainfall events not only bring serious economic losses, but also cause huge casualties [3,4]. According to the data report of the World Resources Institute, the global economic loss caused by flood events was nearly 45.9 billion dollars; as well, 4500 people were killed, accounting for 40% of the global natural disaster deaths in 2019 [5]. The number of casualties caused by floods and the economy will continue to increase in the next decades [6,7]. Numerous studies have shown that record-breaking short-duration rainfall is an important factor causing the increasingly serious urban flood, while the lack of high temporal resolution rainfall records restricts the practices of hydrological engineering and urban flood analysis [8–10]. Zhu et al. [11] and Yu et al. [12] emphasized that hydrologic model-based flood analysis should carefully consider rainfall temporal resolution in the changing complex environment; they found that the simulated peak discharges can be significantly impacted by rainfall with different temporal resolution (e.g., 1-h and 24-h) at the same magnitude. However, most regions lack long-term and high-temporal resolution (sub-daily) rainfall records, especially for developing countries and newly built cities [13]. The available rainfall records show a decrease and non-stationary

9

trend in a changing environment [14,15]. In hydrological practice, however, the length-of-record limitations can limit the traditional methods for calculating the rainfall intensity–frequency–duration relationship.

In order to overcome the lack of rainfall records in urban flood analysis, many researchers have provided various coping methods (e.g., Li et al. [16], Kastridis et al. [17], Papaioannou et al. [18]), which can be categorized into five types. (i) Empirical probability statistics method. Traditional urban flood analysis is often based on frequency statistical methods and empirical assumptions, such as Gumbel, Pearson-III, maximum likelihood estimation, and other probability distribution models for parametric empirical statistical analysis [19]. However, the data time series is highly requisite based on the empirical value hypothesis [20]. Moreover, climate change and human activities lead to non-stationary changes of regional rainfall, making it difficult to ensure the accuracy and rationality of the estimation results [21]. (ii) Hydrologic model-based simulation. With the continuous improvement of hydrological models and hydrological theory, using a hydrological model to simulate urban flooding has become one of the most common methods. To some extent, the scope and application of hydrological data, theory, and tools are improved through the hydrological model. However, it needs detailed basic data to improve its accuracy [22,23]. (iii) Surrogate-data technique. Due to the lack of rainfall datasets, many studies use the rainfall data from adjacent stations to analyze regional flood frequency or calculate hydrological engineering. For example, Mohanty et al. [24] moved the rainfall data of three neighboring rain gauge stations to the study area, which was used for flood analysis. Although the surrogate-data technique can increase the rainfall sample size and make up for the lack of observation data, its accuracy is difficult to guarantee and its uncertainty is high [25]. (iv) Rainfall generator. Rainfall generators are often used to generate more diverse rainfall scenarios or higher spatial and temporal resolution rainfall data to enrich the regional rainfall sample size [26,27]. For example, the meteorological model (e.g., GCM) can simulate more rainfall events and other meteorological elements based on short-record data sets, but it needs strict meteorological data such as temperature and wind speed, and has the disadvantage of requiring complex calculations [26,28]. (v) Remote sensing analysis method. Combined with GIS technology, remote sensing data and the digital elevation model are often used to obtain regional hydrological characteristics and draw flood risk maps for flood analysis [29,30]. It can analyze the distribution of flood risk in a large area with coarse data, but it cannot fully consider the hydrological process [31].

It is undeniable that the above methods can solve the problem of data shortage in flood analysis to a certain extent, but there are still obvious disadvantages in different types of methods [32]. With the increase of high temporal resolution remote sensing rainfall data, there is a new way to do flood analysis in both natural and urban watersheds [33–35]. In recent years, it has become popular to comprehensively analyze floods by coupling remote sensing rainfall data and hydrological models, which solves the shortages of high spatial-temporal resolution rainfall data. For example, Shakti et al. [36] combined remote sensing rainfall data and a distributed hydrological model to analyze inundation. Komi et al. [37] have shown that using relatively rough spatial resolution remote sensing data as inputs to the distributed hydrological model can also roughly predict the flood range in Africa, where topographic and hydrological data are scarce. The coupling of high spatial-temporal resolution remote sensing rainfall data and a hydrological model is used to analyze the regional flood characteristics and widely used by more and more scholars [11,38].

On the other hand, urban flood analysis based on hydrological model mainly focuses on a single factor such as maximum rate, meaning many important indicators are often ignored [39,40]. Zhu et al. [40] emphasized that urban flood analysis should consider not only the maximum rate, but also the flood time, total inundation volume, and other factors. Hereby, urban flood analysis needs to address the high-dimension disaster problem. In order to reflect the characteristics of urban flooding, traditional methods such as the fuzzy comprehensive evaluation method, principal component analysis, and analytic hierarchy process (AHP) are often used for analyzing flood characteristics (e.g., Yang et al. [41];

Nandi eta al. [42]; Sarmah et al. [43]), but most of them have the shortcomings of human-subjective perceptions or being based on an ideal hypothesis [44]. In order to overcome these drawbacks, Zhu et al. [40] used the projection pursuit method to comprehensively analyze urban flood characteristics, and pointed out that this method can objectively evaluate urban flood characteristics.

As stated above, a lack of high-temporal rainfall records is a prominent limitation to flood analysis and hydrological engineering practices [14,45]. Rainfall remote sensing datasets with high temporal-spatial resolution and large coverage can overcome this limitation. This study seeks to propose a modeling framework for urban flood assessments based on short-record remotely sensed rainfall and hydrologic model in ungauged drainage basins. We do so by combining short (2008–2016), hourly remote sensing rainfall data and the RainyDay model to estimate the regional design rainfall under different frequencies. To be consistent with convention [46], the obtained design rainfall is transformed into the Chicago rainfall pattern and put into the SWMM hydrological model to simulate and analyze runoff processes and flood characteristics under different return periods. The projection pursuit method is used to comprehensively analyze flood characteristics based on the outputs of the SWMM hydrological model. It is worth mentioning that this study is not meant to demonstrate the superiority of the proposed framework compared with the traditional methods, but to explore the feasibility of analyzing small ungauged urban drainage basins based on short-term remote sensing rainfall data, and to provide an alternative framework for urban flood assessment.

2. Methodology

The proposed model framework used to analyze urban flooding based on short-record remotely sensed rainfall and hydrologic model includes three parts. (i) Generating extreme rainfall events. A rainfall generator named Rainyday with the short (nine years), gridded (0.1° × 0.1°), and hourly record of remote sensing rainfall is used to generate extreme rainfall events with 20 realizations at 2-, 10-, 20-, 50-, and 100-yr return periods for 2 h, 6 h, 12 h, and 24 h durations. These events are compared to the traditional design rainfall (i.e., intensity-duration-frequency (IDF) formula-based estimates) for rationality analysis. (ii) Simulating runoff under different rainfall return periods and durations. We leverage SWMM to construct a rainfall-runoff model for simulating the runoff under different rainfall return periods and durations, and the time distribution of the design rainfall follows the Chicago rainfall pattern. (iii) Analyzing urban flood. On the basis of analyzing the flood indicators (i.e., flood time, maximum rainfall rate, total maximum rainfall volume) under different rainfall return periods and durations, its comprehensive characteristics are analyzed by projection pursuit method.

2.1. Stochastic Storm Transposition

The traditional estimation methods of design rainfall for urban areas often have some drawbacks, such as a high requirement of rainfall series and a limited scope of application [27]. Many of them cannot meet the requirements of urban flood analysis in areas lacking data [11]. In order to conquer these drawbacks, this study uses RainyDay software with the core technique of stochastic storm transposition (SST) to estimate the design rainfall at different return periods in the area lacking data.

RainyDay is developed by Wright et al. [27] based on Python. The core of this model is to combine SST and remote sensing rainfall products to transpose the spatial location of observed rainfall events. It can effectively lengthen the rainfall record and expand the sample size of observed rainfall events. Figure 1 shows an example of transposing two observed rainfall events to the study area through RainyDay. It is worth mentioning that RainyDay only changes the spatial location of the observed rainfall events, but does not change the temporal distribution. The reader is directed to Zhu et al. [11], Wright et al. [27], Yu et al. [47], and Franchini et al. [48] for more details. The following is a brief introduction to RainyDay.

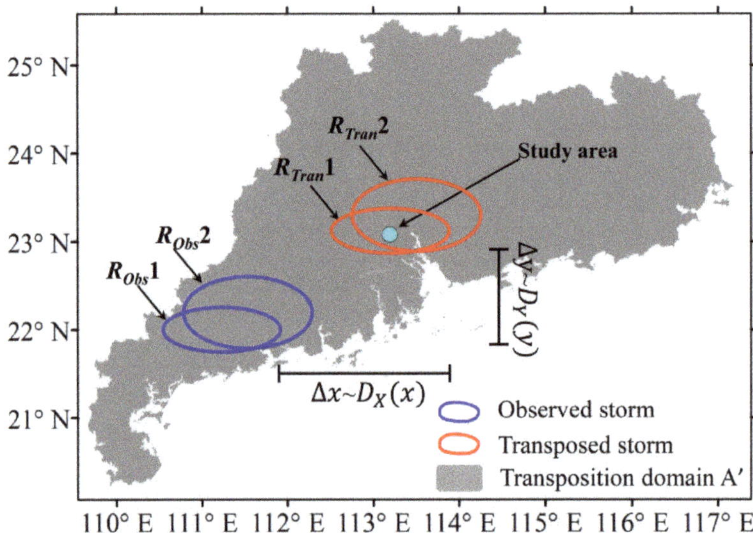

Figure 1. Schematic diagram of rainfall spatial transposition of RainyDay. Where $R_{Obs}1$ and $R_{Obs}2$ are the observed rainfall events in the transposition domain, respectively; $R_{Tran}1$ and $R_{Tran}2$ are the rainfall events after transposition, respectively.

Step 1. Selecting the transposition domain. RainyDay requires that (i) the selected transposition domain should contain the study area; (ii) the selected transposition domain has the same climatic conditions and similar rainfall characteristics as the study area; (iii) the area of the transposition domain is more than 10 times larger than the study area. We selected a typical residential district in Guangzhou as case-study area. Following the requirements of RainyDay, Guangdong Province, which belongs to the same administrative region as the case-study area, is selected as the transposition domain.

Step 2. Identifying the "parent storms". RainyDay selects the m largest t-hour rainfall events that occurred in the transposition domain over n-year record of gridded rainfall dataset, in terms of rainfall accumulation with the same size (i.e., single grid in this study) of study area. The selected rainfall events, which do not occur in the same 24 h, are temporally non-overlapping. That is, RainyDay only selects one t-hour event when there are two or more t-hour events in the top m events occurring in the same 24 h. These selected rainfall events are defined as "parent storms".

Step 3. Calculating the distribution probability of extreme rainfall events. The occurred probability of extreme rainfall events is spatially non-uniform in the transposition domain. RainyDay calculates the probability through the two-dimensional Gaussian kernel according to the storm centers of the "parent storms". The sum of the probability of each grid in the transposition domain is on one.

Step 4. Transposing rainfall events. RainyDay randomly selects k rainfall events from the "parent storms" to generate rainfall events, where k is an integer and indicates a "number of storms per year". Besides, RainyDay assumes that k follows a Poisson distribution with annual occurrence rate λ, where λ represents the ratio of the selected m parent storms to n-year rainfall records, $\lambda = m/n$. More details about Poisson-distributed storm occurrences can be found in Wilson and Foufoula-Georgiou [49]. The selected rainfall events can be transposed to any position in the transposition domain according to the distribution probability of extreme rainfall events, but only the rainfall that occurred in the study area is calculated. RainyDay extracts the t-hour maximum rainfall, and the extracted rainfall is regarded as the maximum t-hour annual rainfall.

Step 5. Generating T_{max} annual maximum rainfall. The T_{max} annual maximum rainfall can be generated through repeating Step 4 T_{max} times. To obtain the intensity-duration-frequency relationships, the maximas are ranked $i = 1 \ldots T_{max}$ from smallest to largest based on rainfall accumulation. Then, the return period P of each these ranks can be calculated as $P_i = 1/(i/T_{max})$. Each return period includes N realizations after repeating Step 4 and this step N times, that is, RainyDay provides the ensemble spread of rainfall accumulation rather than a single estimated value at each return period.

In this study, RainyDay is used to generate 5- to 100-yr design rainfall events with durations of 2 h, 6 h, 12 h and 24 h, respectively. Each return period includes 20 realizations for different durations. For simplicity, we only analyze the mean, minimum, and maximum of 20 realizations, since these results include the ensemble spread of all the realizations. In addition, we compare these results (i.e., RainyDay-based estimates) with IDF formula-based estimates to reflect the reasonability of the proposed framework.

2.2. Constructing Different Rainfall Scenarios

The design rainfall used in urban drainage systems and flood control is often calculated through coupling the IDF formula and the Chicago rainfall pattern [50]. To be consistent with this, the Chicago rainfall pattern is also used to allocate the RainyDay-based estimates at different times. The difference between IDF formula-based and the minimum, maximum and mean in 20 realizations of RainyDay-based estimates are compared. IDF formula is the empirical formula $q = \frac{167 \times A(1 + C \lg P)}{(t+b)^n}$, where q ($L/(s \cdot hm^2)$) indicates the design rainstorm intensity of t-minute duration at return period P (year); A, C, b, and n are the constant parameters that are derived and modified based on long-term rainfall records using the Gauss–Newton iterative algorithm [46]. For the case-study area, the IDF formula is shown in Equation (4).

$$q = \frac{3618.27(1 + 0.438 \lg P)}{(t + 11.259)^{0.750}} \tag{1}$$

In order to analyze the difference between the IDF-based and RainyDay-based estimates impact in urban flood analysis, we combine different return periods (5-, 10-, 20-, 50-, 100-yr), durations (2 h, 6 h, 12 h, 24 h), and estimates (IDF formula-based estimates, and the minimum, maximum, and mean in 20 realizations of RainyDay-based estimates) to generate 80 rainfall scenarios for urban flood analysis. For all rainfall scenarios, the rain peak coefficient is set to 0.375 to be consistent with the design specification for outdoor drainage in China [46].

2.3. Urban Hydrologic Model

In this study, an urban hydrologic model named SWMM is used to simulate and reflect the relationships between rainfall and runoff. SWMM is widely used in urban flood analysis and hydraulic practices, and it has very good simulated performances in both urban and natural basins [51,52]. Since the theory of the SWMM model is introduced in detail in a previous study by Gironás et al. [53], we do not show more details about the SWMM model in this study.

Because the calibrated and verified hydrological model in Zhu et al. [15] is used in this study, the reader is directed to Zhu et al. [15] for more information about case-study area and the performance of the model. In this model, the nonlinear reservoir method is selected to calculate the surface runoff, the Saint-Venant equations are used to calculate the flow, the Horton model is used to calculate the infiltration process, the Manning formula and the approximate continuity equation are used to convert the runoff of each sub basin into the outflow process, and the Newton-Raphson method and finite difference method are used to calculate the time-varying process of runoff. Zhu et al. [54] calibrated and verified the model based on the observed rainfall and runoff data, while the Nash-Sutcliffe efficiency (NSE) index is used to assess the model's performance.

In order to reflect the performance of RainyDay-based estimates for runoff process simulation, we take the time distributions of the RainyDay-based and IDF formula-based estimates as the inputs of the constructed urban hydrologic model and compare their differences. The model used in this study is same as that in Zhu et al. [54] and the calibration and verification results show that the model can be used to simulate the runoff process of the case-study area. The applicability and rationality of the model are demonstrated. More details about the model can be found in Zhu et al. [54].

2.4. Projection Pursuit Algorithm

The projection pursuit algorithm is a robust and powerful algorithm for the exploratory analysis of multivariate high-dimensional data. It is widely used to reduce dimensionality for feature extraction, especially for flood and environment analysis. For instance, Zhi et al. [55] coupled the drainage model, 2D flood simulation model, and projection pursuit algorithm to assess urban flood risk; when Guo et al. [56] proposed an evaluation framework to assess atmospheric environment carrying capacity based on an evaluation index system including 20 indicators, the projection pursuit algorithm was used to reduce dimensionality. The basic theory of the projection pursuit algorithm is to project the data into low-dimensional subspace via projection vectors. It has the advantages of a strong anti-jamming capability and not depending on subjective evaluation criteria. In this study, the projection pursuit algorithm is adopted to analyze the comprehensive characteristics of urban flooding by constructing an evaluation index system. The system includes three indicators, i.e., flood time, maximum rate, and total inundation volume. Zhu et al. [40] demonstrated that flood characteristics could be estimated well based on these indicators. The general steps are summarized as follows; more details are provided in Kruskal and Shepard [57] and Zhu et al. [40].

Step 1: Construct and normalize the evaluation indicator set. Flood time, maximum rate, and total inundation volume are selected as the evaluation indicator set $(X = \{X_{ij} \mid i = 1, 2, 3; j = 1, 2, \ldots, p\})$, where X_{ij} represents the value of the ith evaluation indicator of the jth sample, j and i represent the number of evaluation indicators and sample size, respectively. The normalized set x_{ij} is calculated as follow:

$$x_{ij} = \frac{X_{ij} - X_{j\min}}{X_{j\max} - X_{j\min}} \tag{2}$$

where $X_{j\max}$ and $X_{j\min}$ denote the maximum and minimum of ith evaluation indicator.

Step 2: Establishing the projection indicator function $Q(a)$. The evaluation indicator set is synthesized into a 1×3 vector (i.e., $a = \{a_i \mid i = 1, 2, 3\}$) as the projection direction. Therefore, the projection value of jth sample is calculated as follow:

$$Z_j = \sum_{i=1}^{3} a_i x_{ij} \, (j = 1, 2, \ldots, p) \tag{3}$$

Then, $Q(a)$ can be expressed as:

$$Q(a) = S_Z D_Z \tag{4}$$

$$S_Z = \sqrt{\frac{1}{p-1} \sum_{j=1}^{p} (Z(j) - \overline{Z})^2} \tag{5}$$

$$D_Z = \sum_{i=1}^{3} \sum_{j=1}^{p} (R - R(i,j)) u(R - r(i,j)) \tag{6}$$

where S_Z and D_Z note the interclass distance and local density of Z_j, respectively; $\overline{Z}$ represents the mean of Z_j; $R (R = 0.1 S_Z)$ means the cutoff radius; $u(R - r(i,j))$ is the unit step function, if $R - r(i,j) \geq 0$, $u(R - r(i,j)) = 1$; otherwise, $u(R - r(i,j)) = 0$.

Step 3: Calculating the best projection direction. $Q(a)$ is determined by the projection direction a if the value of the evaluation indicator is given. For the projection direction, the higher the value of $Q(a)$ the better. When the value of $Q(a)$ is at its maximum, the corresponding projection direction is the best. In order to seek the best projection direction, the optimum objective function can be constructed as $\max(Q(a) = S_Z D_Z)$, and the constraint condition is $\sum_{j=1}^{p} a^2(j) = 1$. Seeking the best projection direction is a nonlinear global optimization problem; the particle swarm optimization (PSO) technique is widely used to solve such problems. We also adopt it in this study, and more details are directed to Kennedy and Eberhart [58].

Step 4: Analyzing the comprehensive characteristics of urban flooding. The best projection values can be obtained through putting the best projections direction into Equation (4). The best projection values represent the comprehensive characteristics of urban flooding. The larger the values are, the more severe is the urban flood.

Based on analyzing the runoff processes at the outlet of the case-study area, we focus on the flood characteristics under RainyDay-based and IDF formula-based estimates at the manholes (i.e., junctions) for the case-study area drainage system in this section. Three flood indicators (i.e., flood time, maximum rate, total inundation volume), which are demonstrated to reflect the urban flood characteristics by Zhu et al. [40], are selected to analyze the flood characteristics at each manhole. The comprehensive flood characteristics are analyzed by combining these three indicators with the projection pursuit algorithm.

3. Data and Case-Study Area

3.1. Data

The hourly, 0.1° gauge-adjusted remotely sensed rainfall data (http://www.cma.gov.cn/2011qxfw/2011qsjgx/, accessed date: 15 November 2020) from the China Meteorological Administration merges CMORPH (the Climate Prediction Center Morphing algorithm) and the observations of 30,000 automatic rain gauges. This rainfall product is optimized and verified by the probability density function matching technique and optimal interpolation method. The temporal resolution is coarsened to one hour. Its total error is less than 10%, and the errors for heavy rainfall in the area with sparse ground gauge networks are less than 20%. The accuracy is higher than similar rainfall products and the product has been widely used for precipitation studies [59]. In order to verify the feasibility of estimating the design rainfall based on short-record remote sensing rainfall data, the rainfall data from 2008 to 2016 are selected in this study, where 2008 is the earliest year when data are available.

The rainfall and runoff data used for calibration and verification are observed from the case-study area, where the rainfall data is observed by RainLogger^{TM} rain gauge (RainWise Inc.; USA), and the runoff data is observed by Stingray open channel gauge (Greyline Instruments Inc.; Germany). The observed time steps are 10 min.

3.2. Case-Study Area

In this study, the transposition domain selected for RainyDay is Guangdong Province in the south of China. The latitude ranges from 20.08 to 25.32 °N, and longitude ranges from 109.04 to 117.20 °E (Figure 2). The area belongs to a subtropical monsoon climate, and the rainfall has the characteristics of large amount and high intensity. The annual average rainfall is 1300 to 2500 mm. The rainfall in this area is seasonal, mainly from April to September, and record-breaking rainfall and flood disasters occur frequently during these months.

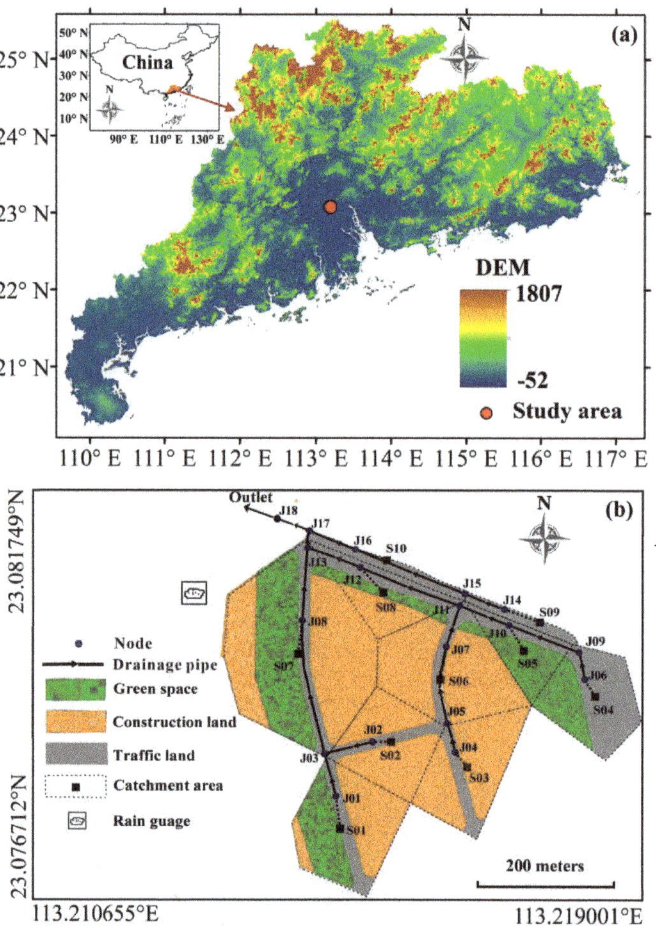

Figure 2. The transposition domain and the location of the case-study area (**a**), and the land use and drainage system of the case-study area (**b**).

In order to verify the rationality of the proposed framework, a highly developed and typical residential area (22.08–23.09 °N, 113.20–113.21 °E) is selected as the case-study area in Guangzhou city (Figure 2). It belongs to the subtropical monsoon climate, and the average annual rainfall is 1675 mm. Extreme rainfall events may occur throughout the year, but are mainly concentrated in April to September. In the past 60 years, the maximum and minimum annual rainfall are 2865 and 1009 mm, respectively. The area of the case-study area is about 1.55×10^5 m^2, and its land use types can be generalized into three types such as building, green space, and road land (Figure 2). The slope ratio of the drainage system is 0.1~1.0%, and the pipe diameter is 600~1650 mm. The drainage system of the case-study area is designed according to rainfall accumulation at 2-yr return period. However, the regional flood problem has become increasingly prominent with increasing record-breaking extreme rainfall events.

According to the generalization theory of SWMM, the case-study area is divided into 10 sub-catchments, while the drainage system is generalized into 18 pipes, 18 manholes, and 1 outlet (Figure 2). More details are referred to in Zhu et al. [54].

4. Results

4.1. Estimating the Design Rainfall

The important assumption of RainyDay for estimating design rainfall is that the storms in the transition domain are likely to occur in the study area. In order to illustrate the rationality of the selected transition domain, this study analyzes the spatial distribution and storm occurrence probability of 200 maximum storms under different durations (2 h, 6 h, 12 h, and 24 h) (Figure 3). The spatial distribution of storms with different durations is basically similar to each other. Generally speaking, the frequency of storms in coastal areas is relatively higher, but its spatial distribution is still relatively random, that is, heavy storms may occur everywhere in the selected transition domain (Figure 3). Similar to the spatial distribution, the spatial probability distribution of storms in the transition domain is relatively uniform, but there are still some differences. The storm occurrence probability decreases from south to north (Figure 3), which is in line with the actual distribution of storms (see Wang et al. [60] for evaluation of rainfall distribution in different precipitation products). The selected transition domain is reasonable since the probability of storm occurrence of 200 maximum storms varies from around 0.0002 to 0.0014 in the transition domain, i.e., the storms in the transition domain are likely to occur in the study area or other regions.

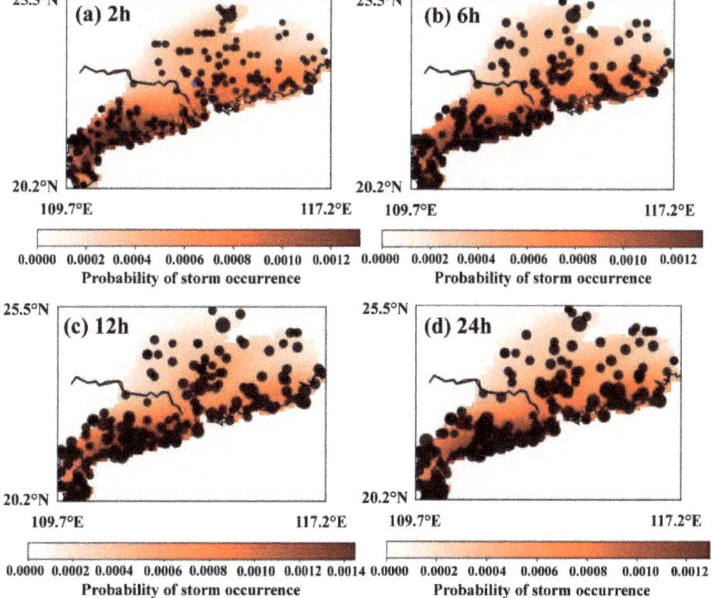

Figure 3. The probability of storm occurrence and spatial distribution of 200 maximum storms under 2 h (**a**), 6 h (**b**), 12 h (**c**), and 24 h (**d**) durations over the transition domain. Shading denotes spatial probability of storm occurrence calculated based on 200 maximum storms. Black dots represent the rainfall centroids of 200 maximum storms, and its size means the relative rainfall depth of each storm.

Figure 4 shows the relationships between IDF formula-based and RainyDay-based estimates for different durations at different return periods. The ensemble spread of 20 realizations is shown as shaded area. Comparing results indicates that RainyDay is generally able to estimate urban extreme rainfall for different durations, but it may relatively underestimate or overestimate the rainfall accumulation. The results show that RainyDay usually underestimates the rainfall accumulation at low return periods or short rainfall durations; the RainyDay-based estimates are usually larger than the IDF formula-based estimates when the rainfall duration is long or the return period is high. Specifically, RainyDay overall underestimates the rainfall accumulation when the rainfall duration is 2 h at different return periods. The degree of underestimation, which varies from 0.4% (100-yr) to 57% (5-yr), decreases with increasing the return period (Table 1). When the duration reaches 6 h, the underestimation is improved. The IDF formula-based estimates, overall, fall within the ensemble spread of RainyDay-based estimates with the increase of duration. At each return period for 6 h or longer durations, the absolute value of the ratio of at least one RainyDay-based estimate (maximum, minimum, or average estimates) to IDF formula-based estimate is less than 10% (Table 1).

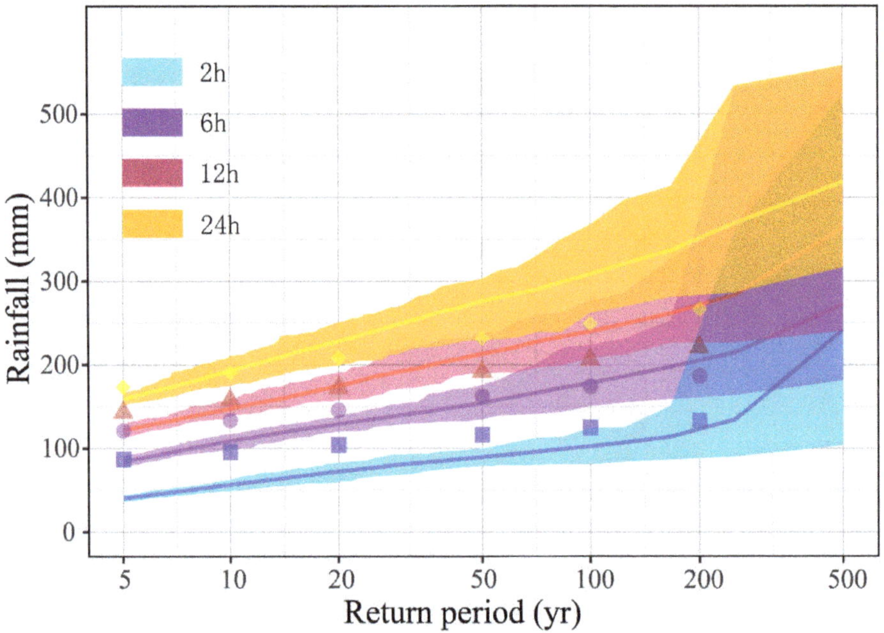

Figure 4. The relationships between IDF formula-based and RainyDay-based estimates for different durations at different return periods. The shaded areas mean the ensemble spread of RainyDay-based estimates. The solid lines denote the ensemble mean for 20 realizations. The symbols of different shapes represent the IDF formula-based estimates.

Table 1. Relative deviations between IDF formula-based and RainyDay-based estimates (%).

Return Period	2 h			6 h			12 h			24 h		
	min	mean	max	min	mean	max	min	mean	max	min	mean	max
5-yr	−57	−53.3	−48.4	−17.2	−9.6	−2.7	−20.6	−15.2	−10.4	−12.7	−8.2	−4.1
10-yr	−48.7	−40.7	−34.9	−18.3	−10.3	−3.4	−12.5	−6.5	1.7	−8.3	0.9	10.4
20-yr	−42.1	−30.6	−20.6	−17.8	−11.3	−4	−8.2	1.5	10.8	−4.5	9.5	20.2
50-yr	−31.9	−22.9	−14.1	−16.8	−4.8	8.6	−1	11.2	24.1	−1	19.1	30.8
100-yr	−34.8	−17.8	−0.4	−13.6	2.4	28.5	1	16.4	33.9	5.2	24	47.6

The IDF formula-based estimates gradually approach to the lower boundary of the shaded area with increasing return period. It indicates that the RainyDay-based estimates basically can reflect the observed design rainfall for long (6 h or longer) durations. To be consistent with the design specification for outdoor drainage in China, the time distributions of the RainyDay-based and IDF formula-based estimates for urban flood simulation are determined by the Chicago rainfall pattern. The time distribution results show that the main difference comes from the rainfall peak. The rainfall peak is underestimated from RainyDay-based estimates at low return periods or short rainfall durations, while it is generally matched or slightly overestimated at high return periods or for long rainfall duration. In order to better explain this fact, the time distributions at different return periods for 6 h duration and at 20-yr return period for different durations are selected as in the below examples (Figures 5 and 6). When the duration is 6 h, the rainfall peak of the RainyDay-based estimates is relatively smaller than the IDF formula-based estimates at 5- and 10-yr return period, but the rainfall peak of IDF formula-based estimates generally falls within the ensemble spread of RainyDay-based estimates, and the average of the ensemble spread is generally matched to the IDF formula-based estimates when the return period reaches 50-yr or higher (Figure 5). On the other hand, when the return period is at 20-yr return period, the time distributions of RainyDay-based and IDF formula-based estimates are essentially coincidental, and the coincidence increases with lengthening rainfall duration (Figure 6). Overall, the RainyDay-based estimates show a good performance for design rainfall analysis. The relationship between the time distributions of RainyDay-based and IDF formula-based estimates at other return periods for other rainfall durations are similar to the above selected rainfall scenarios, so the time distributions of other scenarios are not shown.

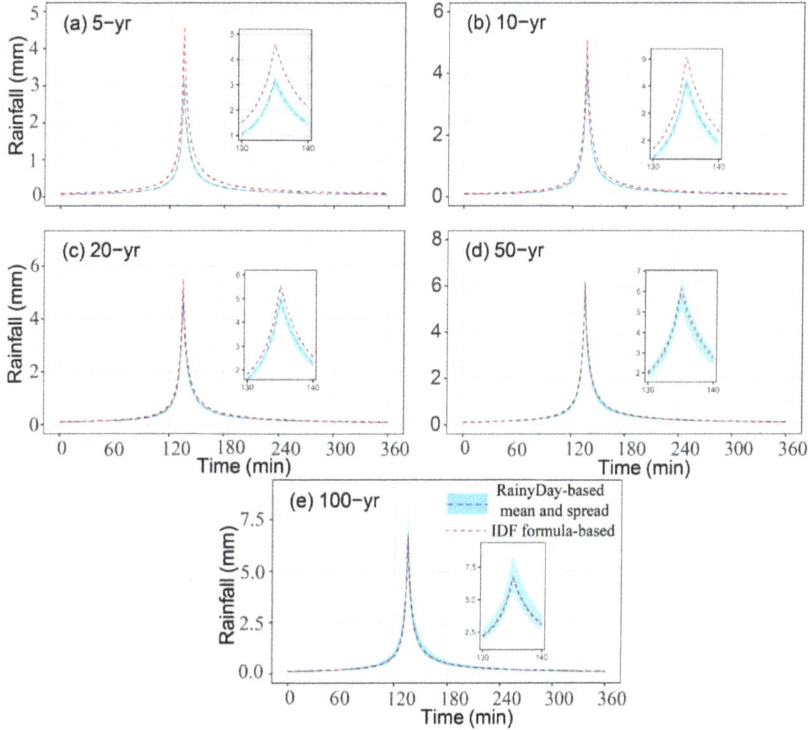

Figure 5. The time distributions of RainyDay-based and IDF formula-based estimates at 5-yr (**a**), 10-yr (**b**), 20-yr (**c**), 50-yr (**d**), and 100-yr (**e**) return periods for 6 h duration.

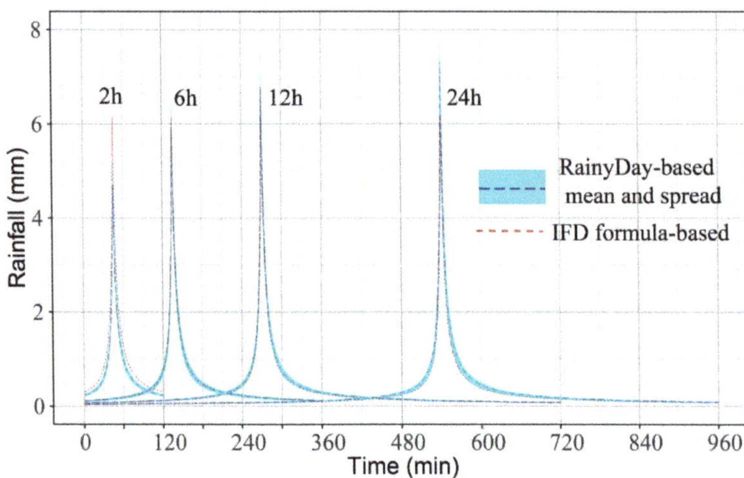

Figure 6. The time distributions of RainyDay-based and IDF formula-based estimates at 20-yr return period for different durations.

4.2. Simulating the Runoff Process Based on RainyDay-Based Estimates

The simulated results show that the RaiyDay-based estimates basically can be used for runoff process simulation. The difference between the runoff processes of RainyDay-based and IDF formula-based estimates is similar to the time distributions of design rainfall, but the difference of peak discharge is smaller than the rainfall peak. Similar to the analysis of time distribution of rainfall estimates, we also take the runoff processes at different return periods for 6 h duration and at 20-yr return period for different durations as in the below example (Figures 7 and 8). The runoff processes of RainyDay-based and IDF formula-based estimates indicate that the difference of runoff process decreases as the rainfall duration lengthens. The difference of peak discharge at high return periods (20-yr or higher) or for long durations (6 h or longer) is very small. For example, the difference of the RainyDay-based and IDF formula-based rainfall peaks is relatively significant, but the differences of peak discharges are very small at 5- and 10-yr return periods (Figures 5 and 7). The RainyDay-based peak discharges become closer and closer, and even approximate overlapping IDF formula-based peak discharges with increasing return period. For the same return period (take 20-yr return period for example), the peak discharge is still slightly underestimated for 2 h duration, but the runoff process is predicted pretty well with the lengthening duration (Figure 8). In addition, we use NSE to evaluate the predicted performance of RainyDay-based estimates, i.e., the difference of runoff processes between RainyDay-based and IDF formula-based estimates. Results show that the values of NSE are generally small for short duration or at low return period (e.g., NSE = 0.53 at 5-yr return period for 6 h duration, NSE = 0.77 at 20-yr return period for 2 h duration); however, the values become larger with increasing rainfall duration or rainfall return period (e.g., NSE = 0.98 at 100-yr return period for 6 h duration, NSE = 0.99 at 20-yr return period for 24 h duration). For long duration (6 h or longer) or high return period (10-yr or higher), the values of NSE are generally above 0.5, i.e., the RainyDay-based estimates of long duration or high return period are satisfied to analyze the runoff process.

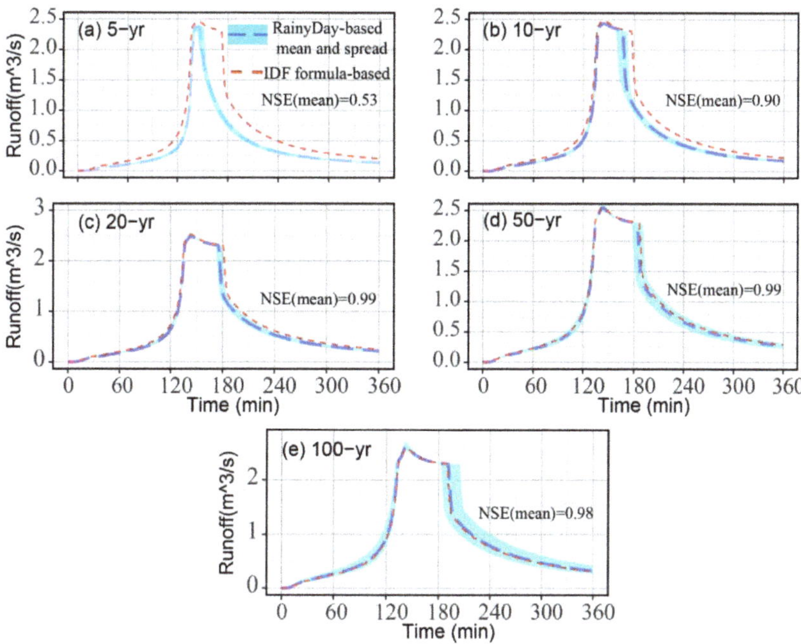

Figure 7. The runoff processes at the outlet of the case-study area at 5-yr (**a**), 10-yr (**b**), 20-yr (**c**), 50-yr (**d**), and 100-yr (**e**) return periods for 6 h duration.

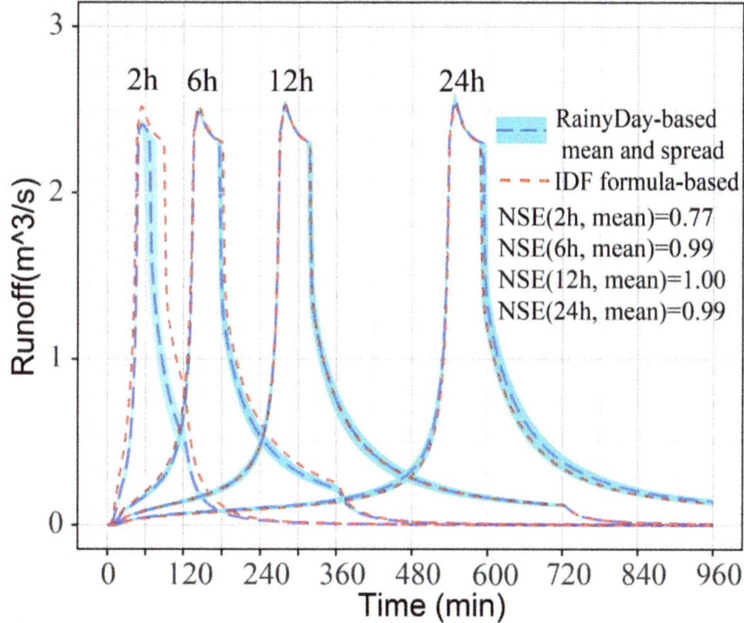

Figure 8. The runoff processes at the outlet of the case-study area at 20-yr return period for different durations.

4.3. Analyzing Flood Characteristics Based on RainyDay-Based Estimates

Results show that the characteristics of urban flooding are generally underestimated based on RainyDay-based estimates at low return periods or short rainfall durations. For short durations or at low return periods, the underestimation of the values of these indicators at each manhole are more significant than runoff processes at the outlet. Specifically, the RainyDay-based estimates underestimate the values of flood time, maximum rate, and total inundation volume when the return period is lower than 10-yr or duration is shorter than 6 h. The underestimation decreases with increasing return period or lengthening rainfall duration. The values of flood time, maximum rate, and total inundation volume simulated based on IDF formula-based estimates generally fall within the ensemble spread of RainyDay-based estimates at high (20-yr or high) return period or long (6 h or longer) duration (Figures 9 and 10). That is to say, the RainyDay-based estimates can be used to assess the flood characteristics at each manhole at relatively high return periods or long rainfall durations.

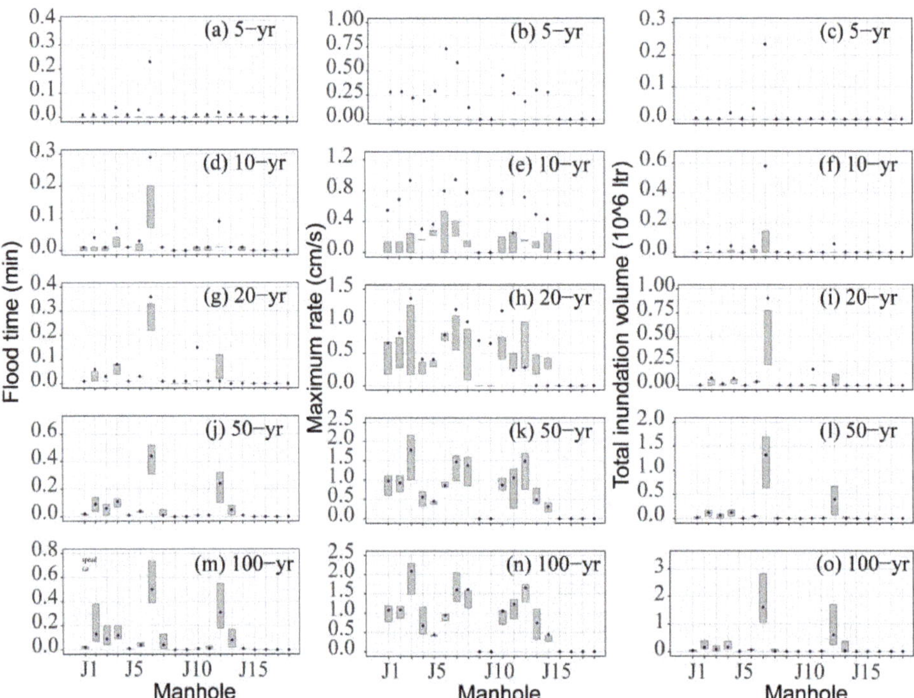

Figure 9. The flood characteristics of each manhole at different return periods for 6 h duration. The flood time at 5-yr, 10-yr, 20-yr, 50-yr, and 100-yr return periods is shown in (**a**), (**d**), (**g**), (**j**), and (**m**), respectively. The maximum rate at 5-yr, 10-yr, 20-yr, 50-yr, and 100-yr return periods is shown in (**b**), (**e**), (**h**), (**k**), and (**n**), respectively. The total inundation volume at 5-yr, 10-yr, 20-yr, 50-yr, and 100-yr return periods is shown in (**c**), (**f**), (**i**), (**l**), and (**o**), respectively. The grey boxes indicate the spread of RainyDay-based, and points represent IDF formula-based, values.

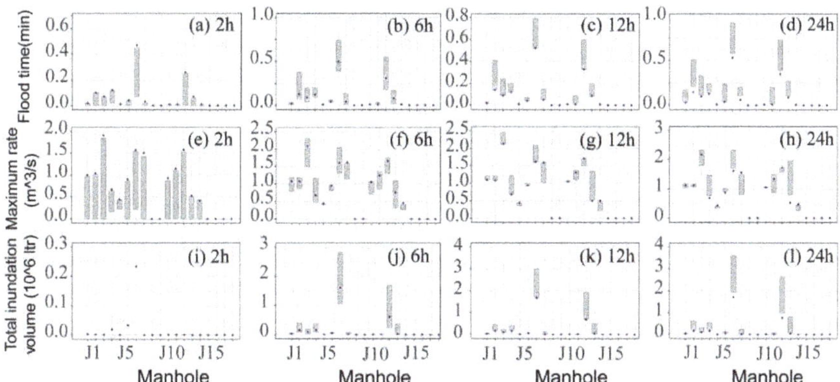

Figure 10. The flood characteristics of each manhole at 20-yr return period for different durations. The flood time for 2 h, 6 h, 12 h, and 24 h durations is shown in (**a–d**), respectively. The maximum rate for 2 h, 6 h, 12 h, and 24 h durations is shown in (**e–h**), respectively. The total inundation volume for 2 h, 6 h, 12 h, and 24 h durations is shown in (**i–l**), respectively. The grey boxes indicate the spread of RainyDay-based, and points represent IDF formula-based, values.

In order to better clarify the changing characteristics of urban flooding at each manhole with rainfall return period or rainfall duration, we also take the flood characteristics of each manhole at 20-yr return period for different durations and at different return periods for 6 h duration as an example. For 6 h rainfall duration, the RainyDay-based estimates significantly underestimate the values of the selected indicators at 5-yr return period; when the return period increases to 10-yr, the RainyDay-based estimates can reflect the flood characteristics at each manhole to a certain extent, but it is still relatively underestimated; while the rainfall return period reaches 20-yr or more, the values of indicators simulated by IDF formula-based estimates basically fall within the ensemble spread of RainyDay-based estimates. On the other hand, when the rainfall return period is 20-yr, the RainyDay-based estimates can basically reflect the flood characteristics of each manhole under different rainfall duration scenarios, especially for long (6 h or longer) rainfall duration. The flood characteristics of some manholes will be slightly overestimated with the increasing rainfall duration.

The results shown in Figures 9 and 10 cannot comprehensively assess the flood characteristics of each manhole, therefore, the projection pursuit algorithm is used to reduce three dimensions (i.e., three indicators) to one dimension. The one-dimension values (i.e., the projection values) indicate the comprehensive characteristics of urban flooding for each manhole. Results show that the flood hotspot manholes are J3, J7, and J13, but they are significant underestimated based on RainyDay-based estimates at low return periods or short rainfall durations (Figures 9 and 10). The changing characteristics of projection values with return periods or duration are similar to the values of the three indicators, but the degree of underestimation for the projection values is larger than the values of indicators (Figures 11 and 12). However, the degree of underestimation decreases with increasing return period or duration. Similar to the values of three indicators, the projection values estimated based on IDF formula-based estimates fall within the RainyDay-based ensemble spread at high (20-yr or higher) return periods or long (6 h or longer) durations. The comprehensive analysis results for urban flooding demonstrates that the RainyDay-based estimates can be used for urban flood analysis, especially for high (20-yr or high) return periods or long (6 h or longer) durations.

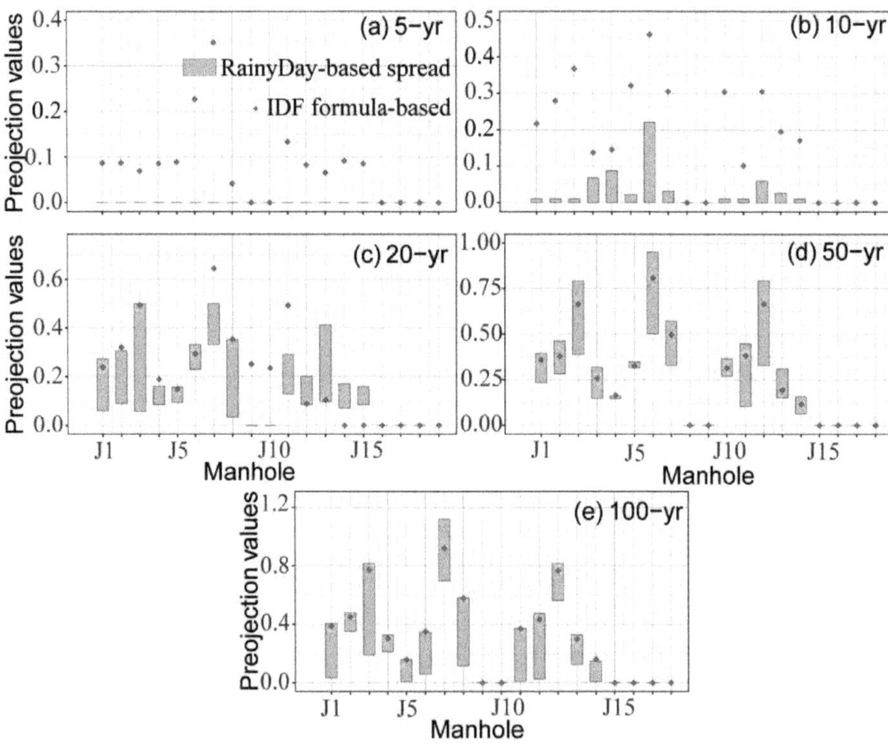

Figure 11. The projection values of each manhole at 5-yr (**a**), 10-yr (**b**), 20-yr (**c**), 50-yr (**d**), and 100-yr (**e**) return periods for 6 h duration.

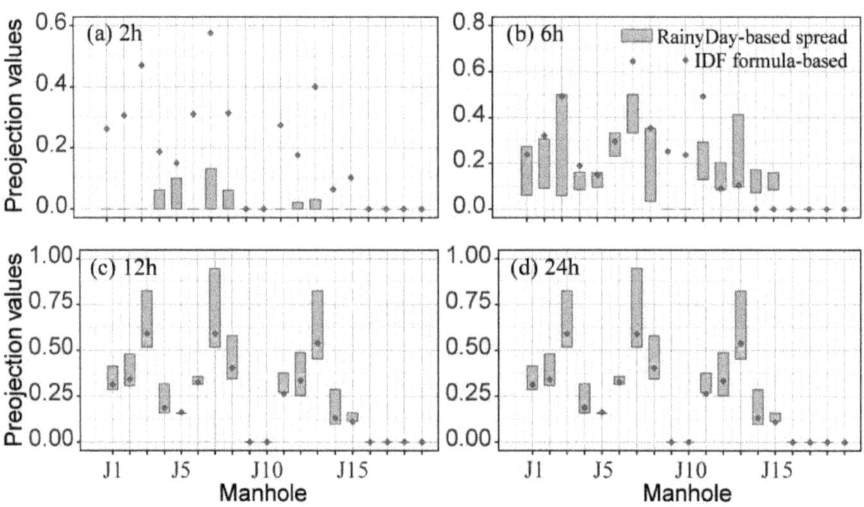

Figure 12. The projection values of each manhole at 20-yr return period for 2 h (**a**), 6 h (**b**), 12 h (**c**), and 24 h (**d**) durations.

5. Discussion

Regarding the limitations of traditional urban flood analysis, lacking high-resolution rainfall records should be one of the primary issues [61]. Although many models and frameworks were proposed to solve this issue, many inherent limitations still exist [16,62]. Therefore, a modeling framework for urban flood analysis is introduced based on short-record rainfall from remote sensing, RainyDay, and urban hydrological model, which effectively overcomes the high-temporal-resolution and long-term rainfall requirements for urban flood analysis. It should be emphasized that this work does not seek to show the proposed framework better than the traditional methods, but rather to provide an alternative framework for urban flood analysis based on short-term remote sensing rainfall records, and discuss its feasibility and rationality.

The results of this study for design rainfall estimates are very similar to Wright et al. [27], though simulated at a much smaller scale (0.155 km^2 vs. 4000 km^2) based on a different time-space resolution (hourly vs. hourly, and 3 h; 0.1° grid vs. 4-km, and 0.25° grid) and length of rainfall records (nine-year vs. 13-year, and 17-year). These two studies show that the design rainfall is generally underestimated with remote sensing data at low return periods or short durations. The underestimation could be explained by the length of rainfall records and spatial resolution (nine-year and 0.1° grid for the remote sensing rainfall record vs. more than 20 years and approximately 0.1 m^2 for the rain gages) in this study. For short duration rainfall, temporal resampling using RainyDay is significantly affected by rainfall detection errors on bias correction and conditional biases [63–66]. Also, this can be attributed to the fundamental structure of RainyDay, i.e., the Poisson distribution is utilized in this study (see Kim and Onof [67] for discussion). Conversely, the slight overestimation of RainyDay-based estimates are showed at high return periods and long durations, but the overestimation is not as severely as the underestimation. This might potentially be attributed to conditional bias for rain rate [68] and the domain area including coastal areas where the typhoon landed. Some existed studies show that the accuracy of the estimates may be improved by higher temporal-spatial resolution remote sensing data, which can better address and understand some rainfall biases [27,69].

The main parts of this study include estimating design rainfall based on nine-year remote sensing rainfall and RainyDay, and revealing the relationship between design rainfall and runoff through hydrological model. Previous studies showed that the design rainfall can be well estimated by RainyDay at different scales (e.g., 14.3 km^2 in Zhou et al. [70], 4400 km^2 in Wright et al. [66]). Though the feasibility is shown varying from small to large scales, the limit on the size of study area can arise since the presence of complex terrain features. The reader is directed to Wright et al. [27] for more discussion. On the other hand, the selected hydrological model (i.e., SWMM) has been widely used for modeling rainfall-driven flood at different scales, especially for urban areas. The proposed modeling framework offers opportunities to analyze urban flooding based on short-record remote sensing rainfall and hydrologic model. However, the size of the case-study area is small, it may cannot represent all the urban flood conditions. We will continue to expand the capabilities of the proposed modeling framework.

Case study shows that the runoff process at the outlet of case-study area and the flood characteristics (i.e., flood time, maximum rate, total inundation volume) of each manhole can be simulated well at relatively high return periods (20-yr or higher) or long durations (6 h or longer) based on the selected rainfall record. But the flood characteristics are more sensitive to the return period and duration of design rainfall than runoff process. The main difference in the rainfall hydrographs between RainyDay-based and IDF formula-based is from the peak rainfall, which can significantly impact the flood characteristics.

Our findings indicate that the rainfall estimates play a key role in flood analysis, similar results are also showed in Peleg et al. [26]. That is, improving the accuracy of the rainfall estimates is the most important in the proposed framework. Lots of studies indicated that rainfall estimates based on historical rainfall records might not be appropriate due to climate change [71]. Doing so would require higher-resolution remote sensing rainfall data and considering climate change [27,70,71]. We are developing frameworks for considering both rainfall space-time structure and climate change based on Regional Climate Model (RCM) simulations for RainyDay-based rainfall estimates.

Despite the proposed framework overcomes some drawbacks (e.g., rainfall records) of traditional approaches for urban flood analysis, there still remain several limitations. (i) Applicability of the proposed framework is insufficient for low return period or short duration rainfall scenarios. The undervaluation of design rainfall and urban flood characteristics are generally showed at these scenarios. The main reason is mentioned above, and the applicability can be improved by utilizing higher resolution and longer rainfall records [70,72]. (ii) The uncertainties from the rainfall data and RainyDay are hard to minimize, which have direct impacts in design rainfall estimates and urban flood analysis. The dominant uncertainty in the input rainfall data comes from the difference between remote sensing rainfall data and ground-based observations [27,73]; and the uncertainty in Rainyday comes from the input requirements (e.g., geographic transposition domain, rainfall record) and its structure [70]. (iii) The proposed framework uses idealized assumption (i.e., Chicago rainfall pattern) to determine the distributions of design rainfall. That is consistent with the guidelines of design rainfall [46]. On the other hand, the rainfall temporal resolution of remote sensing records is general coarser than 30-min. Comparing the relationships between RainyDay-based and IDF formula-based analysis results suggest that the proposed framework is an applicable way for analyzing urban flooding at high return periods (20-yr or higher) or long durations (6 h or longer). Though limitations still remain, we continue to develop its capabilities.

6. Conclusions

Rainfall remote sensing datasets have the advantages of high temporal-spatial resolution and large coverage, which can overcome limitations such as a lack of gauge-based rainfall records. In this study, we propose a modeling framework for urban flood analysis based on short-record remote sensing rainfall and hydrologic model. The framework is largely motivated by the fact that, in spite of increased interest in urban flood analysis using high-temporal remote sensing rainfall data, the inherent limitation of a lack of long-term high-temporal rainfall data still exists. We used RainyDay and a nine-year record of hourly, $0.1°$ remotely sensed rainfall data to generate extreme rainfall events for an urban hydrologic model (SWMM). The rainfall estimates of RainyDay-based and IDF formula-based methods were compared, as well as the corresponding runoff process at 5-, 10-, 20-, 50-, 100-yr return periods for 2 h, 6 h, 12 h, and 24 h durations. In addition, the projection pursuit method was used to reflect the comprehensive characteristics of the urban flooding. A typical urban drainage basin in the south of China was selected as the case-study area. The main conclusions include the following:

1. Combining RainyDay and short-term remotely sensed rainfall data can lengthen the rainfall record through transposing the spatial location of observed rainfall events. It is able to estimate urban extreme rainfall at different return periods (e.g., range in return period from 5- to 100-yr), despite the short (nine-year) observed rainfall record. According to a comparison of the differences between the RainyDay-based and IDF formula-based (a traditional published source of rainfall frequencies) rainfall estimates, RainyDay-based rainfall estimates are basically acceptable for estimating regional design rainfall, especially for relatively high return periods (20-yr or higher) or long durations (6 h or longer).

2. The proposed framework shows a good performance for runoff process simulation at the outlet based on RainyDay-based estimates, especially for high return periods or long durations. In the case study, the difference of runoff process between RainyDay-based and IDF formula-based methods is relatively significant at low return periods or for short durations (e.g., NSE = 0.53 at 5-yr return period for 6 h duration), but the difference decreases with the lengthening rainfall duration or increasing return period. The values of NSE are generally above 0.90 at high return periods or long durations.

3. Contrasting with the flood-simulated results under different return periods and durations, the flood characteristics of urban flooding at each manhole can be generally revealed based on RainyDay-based estimates at relatively high (20-yr and beyond) return periods or long (6 h or longer) durations. Similar to the results of runoff processes, though RainyDay-based estimates basically underestimate the values of flood indicators (i.e., flood time, maximum rainfall rate, total maximum rainfall volume) or the comprehensive characteristics of urban flooding under low return period or short duration scenarios, these values can be well revealed with increasing duration or return period.

4. The proposed modeling framework provides an alternative framework for urban flood analysis in an ungauged drainage basin. This alternative is attractive for the following reasons. First, the proposed framework can produce probabilistic extreme rainfall scenarios based on a very short rainfall record (e.g., nine-year in this study), and it excludes the older rainfall records to eliminate the effect of nonstationarity. Second, the proposed framework provides a way to estimate the ensemble spread of rainfall and flood estimates, rather than a single estimate value; such spread is central to hydrological engineering practices.

Author Contributions: Conceptualization, Z.Z. and Y.C.; methodology, Z.Z. and Y.Y.; software, Z.Z. and Y.Y.; validation, Z.Z., Y.C., and Z.Y.; investigation, Z.Z.; resources, Y.C. and Z.Y.; writing—original draft preparation, Z.Z. and Y.Y.; writing—review and editing, Y.C. and Z.Z.; visualization, Z.Z. and Y.Y.; supervision, Z.Y.; funding acquisition, Z.Y. and Z.Z. All authors have read and agreed to the published version of the manuscript.

Funding: This research was funded by the Local Innovative and Research Teams Project of Guangdong Pearl River Talents Program (2017BT01Z176), Key Special Project for Introduced Talents Team of Southern Marine Science and Engineering Guangdong Laboratory (Guangzhou) (GML2019ZD0403), and National Natural Science Foundation of China (52009021).

Institutional Review Board Statement: Not applicable.

Informed Consent Statement: Not applicable.

Data Availability Statement: The data that support the findings of this study are available from the corresponding author upon reasonable request.

Acknowledgments: The authors wish to express their gratitude to all authors of the numerous technical reports used for this paper. We would also like to thank the editor as well as the reviewers whose constructive criticism contributed greatly to this paper.

Conflicts of Interest: The authors declare no conflict of interest.

References

1. Fowler, H.J.; Lenderink, G.; Prein, A.F.; Westra, S.; Allan, R.P.; Ban, N.; Barbero, R.; Berg, P.; Blenkinsop, S.; Do, H.X.; et al. Anthropogenic Intensification of Short-Duration Rainfall Extremes. *Nat. Rev. Earth Environ.* **2021**, *2*, 107–122. [CrossRef]

2. Liu, B.; Tan, X.; Gan, T.Y.; Chen, X.; Lin, K.; Lu, M.; Liu, Z. Global Atmospheric Moisture Transport Associated with Precipitation Extremes: Mechanisms and Climate Change Impacts. *WIREs Water* **2020**, *7*, e1412. [CrossRef]

3. Kemter, M.; Merz, B.; Marwan, N.; Vorogushyn, S.; Blöschl, G. Joint Trends in Flood Magnitudes and Spatial Extents Across Europe. *Geophys. Res. Lett.* **2020**, *46*, e2020GL087464. [CrossRef]

4. Wu, Z.; Zhou, Y.; Wang, H.; Jiang, Z. Depth Prediction of Urban Flood under Different Rainfall Return Periods Based on Deep Learning and Data Warehouse. *Sci. Total Environ.* **2020**, *716*, 137077. [CrossRef]

5. World Resources Institute RELEASE. New Data Shows Millions of People, Trillions in Property at Risk from Flooding—But Infrastructure Investments Now Can Significantly Lower Flood Risk. Available online: https://www.wri.org/news/2020/04/release-new-data-shows-millions-people-trillions-property-risk-flooding-infrastructure (accessed on 13 April 2021).

6. Du, S.; Scussolini, P.; Ward, P.J.; Zhang, M.; Wen, J.; Wang, L.; Koks, E.; Diaz-Loaiza, A.; Gao, J.; Ke, Q.; et al. Hard or Soft Flood Adaptation? Advantages of a Hybrid Strategy for Shanghai. *Glob. Environ. Chang.* **2020**, *61*, 102037. [CrossRef]

7. Santos, P.P.; Pereira, S.; Zêzere, J.L.; Tavares, A.O.; Reis, E.; Garcia, R.A.C.; Oliveira, S.C. A Comprehensive Approach to Understanding Flood Risk Drivers at the Municipal Level. *J. Environ. Manag.* **2020**, *260*, 110127. [CrossRef] [PubMed]

8. Chen, J.; Wang, Z.; Wu, X.; Lai, C.; Chen, X. Evaluation of TMPA 3B42-V7 Product on Extreme Precipitation Estimates. *Remote Sens.* **2021**, *13*, 209. [CrossRef]

9.   Najibi, N.; Devineni, N. Recent Trends in the Frequency and Duration of Global Floods. *Earth Syst. Dyn.* **2018**, *9*, 757–783. [CrossRef]

10. Zhou, Z.; Smith, J.A.; Yang, L.; Baeck, M.L.; Chaney, M.; Ten Veldhuis, M.-C.; Deng, H.; Liu, S. The Complexities of Urban Flood Response: Flood Frequency Analyses for the Charlotte Metropolitan Region: The complexities of urban flood response. *Water Resour. Res.* **2017**, *53*, 7401–7425. [CrossRef]

11. Zhu, Z.; Wright, D.B.; Yu, G. The Impact of Rainfall Space-Time Structure in Flood Frequency Analysis. *Water Resour. Res.* **2018**, *54*, 8983–8998. [CrossRef]

12. Yu, G.; Wright, D.B.; Holman, K.D. Connecting Hydrometeorological Processes to Low-Probability Floods in the Mountainous Colorado Front Range. *Water Resour. Res.* **2021**, *57*, e2021WR029768. [CrossRef]

13. Sun, Y.; Wendi, D.; Kim, D.E.; Liong, S.-Y. Deriving Intensity–Duration–Frequency (IDF) Curves Using Downscaled in Situ Rainfall Assimilated with Remote Sensing Data. *Geosci. Lett.* **2019**, *6*, 17. [CrossRef]

14. Dai, Q.; Bray, M.; Zhuo, L.; Islam, T.; Han, D. A Scheme for Rain Gauge Network Design Based on Remotely Sensed Rainfall Measurements. *J. Hydrometeorol.* **2017**, *18*, 363–379. [CrossRef]

15. Yu, G.; Wright, D.B.; Zhu, Z.; Smith, C.; Holman, K.D. Process-Based Flood Frequency Analysis in an Agricultural Watershed Exhibiting Nonstationary Flood Seasonality. *Hydrol. Earth Syst. Sci.* **2019**, *23*, 2225–2243. [CrossRef]

16. Li, W.; Lin, K.; Zhao, T.; Lan, T.; Chen, X.; Du, H.; Chen, H. Risk Assessment and Sensitivity Analysis of Flash Floods in Ungauged Basins Using Coupled Hydrologic and Hydrodynamic Models. *J. Hydrol.* **2019**, *572*, 108–120. [CrossRef]

17. Kastridis, A.; Kirkenidis, C.; Sapountzis, M. An Integrated Approach of Flash Flood Analysis in Ungauged Mediterranean Watersheds Using Post-Flood Surveys and Unmanned Aerial Vehicles. *Hydrol. Process.* **2020**, *34*, 4920–4939. [CrossRef]

18. Papaioannou, G.; Vasiliades, L.; Loukas, A.; Aronica, G.T. Probabilistic Flood Inundation Mapping at Ungauged Streams Due to Roughness Coefficient Uncertainty in Hydraulic Modelling. *Adv. Geosci.* **2017**, *44*, 23–34. [CrossRef]

19. Ghorbani, M.A.; Ruskeepää, H.; Singh, V.P.; Sivakumar, B. Flood Frequency Analysis Using Mathematica. *Turk. J. Eng. Environ. Sci.* **2010**, *34*, 171–188. [CrossRef]

20. Salman, A.M.; Li, Y. Flood Risk Assessment, Future Trend Modeling, and Risk Communication: A Review of Ongoing Research. *Nat. Hazards Rev.* **2018**, *19*, 4018011. [CrossRef]

21. François, B.; Schlef, K.E.; Wi, S.; Brown, C.M. Design Considerations for Riverine Floods in a Changing Climate—A Review. *J. Hydrol.* **2019**, *574*, 557–573. [CrossRef]

22. Herman, M.R.; Nejadhashemi, A.P.; Abouali, M.; Hernandez-Suarez, J.S.; Daneshvar, F.; Zhang, Z.; Anderson, M.C.; Sadeghi, A.M.; Hain, C.R.; Sharifi, A. Evaluating the Role of Evapotranspiration Remote Sensing Data in Improving Hydrological Modeling Predictability. *J. Hydrol.* **2018**, *556*, 39–49. [CrossRef]

23. Nguyen, H.D.; Fox, D.; Dang, D.K.; Pham, L.T.; Viet Du, Q.V.; Nguyen, T.H.T.; Dang, T.N.; Tran, V.T.; Vu, P.L.; Nguyen, Q.-H.; et al. Predicting Future Urban Flood Risk Using Land Change and Hydraulic Modeling in a River Watershed in the Central Province of Vietnam. *Remote Sens.* **2021**, *13*, 262. [CrossRef]

24. Mohanty, M.P.; Sherly, M.A.; Karmakar, S.; Ghosh, S. Regionalized Design Rainfall Estimation: An Appraisal of Inundation Mapping for Flood Management Under Data-Scarce Situations. *Water Resour. Manag.* **2018**, *32*, 4725–4746. [CrossRef]

25. Razavi, S.; Tolson, B.A. An Efficient Framework for Hydrologic Model Calibration on Long Data Periods. *Water Resour. Res.* **2013**, *49*, 8418–8431. [CrossRef]

26. Peleg, N.; Fatichi, S.; Paschalis, A.; Molnar, P.; Burlando, P. An Advanced Stochastic Weather Generator for Simulating 2-D High-Resolution Climate Variables: AWE-GEN-2d. *J. Adv. Model. Earth Syst.* **2017**, *9*, 1595–1627. [CrossRef]

27. Wright, D.B.; Mantilla, R.; Peters-Lidard, C.D. A Remote Sensing-Based Tool for Assessing Rainfall-Driven Hazards. *Environ. Model. Softw.* **2017**, *90*, 34–54. [CrossRef] [PubMed]

28. Diaz-Nieto, J.; Wilby, R.L. A comparison of statistical downscaling and climate change factor methods: Impacts on low flows in the river Thames, United Kingdom. *Clim. Change* **2005**, *69*, 245–268. [CrossRef]

29. Cha, S.M.; Lee, S.W. Advanced Hydrological Streamflow Simulation in a Watershed Using Adjusted Radar-Rainfall Estimates as Meteorological Input Data. *J. Environ. Manag.* **2021**, *277*, 111393. [CrossRef]

30. Alahacoon, N.; Matheswaran, K.; Pani, P.; Amarnath, G. A Decadal Historical Satellite Data and Rainfall Trend Analysis (2001–2016) for Flood Hazard Mapping in Sri Lanka. *Remote Sens.* **2018**, *10*, 448. [CrossRef]

31. Notti, D.; Giordan, D.; Caló, F.; Pepe, A.; Zucca, F.; Galve, J.P. Potential and Limitations of Open Satellite Data for Flood Mapping. *Remote Sens.* **2018**, *10*, 1673. [CrossRef]

32. Teng, J.; Jakeman, A.J.; Vaze, J.; Croke, B.F.W.; Dutta, D.; Kim, S. Flood Inundation Modelling: A Review of Methods, Recent Advances and Uncertainty Analysis. *Environ. Model. Softw.* **2017**, *90*, 201–216. [CrossRef]

33. Ding, L.; Ma, L.; Li, L.; Liu, C.; Li, N.; Yang, Z.; Yao, Y.; Lu, H. A Survey of Remote Sensing and Geographic Information System Applications for Flash Floods. *Remote Sens.* **2021**, *13*, 1818. [CrossRef]

34. Getirana, A.; Kirschbaum, D.; Mandarino, F.; Ottoni, M.; Khan, S.; Arsenault, K. Potential of GPM IMERG Precipitation Estimates to Monitor Natural Disaster Triggers in Urban Areas: The Case of Rio de Janeiro, Brazil. *Remote Sens.* **2020**, *12*, 4095. [CrossRef]

35. Li, Y.; Grimaldi, S.; Walker, J.; Pauwels, V. Application of Remote Sensing Data to Constrain Operational Rainfall-Driven Flood Forecasting: A Review. *Remote Sens.* **2016**, *8*, 456. [CrossRef]

36. Shakti, P.C.; Kamimera, H.; Misumi, R. Inundation Analysis of the Oda River Basin in Japan during the Flood Event of 6–7 July 2018 Utilizing Local and Global Hydrographic Data. *Water* **2020**, *12*, 1005. [CrossRef]

37. Komi, K.; Neal, J.; Trigg, M.A.; Diekkrüger, B. Modelling of Flood Hazard Extent in Data Sparse Areas: A Case Study of the Oti River Basin, West Africa. *J. Hydrol. Reg. Stud.* **2017**, *10*, 122–132. [CrossRef]
38. Wright, D.B. Estimating the Frequency of Extreme Rainfall Using Weather Radar and Stochastic Storm Transposition. *J. Hydrol.* **2013**, *488*, 150–165. [CrossRef]
39. D'Oria, M.; Maranzoni, A.; Mazzoleni, M. Probabilistic Assessment of Flood Hazard Due to Levee Breaches Using Fragility Functions. *Water Resour. Res.* **2019**, *55*, 8740–8764. [CrossRef]
40. Zhu, Z.; Chen, Z.; Chen, X.; He, P. Approach for Evaluating Inundation Risks in Urban Drainage Systems. *Sci. Total Environ.* **2016**, *553*, 1–12. [CrossRef]
41. Yang, W.; Xu, K.; Lian, J.; Bin, L.; Ma, C. Multiple Flood Vulnerability Assessment Approach Based on Fuzzy Comprehensive Evaluation Method and Coordinated Development Degree Model. *J. Environ. Manag.* **2018**, *213*, 440–450. [CrossRef]
42. Nandi, A.; Mandal, A.; Wilson, M.; Smith, D. Flood Hazard Mapping in Jamaica Using Principal Component Analysis and Logistic Regression. *Environ. Earth Sci.* **2016**, *75*, 465. [CrossRef]
43. Sarmah, T.; Das, S.; Narendr, A.; Aithal, B.H. Assessing Human Vulnerability to Urban Flood Hazard Using the Analytic Hierarchy Process and Geographic Information System. *Int. J. Disaster Risk Reduct.* **2020**, *50*, 101659. [CrossRef]
44. Liang, L.; Deng, X.; Wang, P.; Wang, Z.; Wang, L. Assessment of the Impact of Climate Change on Cities Livability in China. *Sci. Total Environ.* **2020**, *726*, 138339. [CrossRef]
45. Wu, Z.; Lv, H.; Meng, Y.; Guan, X.; Zang, Y. The Determination of Flood Damage Curve in Areas Lacking Disaster Data Based on the Optimization Principle of Variation Coefficient and Beta Distribution. *Sci. Total Environ.* **2021**, *750*, 142277. [CrossRef]
46. The Ministry of Water Resources of the People's Republic of China Bulletin on China Flood and Drought Disasters [EB/OL]. 2016. Available online: http://Www.Mwr.Gov.Cn (accessed on 7 August 2017).
47. Yu, G.; Wright, D.B.; Li, Z. The Upper Tail of Precipitation in Convection-Permitting Regional Climate Models and Their Utility in Nonstationary Rainfall and Flood Frequency Analysis. *Earths Future* **2020**, *8*, e2020EF001613. [CrossRef]
48. Franchini, M.; Helmlinger, K.R.; Foufoula-Georgiou, E.; Todini, E. Stochastic Storm Transposition Coupled with Rainfall—Runoff Modeling for Estimation of Exceedance Probabilities of Design Floods. *J. Hydrol.* **1996**, *175*, 511–532. [CrossRef]
49. Wilson, L.L.; Foufoula-Georgiou, E. Regional Rainfall Frequency Analysis via Stochastic Storm Transposition. *J. Hydraul. Eng.* **1990**, *116*, 859–880. [CrossRef]
50. Chen, W.; Huang, G.; Zhang, H.; Wang, W. Urban Inundation Response to Rainstorm Patterns with a Coupled Hydrodynamic Model: A Case Study in Haidian Island, China. *J. Hydrol.* **2018**, 1022–1035. [CrossRef]
51. Deng, J.; Yin, H.; Kong, F.; Chen, J.; Dronova, I.; Pu, Y. Determination of Runoff Response to Variation in Overland Flow Area by Flow Routes Using UAV Imagery. *J. Environ. Manag.* **2020**, *265*, 109868. [CrossRef] [PubMed]
52. Shojaeizadeh, A.; Geza, M.; Hogue, T.S. GIP-SWMM: A New Green Infrastructure Placement Tool Coupled with SWMM. *J. Environ. Manag.* **2021**, *277*, 111409. [CrossRef] [PubMed]
53. Gironás, J.; Roesner, L.A.; Rossman, L.A.; Davis, J. A New Applications Manual for the Storm Water Management Model (SWMM). *Environ. Model. Softw.* **2010**, *25*, 813–814. [CrossRef]
54. Zhu, Z.; Chen, Z.; Chen, X.; Yu, G. An Assessment of the Hydrologic Effectiveness of Low Impact Development (LID) Practices for Managing Runoff with Different Objectives. *J. Environ. Manag.* **2019**, *231*, 504–514. [CrossRef]
55. Zhi, G.; Liao, Z.; Tian, W.; Wu, J. Urban Flood Risk Assessment and Analysis with a 3D Visualization Method Coupling the PP-PSO Algorithm and Building Data. *J. Environ. Manag.* **2020**, *268*, 110521. [CrossRef] [PubMed]
56. Guo, Q.; Wang, J.; Yin, H.; Zhang, G. A Comprehensive Evaluation Model of Regional Atmospheric Environment Carrying Capacity_ Model Development and a Case Study in China. *Ecol. Indic.* **2018**, 259–267. [CrossRef]
57. Kruskal, J.; Shepard, R. A Nonmetric Variety of Linear Factor Analysis. *Psychometrika* **1974**, *39*, 123–157. [CrossRef]
58. Kennedy, J.; Eberhart, R. Particle swarm optimization. In Proceedings of the IEEE International Conference on Neural Networks—Conference Proceedings, Perth, WA, Australia, 27 November–1 December 1995; pp. 1942–1948.
59. Shen, Y.; Zhao, P.; Pan, Y.; Yu, J. A High Spatiotemporal Gauge-Satellite Merged Precipitation Analysis over China. *J. Geophys. Res. Atmos.* **2014**, *119*, 3063–3075. [CrossRef]
60. Wang, D.; Wang, X.; Liu, L.; Wang, D.; Huang, H.; Pan, C. Evaluation of TMPA 3B42V7, GPM IMERG and CMPA Precipitation Estimates in Guangdong Province, China. *Int. J. Climatol.* **2019**, *39*, 738–755. [CrossRef]
61. Yalcin, E. Two-Dimensional Hydrodynamic Modelling for Urban Flood Risk Assessment Using Unmanned Aerial Vehicle Imagery: A Case Study of Kirsehir, Turkey. *J. Flood Risk Manag.* **2019**, *12*, e12499. [CrossRef]
62. Anni, A.H.; Cohen, S.; Praskievicz, S. Sensitivity of Urban Flood Simulations to Stormwater Infrastructure and Soil Infiltration. *J. Hydrol.* **2020**, *588*, 125028. [CrossRef]
63. Ciach, G.J.; Morrissey, M.L.; Krajewski, W.F. Conditional Bias in Radar Rainfall Estimation. *J. Appl. Meteorol. Climatol.* **2000**, *39*, 1941–1946. [CrossRef]
64. Ringard, J.; Becker, M.; Seyler, F.; Linguet, L. Temporal and Spatial Assessment of Four Satellite Rainfall Estimates over French Guiana and North Brazil. *Remote Sens.* **2015**, *7*, 16441–16459. [CrossRef]
65. Tian, Y.; Peters-Lidard, C.D. Systematic Anomalies over Inland Water Bodies in Satellite-Based Precipitation Estimates. *Geophys. Res. Lett.* **2007**, *34*, L14403. [CrossRef]
66. Wright, D.B.; Smith, J.A.; Baeck, M.L. Flood Frequency Analysis Using Radar Rainfall Fields and Stochastic Storm Transposition. *Water Resour. Res.* **2014**, *50*, 1592–1615. [CrossRef]

67. Kim, D.; Onof, C. A Stochastic Rainfall Model That Can Reproduce Important Rainfall Properties across the Timescales from Several Minutes to a Decade. *J. Hydrol.* **2020**, *589*, 125150. [CrossRef]
68. Habib, E.; Henschke, A.; Adler, R.F. Evaluation of TMPA Satellite-Based Research and Real-Time Rainfall Estimates during Six Tropical-Related Heavy Rainfall Events over Louisiana, USA. *Atmos. Res.* **2009**, *94*, 373–388. [CrossRef]
69. Bruni, G.; Reinoso, R.; van de Giesen, N.C.; Clemens, F.H.L.R.; ten Veldhuis, J.A.E. On the Sensitivity of Urban Hydrodynamic Modelling to Rainfall Spatial and Temporal Resolution. *Hydrol. Earth Syst. Sci.* **2015**, *19*, 691–709. [CrossRef]
70. Zhou, Z.; Smith, J.A.; Wright, D.B.; Baeck, M.L.; Yang, L.; Liu, S. Storm Catalog-Based Analysis of Rainfall Heterogeneity and Frequency in a Complex Terrain. *Water Resour. Res.* **2019**, *55*, 1871–1889. [CrossRef]
71. Li, J.; Evans, J.; Johnson, F.; Sharma, A. A Comparison of Methods for Estimating Climate Change Impact on Design Rainfall Using a High-Resolution RCM. *J. Hydrol.* **2017**, *547*, 413–427. [CrossRef]
72. Peleg, N.; Blumensaat, F.; Molnar, P.; Fatichi, S.; Burlando, P. Partitioning the Impacts of Spatial and Climatological Rainfall Variability in Urban Drainage Modeling. *Hydrol. Earth Syst. Sci.* **2017**, *21*, 1559–1572. [CrossRef]
73. Tian, Y.; Peters-Lidard, C.D.; Eylander, J.B.; Joyce, R.J.; Huffman, G.J.; Adler, R.F.; Hsu, K.; Turk, F.J.; Garcia, M.; Zeng, J. Component Analysis of Errors in Satellite-Based Precipitation Estimates. *J. Geophys. Res. Atmos.* **2009**, *114*, D24101. [CrossRef]

Article

Effect of the Assimilation Frequency of Radar Reflectivity on Rain Storm Prediction by Using WRF-3DVAR

Yuchen Liu [1], Jia Liu [1,*], Chuanzhe Li [1], Fuliang Yu [1] and Wei Wang [1,2]

[1] State Key Laboratory of Simulation and Regulation of Water Cycle in River Basin, China Institute of Water Resources and Hydropower Research, Beijing 100038, China; melodieyu@163.com (Y.L.); lichuanzhe@iwhr.com (C.L.); yufl@iwhr.com (F.Y.); wangwei_hydro@163.com (W.W.)

[2] College of Hydrology and Water Resources, Hohai University, Nanjing 210098, China

* Correspondence: jia.liu@iwhr.com; Tel.: +86-150-1044-3860

Abstract: An attempt was made to evaluate the impact of assimilating Doppler Weather Radar (DWR) reflectivity together with Global Telecommunication System (GTS) data in the three-dimensional variational data assimilation (3DVAR) system of the Weather Research Forecast (WRF) model on rain storm prediction in Daqinghe basin of northern China. The aim of this study was to explore the potential effects of data assimilation frequency and to evaluate the outputs from different domain resolutions in improving the meso-scale NWP rainfall products. In this study, four numerical experiments (no assimilation, 1 and 6 h assimilation time interval with DWR and GTS at 1 km horizontal resolution, 6 h assimilation time interval with radar reflectivity, and GTS data at 3 km horizontal resolution) are carried out to evaluate the impact of data assimilation on prediction of convective rain storms. The results show that the assimilation of radar reflectivity and GTS data collectively enhanced the performance of the WRF-3DVAR system over the Beijing-Tianjin-Hebei region of northern China. It is indicated by the experimental results that the rapid update assimilation has a positive impact on the prediction of the location, tendency, and development of rain storms associated with the study area. In order to explore the influence of data assimilation in the outer domain on the output of the inner domain, the rainfall outputs of 3 and 1 km resolution are compared. The results show that the data assimilation in the outer domain has a positive effect on the output of the inner domain. Since the 3DVAR system is able to analyze certain small-scale and convective-scale features through the incorporation of radar observations, hourly assimilation time interval does not always significantly improve precipitation forecasts because of the inaccurate radar reflectivity observations. Therefore, before data assimilation, the validity of assimilation data should be judged as far as possible in advance, which can not only improve the prediction accuracy, but also improve the assimilation efficiency.

Keywords: assimilation frequency; data assimilation; WRF-3DAVR; radar reflectivity; rainfall forecast

Citation: Liu, Y.; Liu, J.; Li, C.; Yu, F.; Wang, W. Effect of the Assimilation Frequency of Radar Reflectivity on Rain Storm Prediction by Using WRF-3DVAR. *Remote Sens.* **2021**, *13*, 2103. https://doi.org/10.3390/rs13112103

Academic Editor: Yongqiang Zhang

Received: 23 April 2021
Accepted: 24 May 2021
Published: 27 May 2021

Publisher's Note: MDPI stays neutral with regard to jurisdictional claims in published maps and institutional affiliations.

1. Introduction

There have been higher requirements put forward for the prediction of convective systems precipitation and its related disasters in recent years. Improving the accuracy of precipitation forecast has long been a challenge for Numerical Weather Prediction (NWP) model researchers and operational communities [1,2]. Kryza and Werner et al. [3] forecasted several short and intensive rainfalls over the SW area of Poland using the Weather Research and Forecasting Model (WRF) with different parameterization and spatial resolution. The results show that none of the experimented model configurations was able to reproduce a local intensive rainfall properly. Hamill and Thomas [4] applied ensemble prediction systems to describe the performance of the WRF precipitation forecasts. Although ensemble forecast can reflect the predictability or reliability of real atmosphere to some extent, it cannot improve the physical mechanism of the model. Among many considerable causes that could lead to the inherent low predictability of convective precipitation forecasting using an NWP system, ambiguous description of the atmosphere initial state is one of them. Rainfall features are always generated in an inaccurate manner, regarding location, initiation, timing, and intensity, especially of convective storms [5]. As a result, accurately obtaining the initial state of a storm for the regional model is the key issue for a successful prediction of the convective system.

Several studies have confirmed it is possible to remedy defects through data assimilation into the NWP model. Among the numerous data assimilation methods, those commonly used by researchers include the optimal interpolation methods, the three-dimensional variational (3DVAR) [6,7], the four-dimensional variational (4DVAR) [8,9], and the Kalman filter [10] approaches. Thereinto, 3DVAR is practicable in terms of computational efficiency, thus is adopted the most frequently [5]. Variational data analysis system was first developed by Sun and Juanzhen et al. [11], which is called Variational Doppler Radar Analysis System (VDRAS) and began this pioneering work. Subsequently, it was expanded by Sun and Crook [12] to use for short-term forecast initialization during convective precipitation. Yang and Duan et al. [13] discussed if 3DVar data assimilation could potentially improve the rainfall forecast in the WRF model, an obvious improvement of the event with even rainfall in temporal distribution was found regarding the northeastern Tibetan Plateau area. Physical initialization combined with the three-dimensional variational data assimilation method (PI3DVAR_rh) was designed by Gan and Yang et al. [14], through which the spatial pattern forecasting of radar reflectivity and precipitation were improved based on WRF. Vedrasco and Sun et al. [15] certified that the 3DVAR analysis with the constraint introduced into WRF could improve the initial state of the model and guide the convective characteristics exhibited by six summer convection cases across Brazil. These studies show that 3DAVR plays a positive role in rainfall forecast, and this system, which is widely used in data assimilation, is also adopted in this study.

With high-resolution observation data becoming increasingly rich, it has been a hot research topic worldwide in recent years to improve the short-term quantitative precipitation prediction level by using high-resolution observation. High-resolution observations like Doppler radar and GTS data, can provide huge amounts of detailed information with high spatial and temporal resolution, which makes it possible to further improve the convective rainstorm forecast. Doppler radar can make millions of measurements of precipitation with a spatial resolution of a few kilometers and a temporal resolution of a few minutes [5]. Compared with existing meso-scale measure platforms, such a significant advantage of spatial and temporal resolution has great potential for improving small-scale and short-term rainfall prediction.

Many researchers combine high-resolution observational data with numerical models, which plays a positive role in promoting the development of convective rainfall prediction, improving the initial state of numerical models, as well as alleviating the imbalance caused by interpolation [16–19]. Routray and Mohanty [20] believed that the assimilation of radar data (radial velocity and reflectivity factor) has a positive impact on enhancing performance of the WRF-3DVAR (WRF-three-dimensional variational) system for the Indian region. Gvindankutty and Chandrasekar et al. [21] investigated how the 3DVAR assimilation of DWR radial wind and GTS data positively affected the precipitation intensity as well as spatial distribution. Abhilash and Sahai et al. [22] demonstrated improvement in spatial pattern of rainfall of convective systems precipitation by assimilating relevant parameter obtained by the WRF-3DVAR system, including DWR radial velocity and reflectivity as well as GTS data. Osuri and Mohanty et al. [23] used the WRF-3DVAR system to assimilate DWR with GTS, thus conducting 24 cases to predict tropical cyclones of Bay of Bengal, and the results show that this method helps to provide a positive impact on the credibility of prediction. Sugimoto and Crook et al. [5] indicated that in the 3DVAR framework and the storm case, assimilating the radial velocity as well as reflectivity can achieve the best performance, applied on short-range precipitation forecasting. A cycling data assimilation improves the regional models' initial state on the one hand, and meanwhile introduces observation data to mitigate the imbalance caused by interpolation of prediction on the other hand [24]. The data assimilation system allows higher-resolution observations to be used as the background through update-cycling procedure [25,26], which makes the data assimilation system rely on a specified combination of error statistics to obtain the optimal short-term forecast analysis.

Although data assimilation is capable of improving the NWP (i.e., WRF) convective precipitation in an effectively manner, the quality of data assimilation creates uncertainty about results. Some studies have shown that the quality of assimilated data is critical to forecast results. Liu and Tian et al. [27] indicated that data assimilation via WRF-3DVar could potentially improve the rainfall forecasting in northern China, with GTS data, radar reflectivity, and radial velocity assimilated every 6 h. From their study, it is concluded that it is the effective information in the assimilated data that exhibited more significant results rather than the volume of data. Tian and Liu et al. [28] further explored the effect of assimilating radar data from different height layers on the improvement of the NWP rainfall accuracy. The results showed that the accuracy of the forecasted rainfall deteriorated with the rise of the height of the assimilated radar reflectivity. In the process of data assimilation, the frequency of assimilation determines the amount of effective data assimilated. Considering the time cost, most operation departments choose a 6 h assimilation interval [3,27,29]. In later developments, many researchers and principal operational centers upgraded the regional NWP systems with a 3 h assimilation time interval [30]. For highly convective storms, more frequent data assimilation with a shorter time interval is found to be more effective to produce reliable predictions [31]. Li and Wang et al. [32] demonstrated that the assimilation of radial velocity every half an hour could enhance the intensity analyses and forecasts of rainfall compared to results without assimilating radar data. Kawabata and Seko et al. [33] applied four-dimensional variational (4DVAR) with a horizontal resolution of 2 km and 1 h length of the assimilation window to forecast heavy rainfall at the central part of Tokyo. In most cases, researchers tend to assimilate observations as they are originally obtained, rather than choose an appropriate assimilation frequency or time interval.

Despite the wide application of data assimilation in enhancing precipitation forecasts, the sensitivity of data assimilation frequency has not yet gained enough attention. This study mainly aims at evaluating how the WRF-3DVAR with different assimilation frequencies affects the accuracy of the forecast precipitation. Four typical rain storms that occurred in semi-humid and semi-arid area of northern China were chosen as study objects. Doppler radar and GTS data were assimilated in four designed experiments with the time intervals of 6 and 1 h by WRF-3DVAR. The results would be helpful to improving data assimilation efficiency with WRF-3DVAR, and provide guidance for the development of a similar basin rainstorm forecast system. At the same time, in order to study the sensitivity of data assimilation to rainfall forecast, the quality of radar data is also analyzed.

2. Methodology

To explore the data assimilation frequency of the short-range precipitation forecasting, the numerical model WRF is chosen for this study [7] and its assimilating extension WRF-3DAVR is used. WRF version 3.7 is used for all experiments. The WRF model is a fully compressible, non-hydrostatic, mesoscale NWP, and atmospheric simulation system. In addition, WRF-3DAVR can further influence the initial state of the WRF model by assimilating different high-resolution observations. A brief overview of the basic model settings and the system description can be found in the following sections.

2.1. A Brief Description of WRF-3DVAR

The fundamental objective of the 3DVAR system is to seek an optimal estimate of initialization at parsing time through an iterative solution of a pre-determined function, including observations, background forecast from the NWP system, etc. [34].

$$J(x) = J^b + J^o = \frac{1}{2}\left(x - x^b\right)^T B^{-1}\left(x - x^b\right) + \frac{1}{2}(y - y^o)^T (E + F)^{-1}(y - y^o) \tag{1}$$

Iterative solutions to Equation (1) can summarize the Dimensional Variational Assimilation problem, to seek the analysis state variable x to minimize $J(x)$. In the formula, x is the variable that represents the surface and atmospheric surface state, x^b is the first guess (or background) acquired from the previous forecast, and y^0 is the assimilated observation. $Y = H(x)$ is the model-derived observation space, which is transformed from gridded analysis x by the observation operator H for comparison against y^0. The individual data points are fitted by the weight of its error estimate: B, E, and F are the background error covariance matrix, observation error covariance matrix, and representativity error covariance matrices, respectively. This solution represents a minimum variance estimate of the true state of the atmosphere given the two sources of a priori data: the first guess x^b and the observation y^0 [35].

The WRF-3DVar is a variational data assimilation system designed and built in the WRF model, and is used to assimilate radar reflectivity and GTS in this study [6,36]. The background error covariance CV3 created by the National Meteorological Center (NMC) was employed, which has the advantage of wide applicability [37].

2.2. WRF Model Configuration

The primary focus of this study is Fuping and Zijingguan catchments covered by the innermost domain of a three nested domain. For the study area, the center of the domain is at 39°26′00″N and 114°46′00″E, and from the outermost to the innermost the nested domain sizes are 1260 × 1260, 450 × 360, and 145 × 115 km². Considering the diversity in assimilation variables, the horizontal grid spacing of the outermost domain is set at 9 km and the grid size of the innermost layer is set to be 1 km, where the downscaling radio is set to be 1:3 [38,39]. The locations of the nested domain and radar coverage are shown in Figure 1. Forty vertical pressure levels are considered for the three nested domains with a model top at 50 hPa [40,41].

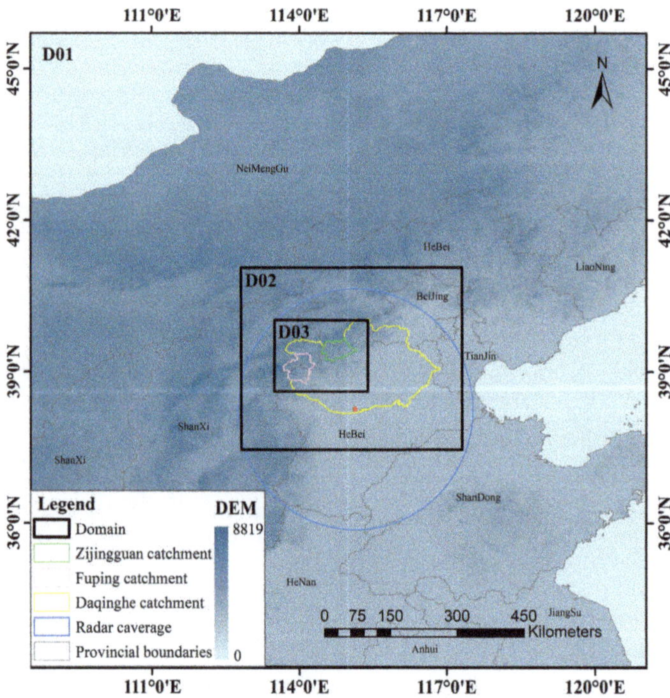

Figure 1. The study area with nested configuration of WRF domains at 9, 3, and 1 km resolution.

The $1° \times 1°$ spatial resolution of Global Forecast System (GFS) forecast data provide the initial and lateral boundary conditions for the simulations. Considering the resolution of the GFS data, it would be more reasonable to add another domain with 27 km horizontal grid spacing as the outermost one. Unfortunately, experimental runs are carried out and the effects of data assimilation are found to be similar with the current domain settings. Since rainstorm forecasting has a high requirement of effectiveness for a given period of time, in practical application it is beneficial to obtain the forecasting information as soon as possible. Therefore, the 27 km grid is not applied in this study. The integration time-step of the WRF model and WRF-3DAVR system is 6 s and the time interval for the output is set to be 1 h. The GFS data, which is updated every six hours, is a real-time product provided by National Centers for Environmental Prediction (NCEP). Because GFS has not been processed by assimilation analysis, it has been used in many studies to forecast the storm events in WRF model and WRF-3DAVR [3,5,31].

The performance of the model is highly dependent on the parameterized scheme, and the scheme selected in the study may be suitable for one storm event in one region, but not necessarily for others [42]. Since it is difficult to judge which scheme is most suitable for future storm events, parameterized schemes are usually determined in advance for practical application [43]. Based on relevant experimental research on the selection of sensitive parameterization schemes for ensemble rainfall forecasting [44,45], the physical parameters with the best applicability in the study area were adopted in this study. Details of the parameterizations that have significant impacts on the generation of rainfall used in the assimilation experiments are shown in Table 1. It is worth noting that the cumulus scheme is switched off for the 3 and 1 km domain.

Table 1. Details of the parameterizations used in the assimilation experiments.

Parameterization	Chosen Option	Reference
Microphysics scheme	WSM6	[46]
Longwave radiation	Rapid Radiative Transfer Model (RRTM)	[47]
Shortwave radiation	Dudhia	[48]
Land surface scheme	Noah	[49]
Planetary boundary layer	Mellor-Yama-da-Janjic (MYJ)	[50]
Cumulus convection	Kain-Fritsch (KF)	[51]

3. Case Study and Data

3.1. Study Area and Storm Events

The main focus of this study is the Fuping and Zijingguan watershed covered by Domain 2 and Domain 3 of the WRF model. According to the temporal and spatial distribution characteristics of rainfall and the representatives of rainfall-runoff generation characteristics, the case used in this study is four 24-h rainstorm events, which occurred in Fuping and Zijingguan watershed. Fuping catchment is located at 39°22'N~38°47'N and 113°40'E~114°18'E with a drainage area of 2210 km^2 and Zijingguan catchment is located at 39°13'~39°40'N and 114°28'~115°11'E with a drainage area of 1760 km^2. They belong to the south and the north branch of the Daqinghe basin respectively, located in northern China, having a warm temperate continental monsoon climate. Terrain elevation of the study area varies from 2286 m in the northwest part to 200 m in the southeast mountains. The average annual rainfall in the study catchments is approximately 600 mm, the short episodes of intensive precipitation are more frequent in late May and early September. The steep terrain leads to a short confluence time of the flood, which together with high-intensity, short-duration precipitation is prone to cause severe flood disasters. They are concentrated in areas with thin soil layer and low vegetation coverage especially. In particular, on 21 July 2012, the 24 h rainstorm event occurred in the Beijing-Tianjin-Hebei region, which has been widely concerned because of the heavy rainfall intensity and large losses. The duration of the four storm events and the accumulative rainfall amounts are shown in Table 2. Among the four storms, event 4 is the heaviest one trigged by a severe convective system, concentrated in a small area of the watershed and with high rainfall intensity in a short time period. The other three storms are typical stratiform rains with moderate intensity but various distributions in space and time. The measured rainfall data are rainfall stations and the time interval is 1 h. The 24 h rainfall accumulation is computed by Thiessen polygon method, which averages the observations from the rain gauges. Figure 2 demonstrates a detailed map of the study area, such as the rain gauges, watershed boundary, catchment outlets topography, and major rivers.

Table 2. Duration, accumulated rainfall, and maximum stream flow of the four selected 24 h storm events.

Event ID	Catchment	Storm Start Time	Storm End Time	Accumulated Rainfall (mm)
I	Fuping	29/07/2007 20:00	30/07/2007 20:00	63.38
II	Fuping	30/07/2012 10:00	31/07/2012 10:00	50.48
III	Fuping	11/08/2013 07:00	12/08/2013 07:00	30.82
IV	Zijingguan	21/07/2012 04:00	22/07/2012 04:00	155.43

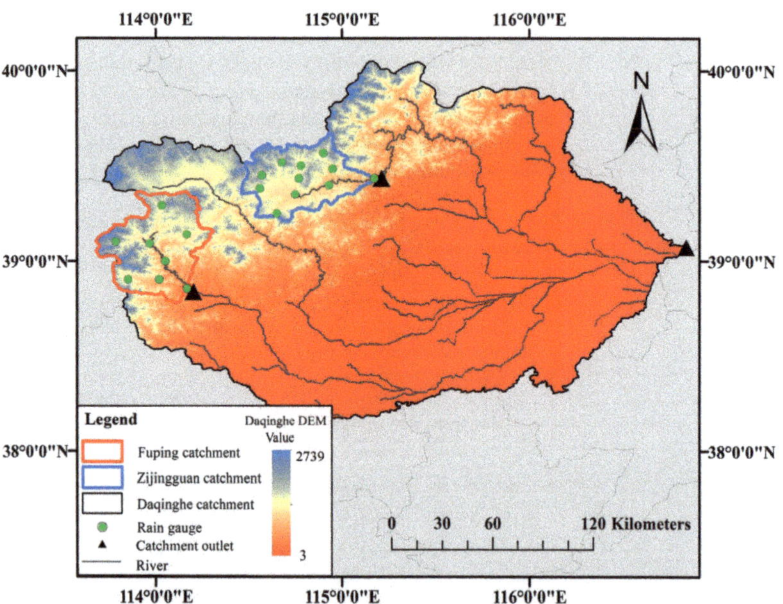

Figure 2. Location map and two study sites in the Daqinghe catchment.

In reality, the precipitation in northern China is not absolutely even in time and space, which is different from that in southern China. To study the spatial and temporal evenness of the precipitation in study catchments, both spatial and temporal variation coefficient (Cv) of four storm events from 1985 to 2015 are calculated. A threshold of 5% was employed to separate even and uneven rainfall events. The evenness of the rainstorm events is quantitatively evaluated in the spatial and temporal dimensions by variation coefficient Cv in this study. The lower Cv is, the more even the rainfall is. Based on the thresholds selected, we found two critical values: 0.4 for spatial Cv and 0.6 for temporal Cv. When the rainfall Cv values were less than the threshold Cv values, storm events have a relatively even distribution of rainfall over the respective catchment. Table 3 shows the Cv values of the four storm events in the both spatial and temporal dimensions. It can be found that the rainfall is more uneven in time than in space, which is helpful to analyze the spatiotemporal distribution of rainfall forecast results. It can be seen from Table 3 that the four rainstorms have different spatial and temporal evenness, and the selection of rainfall events can provide reference for different types of rainfall in northern China.

$$Cv = \sqrt{\frac{\sum_{j=1}^{N}\left(\frac{x_j}{\bar{x}} - 1\right)^2}{N}} \tag{2}$$

In the spatial dimension, x_j is the accumulated 24 h rainfall at the rain gauge j, and $\bar{x}$ is the average of x_j. N is the number of rain gauges. In the temporal dimension, x_j is average hourly rainfall from all the rain gauges at time j, and N is the number of hours. The higher Cv is, the more uneven the rainfall distribution is.

Table 3. Rainfall evenness of the four selected 24-h storm events in space and time.

Event ID	I	II	III	IV
Spatial Cv	0.3975	0.1927	0.7400	0.6098
Temporal Cv	0.6011	1.0823	2.3925	1.8865

3.2. Data Assimilation Experiments

In this study the radar reflectivity and GTS data observations forecasting are assimilated into WRF and WRF-3DVAR system for 24 h typical storm events.

3.2.1. Weather Radar Data

The Doppler radar used in this study is the S-band radar located at Shijiazhuang city, which is provided by the National Meteorological Administration and covers a radius of 250 km and can completely cover the two study areas. The radar completed a volume sweep every 6 min, with 9 beam angles. By using a data format conversion program, the binary base data file of Shijiazhuang Doppler radar is directly written into "ob.radar" file as the observation field input file of wrfvar.exe. The format of the ob.radar file can be found in the WRFDA user guide. The ob.radar file includes the latitude and longitude of the pixel center and the height of the radar beam above that pixel. The technique details of the radar data are provided in Table 4.

Table 4. Basic parameters of S-band radar in Shijiazhuang.

Parameters	Information
Location	38.5°, 114.68°
Administrative location	Shijiazhuang
Antenna diameter	1.3 m
Emission frequency	2.7~3.0 GHz
Observation radius	250 km
Effective radius of observation	230 km
Spatial resolution	1km
Sweep time	6 min
Beam angles	0.5°, 1.5°, 2.4°, 3.4°, 4.3°, 6.0°, 9.9°, 14.6°, 19.5°

The radar reflectivity is conducted through the quality control chain to improve the quality of the measurements before importing the assimilation process [52]. Therefore, before being assimilated by WRF-3DVAR, radar data are supported by China Integrated Meteorological Information Service System (CIMISS) of China Meteorological Administration to remove artefacts such as ground clutter, radial interference echo, speckles through quality control. For convenience, dBZ is often used to represent radar reflectivity:

$$dBZ = 10\,lg^Z/Z_0 \tag{3}$$

where, $Z_0 = 1\ \text{mm}^{6/}\text{m}^3$, is a constant, and Z is reflectivity.

According to Sun and Crook [12], the radar data are assimilated through an observational operator that describes the relationship between radar reflectivity and rainwater mixing ratio by

$$Z = 43.1 + 17.5 log \left(\rho q_r \right) \tag{4}$$

where Z is the reflectivity, ρ is the air density in kg/m^3, and q_r is the rainwater mixing ratio. This relation is derived analytically by assuming the Marshall–Palmer distribution of raindrop size. At the same time, the pixel-based radar reflectivity is assimilated directly in WRF-3DVar, by stating the latitude and longitude of the pixel center and the height of the radar beam above that pixel.

3.2.2. GTS Data

The downloaded GTS data is a collection of all kinds of meteorological observation data, including surface weather station, ship, buoy, pilot balloon, sonde, aircraft, and satellite observations, and its format is easily recognized by WRF-3DVAR. The decoded data is converted into a suitable Little_R format by shell script, which is used for data assimilation of WRF-3DVAR. Five GTS datasets, including Sound, Synop, Pilot, AIREP, and Metar, were incorporated into the WRF model at 6 h assimilation time interval in this study, the number of observations were 2718, 4217, 733, 201, 612, respectively. The assimilated GTS data was directly interpolated into the background field of the model, and the background field was corrected by a certain algorithm.

GTS data has the characteristics of wide coverage and small spatial density, which is suitable for assimilation on a large scale. On the other hand, radar data coverage is relatively restricted, but the data spatial density is intense, and therefore is more suitable for assimilation in small scale. Since the scanning radius of the radar is 250 km, thus the coverage range is similar to that of Domain 2, which is much smaller than that of the outer domain and larger than the innermost domain. Therefore, in this study the radar data is assimilated only in Domain 2, whereas the GTS data is assimilated in the outer domain (Domain 1). Since the GTS data is released every six hours, in the hourly assimilation scheme, GTS is only assimilated at the 6th, 12th, 18th, and 24th hour from the start of the storm.

Four experiments were conducted at different horizontal resolutions while keeping all physical settings the same. In scheme "NA_1km", Global Forecast System (GFS) forecast data is interpolated to the model grid as the initial conditions for the 24 h forecast without data assimilation in all the three nested domains. In scheme "DA_1h_1km", 1 h data assimilation is carried out with rainfall forecasts output from the 1 km domain (Domain 3). The settings of Scheme "DA_1h_3km" are the same as DA_1h_1km, except that rainfall forecasts are output from the 3 km domain (Domain 2). In the scheme "DA_6h_3km", the 1 km innermost domain is removed with only Domain 1 and Domain 2 left. Radar and GTS data are assimilated every 6 h and rainfall forecasts are output from Domain 2. Detailed explanations of the experimental design and the settings of the four schemes are shown in Table 5. Before the data assimilation cycle starts, the WRF model spins up for 30 h in all schemes with the initial condition from GFS. For cycling data assimilation, the prediction of the previous assimilation run serves as the background for the next run.

The root mean square error (*RMSE*), the mean bias error (*MBE*), and the critical success index (*CSI*) are used to evaluate the simulated precipitation of the WRF and WRF-3DAVR model. After the analysis of indices, *CSI/RMSE* is used as the comprehensive evaluation index to explore the more intuitive response of different rainfall types to forecast errors in temporal and spatial scales.

Table 5. Experiment description.

Experiments	Domain Resolutions	Data Assimilation	Radar Data Assimilation Time Interval	GTS Data Assimilation Time Interval	Output Resolutions
NA_1km (=no assimilation)	Domain 1 (9 km) Domain 2 (3 km) Domain 3 (1 km)	no	no	no	1 km (Domain 3)
DA_1h_1km (=data assimilation with 1 h interval and output from 1 km domain)	Domain 1 (9 km) Domain 2 (3 km) Domain 3 (1 km)	Radar reflectivity in Domain 2 + GTS in Domain 1	1 h	6 h	1 km (Domain 3)
DA_1h_3km (=data assimilation with 1 h interval and output from 3 km domain)	Domain 1 (9 km) Domain 2 (3 km) Domain 3 (1 km)	Radar reflectivity in Domain 2 + GTS in Domain 1	1 h	6 h	3 km (Domain 2)
DA_6h_3km (=data assimilation with 6 h interval and output from 3 km domain)	Domain 1 (9 km) Domain 2 (3 km)	Radar reflectivity in Domain 2 + GTS in Domain 1	6 h	6 h	3 km (Domain 2)

CSI [53] denotes the percentage of correct simulation between the forecast and observations, and the perfect score is 1. According to Equation (6), the calculation of *CSI* depends on whether it rains or not. Since the essence of WRF model simulation is to solve equations, it is inevitable that the rainfall calculation result is close to 0. In order to avoid the light rain in the prediction being included in H or R, hourly rain rate less than 0.01 mm is considered as no rain, and *CSI* is based on the method in Table 6 to classify the rainfall simulation results.

Table 6. Rain/no rain contingency table for the WRF simulation against observation.

Prediction/Observation	Yes (>0.01 mm)	No
Yes	*hits (H)*	*misreports (R)*
No	*misses (S)*	/

Classified variables H, R and S represent whether the predicted and observed values in a certain observation period or observation position are greater than 0.01. If both predicted and observed values are greater than 0.01, that is, rainfall is captured in model, $H+1$; If the predicted value is greater than 0.01 and the observed value is less than or equal to 0.01, that is, the model misreports rainfall, then $R+1$; If the predicted value is less than or equal to 0.01 and the observed value is greater than 0.01, that is, the rainfall is missed in model, then $S+1$. If both the predicted value and the observed value are equal to 0.01, that is, the model accurately predicts the scenario without rainfall.

$$RMSE = \sqrt{\frac{1}{M}\sum_{i=1}^{M}\left(Q_i' - Q_i\right)^2} \tag{5}$$

$$CSI = \frac{1}{M}\sum_{i=1}^{M}\frac{H_i}{H_i+S_i+R_i} \tag{6}$$

$$MBE = \frac{1}{M}\sum_{i=1}^{M}\left(Q_i' - Q_i\right) \tag{7}$$

For the spatial dimension, Q_i' and Q_i denote the observation and prediction of 24 h rainfall accumulations at each rain gauge i. M is the number of the rain gauges, which is 8 for the Fuping catchment and 11 for the Zijingguan catchment. For the temporal dimension, $Q_i{'}$ and Q_i are the average areal rainfall of observation and prediction at each time step i. This time M is 24, which represents the number of the time steps. The specific meaning can be seen in Table 7.

Table 7. The meaning of letter in formulas, including $Q_i{'}$, Q_i, i, and M.

Letter	For the Spatial Dimension	For the Temporal Dimension
$Q_i{'}$	Observation of 24 h rainfall accumulations at each rain gauge	average areal rainfall of observation
Q_i	Prediction of 24 h rainfall accumulations at each rain gauge	average areal rainfall of prediction
i	Rain gauge ID	each time step
M	Total numbers of rain gauges	24 h

The calculation of *CSI* is based on the rain or no rain contingency table (Table 7). For the spatial dimension, the predicted rainfall was compared with observations at rain gauge locations to calculate the indices of H, R, and S at the time i, and then the values of the indices at all times are averaged to obtain *CSI* according to Equation (6). M is the number of total times. In this study, the time i is consistent with output frequency of the model, which is 1 h. That is, *CSI* is the average value of the $H/(H + R + S)$ for each hour of the 24 h rainfall duration. Similarly, for the temporal dimension, the indices in Table 7 are calculated based on the time series data obtained for the simulated and observed areal rainfall at the rain gauge i. The values of the indices at all rain gauges are then averaged to produce the final *CSI* value based on Equation (6). In this case, M refers to the total number of rain gauges rather than the simulation time.

4. Results

4.1. Effect of Data Assimilation on Temporal Rainfall Distributions

As illustrated by Figure 3, it shows the results of different assimilation frequencies. The first guess file generated from the previous run will provide the initial conditions for the next run. "DA_1h_1km" and "DA_1h_3km" are assimilated data with an interval of 1 h, and "DA_6h_3km" is 6 h. The forecasted accumulative rainfall is calculated from the average value of rainfall at each grid point in the study area. When the area of the grid within the watershed boundary accounts for more than 50% of the grid area, the rainfall value of the grid point participates in the calculation of the rainfall accumulations. As for the observations, the observed accumulative rainfall is calculated by averaging rain gauge observations using the Thiessen polygon method.

When the different assimilation frequencies are chosen in the model, the curve structure in the rainfall forecasting is significantly altered ("DA_1h_3km" and "DA_6h_3km"). The evolution of "DA_1h_3km" and "DA_6h_3km" in WRF-3DAVR system shows similar patterns with higher differences in the rainfall peak. The improvement of assimilation frequency led to a significant increase in precipitation. In WRF-3DVAR system, the operation with high assimilation frequency will produce incremental adjustment, which makes the prediction closer to the observation. In addition, through evaluating the outputs from different domain resolution on rainfall prediction, the improvement of WRF-3DVAR system domain output resolution is less obvious on the accumulative rainfall ("DA_1h_1km" and "DA_1h_3km"). In other words, the data assimilation of the outer domain has a positive effect on the output of the inner domain, but the improvement is not obvious. This may be due to the fact that no data is assimilated on the 1 km horizontal resolution domain. The larger the volume of assimilation data import to model, the longer time it will take to forecast the rainstorm. However, rainstorm forecasting has a high requirement for effectiveness for a given period of time, so in practical application it is beneficial to obtain the effective information of rainfall as soon as possible. In order to balance the accuracy and timeliness of rainstorm forecasting, data assimilation is not carried out in the innermost domain.

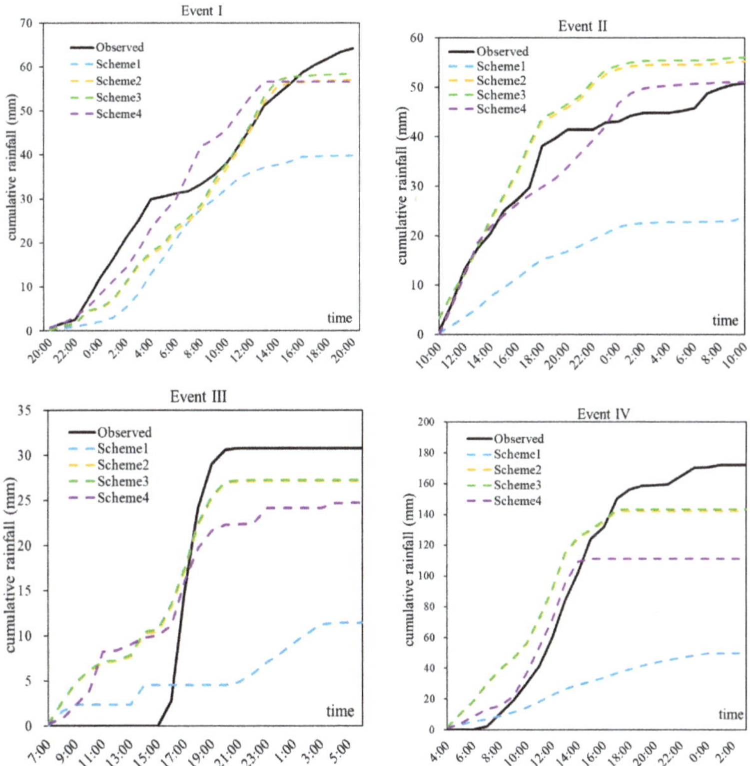

Figure 3. Rainfall accumulation of rain gauge and four data assimilation schemes for Event I, II, III, IV.

4.2. Effect of Data Assimilation on Spatial Rainfall Distributions

The spatial distributions of predicted precipitation with and without assimilation are shown in Figure 4. It is not difficult to see that after the data assimilation, the spatial temporal distribution of rainfall forecast has been improved in varying degrees compared with without assimilation. "DA_1h_1km", "DA_1h_3km", and "DA_6h_3km" performed much better as rainfall forecasts, respectively, than "NA_1km" in spatial distributions. The results show that the WRF-3DAVR system can obtain the major rain band located around the east and south border of the study area while the same rain band is disorganized in the WRF model. That is, the improvements to certain extents in spatial distributions after data assimilation.

Figure 4 shows spatial distributions of the 24 h accumulative rainfall for the four storm events in the Fuping and Zijingguan catchment. It can be intuitively seen from the spatial variations in Event II-IV that the rainfall forecast after data assimilation is significantly larger in numerical value than before assimilation. The storm centers of events were captured relatively well by WRF-3DAVR; however, some parts of the catchments with high rainfall accumulations were missed by WRF-3DAVR, such as the northern rainband of Event III and the western rainband of Event IV. By analyzing the evenness of storm events, the temporal and spatial distribution of Event I is more even, Event II is uneven in time, and Event III-IV is uneven in space and time. The results show that the rainfall with even spatial distribution has the best predicted results on the spatial scale, while the rainfall with uneven spatial distribution has the worst predicted results on the spatial scale.

WRF-3DAVR is easier to accurately simulate or forecast rainfall with more uniform spatial distribution, while WRF-3DAVR is more difficult to accurately forecast rainfall area with uneven spatial distribution. It can also be obtained from the index analysis of each rainfall forecast result in Table 8. In addition, whether the rainfall is evenly distributed on the time scale has a certain influence on the forecast results of rainfall on the spatial scale but does not play a decisive role. From "DA_1h_1km" and "DA_1h_3km", simply increasing the resolution of domain has no significant improvement in the spatial dimension of the rainfall simulations, this is probably because the innermost domain does not assimilate data. Furthermore, Table 8 in Section 4.3 provides the root mean square error (*RMSE*), mean bias error (*MBE*), critical success index (*CSI*), and *CSI/RMSE* of 24 h rainfall accumulation values using WRF and the different WRF-3DAVR schemes.

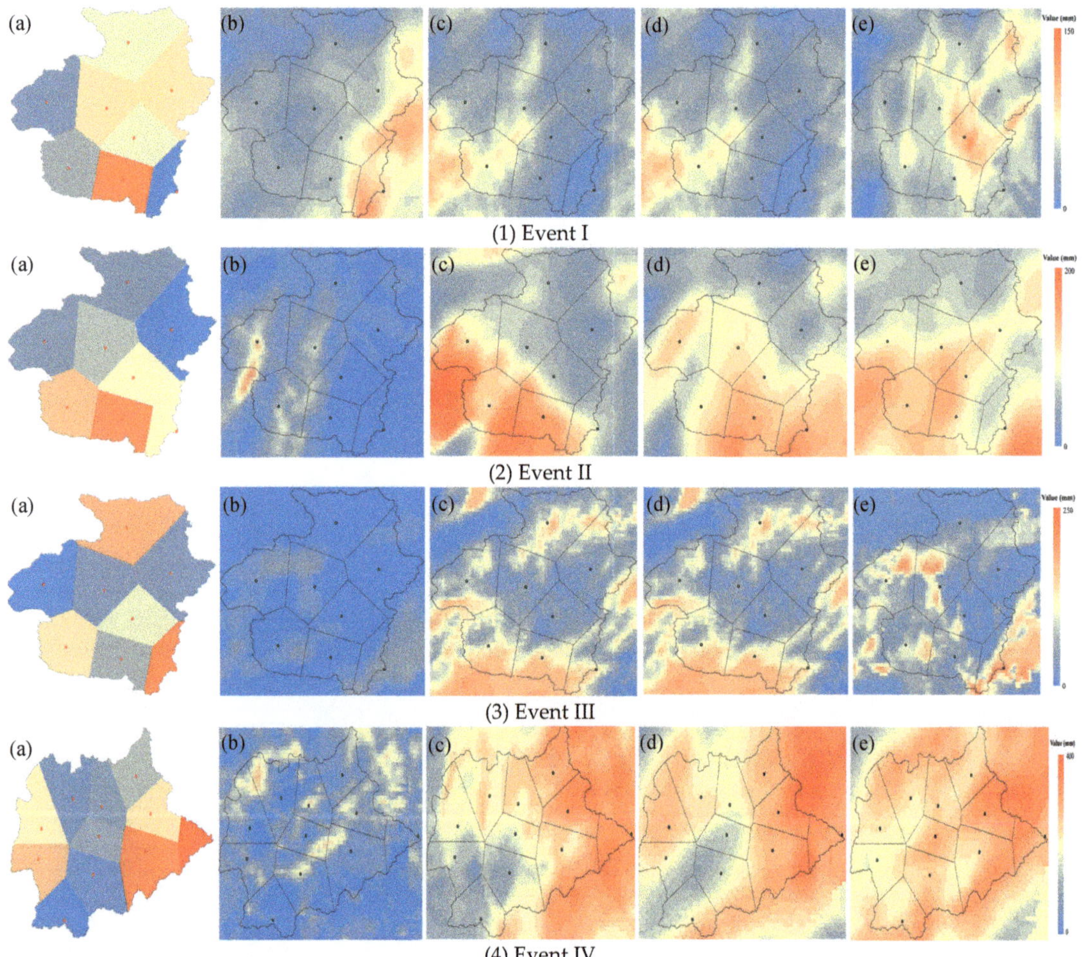

Figure 4. Spatial rainfall distributions of gauge observations and forecasts from four data assimilation schemes for Event I, II, III, IV, from left to right: (**a**) observation; (**b**) NA_1km; (**c**) DA_1h_1km; (**d**) DA_1h_3km; (**e**) DA_6h_3km.

4.3. Evaluation on the Storm Process Improvements

The evaluation scores for the four 24-h rainstorm forecasts from 1 and 6 h time intervals with and without assimilation are shown in Table 8. In evaluations in the spatial scale, the model forecasts are interpolated to the rain gauge locations for comparisons with the observations. Firstly, compared with "NA_1km", the rainfall forecasts at different assimilation schemes are improved significantly. For example, *RMSE* of "NA_1km" without assimilation in Event I on the temporal dimension was 2.3393, while the highest *RMSE* with data assimilation is 2.1320 (DA_6h_3km) and the lowest *RMSE* is 1.7816 (DA_1h_1km). Possibly it is because Event I has a relatively even distribution in both time and space (0.3975 for spatial *Cv* and 0.6011 for temporal *Cv*), and data assimilation has no large improvement on the rainfall forecasting with even spatial-temporal distribution. Similarly, on the spatial dimension, *RMSE* of the assimilated schemes is better than that without assimilation. In the spatial dimension, for example, *RMSE* of "NA_1km" in Event III is 2.8849, which is no assimilation, while the worst *RMSE* with data assimilation in Event III is 1.4068 (DA_6h_3km). Therefore, the good performance for selected typical precipitation events shown by the cycling data assimilation gradually improves not only temporal but also spatial variability.

Secondly, the performance of the 1 h assimilation time interval with respect to its 6 h counterpart with data assimilation is examined. It is shown (Table 8) that experiment hourly assimilation time interval has lower *RMSE* than its 6-hourly counterpart in most events in both the temporal and spatial dimension, indicating the potential for a better forecast. Event I, Event III, and Event IV show positive effect in rapid update assimilation, and much precipitation rises in WRF-3DAVR compared to low assimilation frequency ("DA_1h_3km" and "DA_6h_3km"). In the case of Event IV in the spatial dimension, for example, a decrease in *RMSE* from 12.6979 after 6 h assimilation time interval to 8.7782 occurred after hourly assimilation frequency; a similar trend was noted during Event I and Event III. In the meantime, hourly assimilation frequency has a lower MBE than 6-hourly assimilation time interval for most events. This might be the result that the regional approach with higher-resolution observations and closing to actual atmospheric boundary conditions may improve the assimilation effect and help offset temporal and spatial information lost by WRF. For the study area with small-scale, the assimilation time interval of 6 h is too long, and the model background field is not corrected in time. As time goes on, the observation error of radar is constantly amplified in the model background field, which reduces the effect of rainfall forecast.

However, WRF-3DAVR and high assimilation frequency are mixed. In Event II, experiment hourly cycled configuration had slightly lower scores than those of the 6-hourly counterpart in the both time dimension and spatial distribution for Event II. Although the low assimilation frequency appears to be slightly better than the high assimilation frequency for Event II, this does not seem to pose a threat to the hourly assimilation frequency. But it also reflects the disadvantages of spreading too much radar information to places where the radar data are not available.

In addition, the influence of data assimilation of outer domain on the output of inner domain is discussed, and the precipitation outputs of 3 and 1 km domain are compared. The results show that although the data assimilation of outside domain has a positive impact on the output of the inside domain, inside domain generated very small helpful increments, especially in the time scale.

All the schemes show different amounts of false precipitation in study areas from *CSI*. The high *CSI* are Event I and II, indicating that the model basically captures the occurrence time and rainstorm area of Event I and II, the low *CSI* are Event III and IV, and the lowest is Event III, indicating that the simulation results of these rainfall fields are poor in terms of time and space. In order to further evaluate the forecasting results of WRF-3DAVR system for each rainfall type on the temporal scale, *CSI/RMSE* was taken as a comprehensive index to evaluate the forecasting results. In the temporal dimension, the forecasting results of Event I and Event II are the best, with the value range of four schemes of *CSI/RMSE* being 0.2412 to 0.4538, while the forecasting results of Event III and Event IV are the worst, value range of four schemes of *CSI/RMSE* being 0.0345 to 0.0948. In the spatial distribution, the same law is presented, that is, WRF-3DAVR system is easier to accurately simulate or forecast rainfall with more uniform time distribution, but more difficult to simulate or forecast rainfall with short duration and concentrated rainfall. Combined with the calculation of the spatiotemporal variation coefficient (Cv) of precipitation events, it shows that Event II–IV are more uneven in time and space than Event I. In Event III, for example, the temporal Cv value of 2.3925 and the spatial Cv value of 0.7400 are much higher than those of Event I (0.6011 and 0.3975). This may explain the increased bias in assimilation, since the improved effectiveness of the rainfall forecast after assimilation is determined by the amount of effective information contained in the data. It is easier for radar reflectivity and GTS data to capture data during periods of rainfall that is homogeneously distributed in space and time.

Table 8. Temporal and spatial values of the four assessment indices for Event I, II, III, IV with four schemes.

Events	Experience Scheme	Temporal Dimension				Spatial Dimension			
		RMSE	MBE	CSI	CSI/RMSE	RMSE	MBE	CSI	CSI/RMSE
I	NA_1km	2.3393	−1.9322	0.7519	0.3214	1.7908	−1.7003	0.7478	0.4176
	DA_1h_1km	1.7816	−1.4615	0.6820	0.3828	1.1295	−1.0476	0.7835	0.6936
	DA_1h_3km	1.7967	−1.4837	0.8153	0.4538	1.1343	−0.9890	0.8155	0.7189
	DA_6h_3km	2.1320	−1.6189	0.7872	0.3692	1.7171	−1.6280	0.6719	0.3913
II	NA_1km	2.3752	−1.5909	0.5729	0.2412	2.5884	−2.5600	0.5729	0.2213
	DA_1h_1km	1.9468	1.3097	0.5791	0.2974	0.9473	0.9425	0.5744	0.6063
	DA_1h_3km	1.9584	1.3263	0.5759	0.2940	0.9523	0.8136	0.5729	0.6016
	DA_6h_3km	1.9360	1.2940	0.5791	0.2991	0.9047	0.6404	0.5744	0.6349
III	NA_1km	3.4189	−1.7446	0.1180	0.0345	2.8849	−2.4168	0.1910	0.0662
	DA_1h_1km	2.0987	−1.0273	0.1038	0.0495	1.0594	−1.0353	0.1875	0.1770
	DA_1h_3km	2.1185	−1.0381	0.1676	0.0791	1.2250	−1.1401	0.1667	0.1361
	DA_6h_3km	2.2778	−1.2869	0.2004	0.0880	1.4068	−1.2895	0.1806	0.1283
IV	NA_1km	8.5700	−5.8656	0.6601	0.0770	12.6979	−10.4946	0.5524	0.0435
	DA_1h_1km	5.9530	−4.0378	0.5449	0.0915	8.7782	−3.5646	0.5524	0.0629
	DA_1h_3km	5.9566	−4.0304	0.5648	0.0948	8.8033	−3.6767	0.5131	0.0583
	DA_6h_3km	6.6525	−4.3429	0.6601	0.0992	9.3979	−5.4607	0.4270	0.0454

Assimilation of all possible data with high assimilation frequency may not be the most effective method in precipitation forecasting. Especially, the influence of assimilation frequency on rainfall forecast is rather small in Event II; the precipitation forecasts with 1 h cycle do not have much difference from those with 6 h cycle. That may indicate the impact of false rainfall forecasting fields is enlarged because of the inaccurate radar observed data. One may wonder whether the results of assimilation have anything to do with the quality of the assimilated data, such as radar data. To answer that question, the ability of Doppler radar to retrieve precipitation is plotted (Figure 5). In each single subfigure, the black bars and yellow solid curve indicate measure rainfall and accumulative rainfall from rain gauge, respectively. Green bars and pink curve indicate rainfall and accumulative rainfall from radar observed reflectivity inverse calculation. The left y-axis is the cumulative rainfall value, corresponding to the curve, and the y-axis on the right is the value of hourly rainfall, which corresponds to the bar graph.

As can be seen from Figure 5, the radar precipitation estimation of Event I was closer to the observation accumulation curve, and at the same time, assimilation effect of Event I was also the best in all events. In addition, there was substantial rainfall growth during the first nine hours of the rainfall after the radar data assimilation for Event III, and the accumulated rainfall increased abruptly. Additionally, we found that the radar measures rainfall from the 1st hour to the 9th hour as much larger than the observed rainfall (Figure 5), as revealed in many previous studies [54]. Therefore, in our storm events selected, the accuracy of radar reflectivity is of primary importance in improving the quality of precipitation forecast within the time range of forecasting [5].

When WRF-3DVAR technology is applied, a matter of effective radar data assimilation could be tackled by using shorter assimilation time interval to achieve greater information assimilation. Although the assimilated radar data can help WRF model to forecast precipitation effectively, it increases the conflict between radar data and domain. In the process of data assimilation, the validity of assimilated data should be judged as far as possible in advance, which can not only improve the prediction accuracy of WRF model, but also improve the assimilation efficiency. There are many factors affecting assimilation, and radar data may be only a part of them, and more factors need to be further explored, such as the resolution of GFS data, nested boundary conditions, the dynamic structure of the model, numerical discretization, etc.

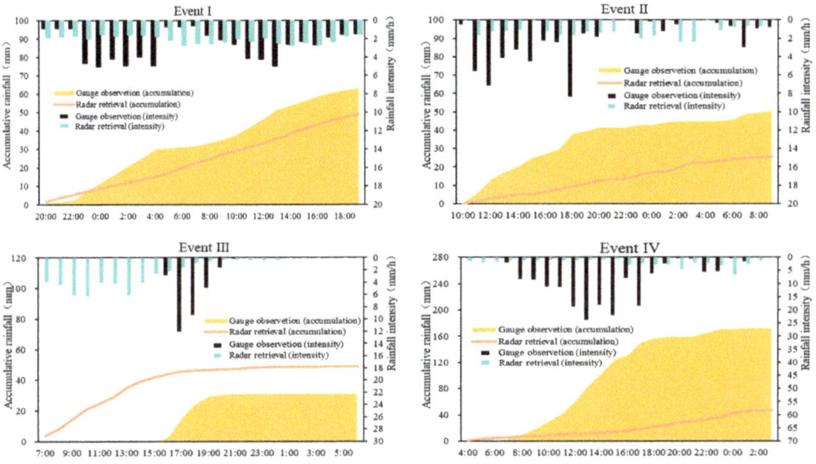

Figure 5. Observations and radar measurements of 24 h rainfall accumulations for Event I, II, III, IV.

5. Discussion

By carrying out a comparative analysis on the output results of different assimilation frequencies, it can be proved that by reducing the length of the assimilation window, the rainfall forecasting can be closer to the observations and can better present storm center. The hourly data assimilation frequency in the WRF-3DVAR allows for the addition of useful information and results in improved performance of the rainfall forecasting. However, misestimates may be caused when applying high assimilation frequency.

First, the quality of the output analysis depends on the error in the model initialization when inaccurate boundary field and background information is inputted into the assimilation system. As is widely understood, accurate initial condition is crucial to data assimilation and prediction of numerical weather prediction (NWP) system. Data assimilation system can combine all useful information about atmospheric conditions in the given time window, and obtain estimated value of the valid atmospheric conditions in given analysis time. Where, information used for model calculations is sourced from the observations, background, previous estimate of the atmospheric state, as well as their specific inherent errors. Compared with other error sources (e.g., physical parameterization, boundary conditions, and predicting dynamics), the relative importance of prediction errors caused by initial condition errors depend on many factors, including resolution, domain, data density, orography as well as the forecast product of interest [55]. The assimilation data of the WRF-3DVAR system is dedicated to providing initial conditions and further improving the WRF predictions, and then applied to creating forecasting of regional climates in the future, which is excitingly possible.

Especially, the radar data as one of the assimilation data sources, which represent the small-scale and rapid evolution in the boundary layer, are often unsatisfactory, especially in the events of extreme rainfall intensity. This is because the rainfall estimate of radar is not directly measured, but is indirectly obtained by the measured radar reflectivity. The measurement of radar reflectivity and the conversion process from reflectivity to intensity are affected by many error sources. In order to improve the accuracy of the radar data used for initial conditions of mode, it can be adjusted according to the rain gauge measurements. In this way, it can combine the advantages of the precise measurement of rainfall by the rain gauge at the point and the better performance of the radar in the spatial distribution, and overcome the disadvantages of the rainfall deviation caused by the uncertainty of the radar in the rain measurement. Studies are being conducted to find the optimal technique for radar observations to improve the initial state of the model.

Second, the error in the actual numerical prediction system may be highly non-linear, although the variational method includes linearized dynamic and physical processes, which limits the practicability of variational data assimilation in highly non-linear regions (such as convective scale or tropical region). Therefore, it is necessary to develop a more objective method, such as a method based on ensemble prediction for estimating the uncertainty of a flow-related prediction background [5]. Nevertheless, despite these limitations, this study provides a prototype for short-term practical prediction of a local convective weather system. It is hoped that these research topics can be discussed in the future research and application of the 3DVAR system.

Furthermore, it is shown that the WRF model can reasonably predict a low-intensity and long-lasting rainfall event. However, the result from this study indicates that this model often leaves out rare small-scale and short-term rainfall prediction events or underestimates the precipitation intensity, because small-scale interference is filtered when large-scale analysis keeps a better balance [15]. The main reason for this is that convection is a small-scale phenomenon, and false estimates may be caused if the increment in diffusion spreads too far. This is exactly what should be improved in the predictions, by applying different model parameterization technology [56] or data assimilation technology for instance [3].

For mesoscale catchments, the spatial distribution of rainfall is also important due to its significant impact on the flood volume, flood peak, and time to peak [28]. In addition, approaches to improve the spatial accuracy of precipitation predictions after data assimilation are also worth exploring. It is necessary to analyze more storm events in different survey regions in order to find more general radar data assimilation criteria and thus facilitate numerical prediction.

6. Conclusions

This study explores the effect of radar reflectivity and GTS data assimilation from assimilation frequency using WRF-3DVar for rainfall forecasting. Four heavy storm events at the Daqinghe catchment in the Beijing Tianjin Hebei region of northern China are selected to be regenerated by the WRF model. We employed three nested domains, and adopted the GFS data for driving the WRF model. From two aspects of cumulative rainfall and spatial distribution of rainfall, two observational data types (radar reflectivity and GTS data) assisted in investigating how WRF rainfall forecasts were potentially improved in space and time through data assimilation. We designed four data assimilation schemes considering various possible combinations of the two data assimilation frequency types in the three nested domains. We compared the analysis with data assimilation and that without data assimilation, finding that the assimilation results partly fit observations in the case and that WRF-3DAVR with radar reflectivity and GTS data better represents the rainfall forecasts in space and time.

Precipitation simulated by the WRF model is always much lower than observed rainfall, but assimilation systems can increase rainfall. The improved initial conditions in WRF-3DVAR system via radar data assimilation and GTS data achieved better short-term and convective strong precipitation in the both temporal dimension and spatial dimension. The high assimilation frequency significantly helps to trigger and maintain the convective activities in the 3DVAR framework as well as the storm case applied. Forecasts of events indicate that the temporal rainfall distributions of convective storms can be much better predicted with high assimilation frequency, compared with the 6 h assimilation time interval run. At the same time, employing the high assimilation frequency to the assimilation showed improved skill of precipitation forecasting in WRF-3DVAR on spatial rainfall distributions. The only exception happened in Event II. In this case, the impact of false rainfall forecasting fields is enlarged because of the inaccurate radar observed data, so that the negative impact was found after assimilation. However, this does not seem to pose a threat to the hourly assimilation frequency. In addition, the data assimilation of outside domain has small impact on output of inside domain in not only temporal but also spatial dimension. In general, the hourly data assimilation frequency together with strict outputs from domain resolutions is closer to actual precipitation. In this study, the assimilation by combining the radar reflectivity and Global Tele-communication System (GTS) data with high assimilation frequency is helpful for further enhancing the temporal and spatial distribution of the short-term precipitation forecast. The results can be used as a reference for areas with similar climatic conditions as well as rainfall characteristics. The methodology is of guiding significance for WRF-3DAVR rainfall forecasting. In this case, in order to explore universally applicable data assimilation guidelines for rainfall forecasting, research should be conducted over more storm events in different study areas.

Author Contributions: Conceptualization, Y.L. and J.L.; Methodology, Y.L., J.L. and F.Y.; Software, Y.L. and Y.L.; Validation, Y.L. and W.W.; Formal analysis, Y.L. and J.L.; Investigation, W.W., C.L. and F.Y.; Writing-original draft preparation, Y.L. and J.L.; Writing-review and editing, Y.L., J.L. and C.L.; Visualization, Y.L.; Funding acquisition, J.L. and C.L. All authors have read and agreed to the published version of the manuscript.

Funding: This research was funded by the Major Science and Technology Program for Water Pollution Control and Treatment (2018ZX07110001), the National Natural Science Foundation of China (51822906), the National Key Research and Development Project (2017YFC1502405), and the IWHR Research & Development Support Program (WR0145B732017).

Conflicts of Interest: The authors declare no conflict of interest.

References

1. Fritsch, J.M.; Carbone, R.E. Improving Quantitative Precipitation Forecasts in the Warm Season: A USWRP Research and Development Strategy. *Bull. Am. Meteorol. Soc.* **2004**, *85*, 955–966. [CrossRef]
2. Fritsch, J.M.; Houze, R.A.; Adler, R.; Bluestein, H.; Bosart, L.; Brown, J.; Carr, F.; Davis, C.; Johnson, R.H.; Junker, N.; et al. Quantitative Precipitation Forecasting: Report of the Eighth Prospectus Development Team, U.S. Weather Research Program. *Bull. Am. Meteorol. Soc.* **1998**, *79*, 285–299. [CrossRef]
3. Kryza, M.; Werner, M.; Waszek, K.; Dore, A.J. Application and evaluation of the WRF model for high-resolution forecasting of rainfall—A case study of SW Poland. *Meteorol. Z.* **2013**, *22*, 595–601. [CrossRef]
4. Hamill; Thomas, M. Performance of Operational Model Precipitation Forecast Guidance during the 2013 Colorado Front-Range Floods. *Mon. Weather Rev.* **2012**, *142*, 2609–2618. [CrossRef]
5. Sugimoto, S.; Crook, N.A.; Sun, J.; Xiao, Q.; Barker, D.M. An Examination of WRF 3DVAR Radar Data Assimilation on Its Capability in Retrieving Unobserved Variables and Forecasting Precipitation through Observing System Simulation Experiments. *Mon. Weather Rev.* **2009**, *137*, 4011–4029. [CrossRef]
6. Barker, D.M.; Huang, W.; Guo, Y.R.; Bourgeois, A.J.; Xiao, Q.N. A Three-Dimensional Variational Data Assimilation System for MM5: Implementation and Initial Results. *Mon. Weather Rev.* **2004**, *132*, 897–914. [CrossRef]
7. Skamarock, W.C.; Klemp, J.B. A time-split nonhydrostatic atmospheric model for weather research and forecasting applications. *J. Comput. Phys.* **2008**, *227*, 3465–3485. [CrossRef]
8. Le Dimet, F.O.X.; Talagrand, O. Variational algorithms for analysis and assimilation of meteorological observations: Theoretical aspects. *Tellus Dyn. Meteorol. Oceanogr.* **1986**, *38*, 97–110. [CrossRef]
9. Lewis, J.M.; Derber, J.C. The use of adjoint equations to solve a variational adjustment problem with advective constraints. *Tellus* **1985**, *37A*, 309–322. [CrossRef]
10. Evensen, G. Sequential data assimilation with a nonlinear quasi-geostrophic model using Monte Carlo methods to forecast error statistics. *J. Geophys. Res. Ocean.* **1994**, *99*, 10143–10162. [CrossRef]
11. Sun, J.; Flicker, D.W.; Lilly, D.K. Recovery of Three-Dimensional Wind and Temperature Fields from Simulated Single-Doppler Radar Data. *J. Atmos. Sci.* **1991**, *48*, 876–890. [CrossRef]
12. Sun, J.; Crook, N.A. Dynamical and Microphysical Retrieval from Doppler Radar Observations Using a Cloud Model and Its Adjoint. Part I: Model Development and Simulated Data Experiments. *J. Atmos.* **1998**, *55*, 835–852. [CrossRef]
13. Yang, J.; Duan, K.; Wu, J.; Qin, X.; Shi, P.; Liu, H.; Xie, X.; Zhang, X.; Sun, J. Effect of Data Assimilation Using WRF-3DVAR for Heavy Rain Prediction on the Northeastern Edge of the Tibetan Plateau. *Adv. Meteorol.* **2015**, *2015*, 1–14. [CrossRef]

14. Gan, R.; Yang, Y.; Xie, Q.; Lin, E.; Wang, Y.; Liu, P. Assimilation of Radar and Cloud-to-Ground Lightning Data Using WRF-3DVAR Combined with the Physical Initialization Method—A Case Study of Mesoscale Convective System. *J. Meteorol. Res.* **2020**, *34*, 1–14. [CrossRef]

15. Vendrasco, E.P.; Sun, J.; Herdies, D.L.; Carlos, F.D.A. Constraining a 3DVAR Radar Data Assimilation System with Large-Scale Analysis to Improve Short-Range Precipitation Forecasts. *J. Appl. Meteorol. Clim.* **2016**, *55*, 673–690. [CrossRef]

16. Chung, K.S.; Zawadzki, I.; Yau, M.K.; Fillion, L. Short-Term Forecasting of a Midlatitude Convective Storm by the Assimilation of Single-Doppler Radar Observations. *Mon. Weather Rev.* **2009**, *137*, 4115–4135. [CrossRef]

17. Hu, M.; Xue, M.; Gao, J.; Brewster; Keith. 3DVAR and Cloud Analysis with WSR-88D Level-II Data for the Prediction of the Fort Worth, Texas, Tornadic Thunderstorms. Part II: Impact of Radial Velocity Analysis via 3DVAR. *Mon. Weather Rev.* **2004**, *134*, 699–721. [CrossRef]

18. Rennie, S.J.; Dance, S.L.; Illingworth, A.J.; Ballard, S.P.; Simonin, D. 3D-Var Assimilation of Insect-Derived Doppler Radar Radial Winds in Convective Cases Using a High-Resolution Model. *Mon. Weather Rev.* **2011**, *139*, 1148–1163. [CrossRef]

19. Tai, S.L.; Liou, Y.C.; Sun, J.; Chang, S.F.; Kuo, M.C. Precipitation Forecasting Using Doppler Radar Data, a Cloud Model with Adjoint, and the Weather Research and Forecasting Model: Real Case Studies during SoWMEX in Taiwan. *Weather Forecast.* **2011**, *26*, 975–992. [CrossRef]

20. Routray, A.; Mohanty, U.C.; Osuri, K.K. Improvement of Monsoon Depressions Forecast with Assimilation of Indian DWR Data Using WRF-3DVAR Analysis System. *Pure Appl. Geophys.* **2013**, *170*, 2329–2350. [CrossRef]

21. Govindankutty, M.; Chandrasekar, A.; Pradhan, D. Impact of 3DVAR assimilation of Doppler Weather Radar wind data and IMD observation for the prediction of a tropical cyclone. *Int. J. Remote Sens.* **2010**, *31*, 6327–6345. [CrossRef]

22. Abhilash, S.; Sahai, A.K.; Mohankumar, K.; George, J.P.; Das, S. Assimilation of Doppler Weather Radar Radial Velocity and Reflectivity Observations in WRF-3DVAR System for Short-Range Forecasting of Convective Storms. *Pure Appl. Geophys.* **2012**, *169*, 2047–2070. [CrossRef]

23. Osuri, K.K.; Mohanty, U.C.; Routray, A.; Niyogi, D. Improved Prediction of Bay of Bengal Tropical Cyclones through Assimilation of Doppler Weather Radar Observations. *Mon. Weather Rev.* **2015**, *143*, 1171093873. [CrossRef]

24. Hsiao, L.; Chen, D.; Kuo, Y.; Guo, Y.; Yeh, T.; Hong, J.; Fong, C.; Lee, C. Application of WRF 3DVAR to Operational Typhoon Prediction in Taiwan: Impact of Outer Loop and Partial Cycling Approaches. *Weather Forecast.* **2012**, *27*, 1249–1263. [CrossRef]

25. Hodur, R.M. Evaluation of a Regional Model with an Update Cycle. *Mon. Weather Rev.* **1987**, *115*, 2707–2718. [CrossRef]

26. Xiao, Q.; Kuo, Y.H.; Sun, J.; Lee, W.C.; Lim, E.; Guo, Y.R.; Barker, D.M. Assimilation of Doppler Radar Observations with a Regional 3DVAR System: Impact of Doppler Velocities on Forecasts of a Heavy Rainfall Case. *J. Appl. Meteorol.* **2005**, *44*, 768–788. [CrossRef]

27. Liu, J.; Tian, J.; Yan, D.; Li, C.; Yu, F.; Shen, F. Evaluation of Doppler radar and GTS data assimilation for NWP rainfall prediction of an extreme summer storm in northern China: From the hydrological perspective. *Hydrol. Earth Syst. Sci.* **2018**, *22*, 4329–4348. [CrossRef]

28. Tian, J.; Liu, J.; Yan, D.; Li, C.; Chu, Z.; Yu, F. An assimilation test of Doppler radar reflectivity and radial velocity from different height layers in improving the WRF rainfall forecasts. *Atmos. Res.* **2017**, *198*, 132–144. [CrossRef]

29. Thiruvengadam, P.; Indu, J.; Ghosh, S. Assimilation of Doppler Weather Radar data with a regional WRF-3DVAR system: Impact of control variables on forecasts of a heavy rainfall case. *Adv. Water Resour.* **2019**, *126*, 24–39. [CrossRef]

30. Sun, J.; Xue, M.; Wilson, J.W.; Zawadzki, I.; Ballard, S.P.; Onvlee-Hooimeyer, J.; Joe, P.; Barker, D.M.; Li, P.W.; Golding, B. Use of NWP for Nowcasting Convective Precipitation: Recent Progress and Challenges. *Bull. Am. Meteorol. Soc.* **2014**, *95*, 409–426. [CrossRef]

31. Tong, W.; Li, G.; Sun, J.; Tang, X.; Zhang, Y. Design Strategies of an Hourly Update 3DVAR Data Assimilation System for Improved Convective Forecasting. *Weather Forecast.* **2016**, *31*, 1673–1695. [CrossRef]

32. Li, Y.; Wang, X.; Xue, M. Assimilation of Radar Radial Velocity Data with the WRF Hybrid Ensemble-3DVAR System for the Prediction of Hurricane Ike (2008). *Mon. Weather Rev.* **2012**, *140*, 3507–3524. [CrossRef]

33. Kawabata, T.; Seko, H.; Saito, K.; Kuroda, T.; Wakazuki, Y. An Assimilation and Forecasting Experiment of the Nerima Heavy Rainfall with a Cloud-Resolving Nonhydrostatic 4-Dimensional Variational Data Assimilation System. *J. Meteorol. Soc. Jpn. Ser. II* **2007**, *85*, 255–276. [CrossRef]

34. Ide, K.; Courtier, P.; Ghil, M.; Lorenc, A.C. Unified Notation for Data Assimilation: Operational, Sequential and Variational (gtSpecial IssueltData Assimilation in Meteology and Oceanography: Theory and Practice). *J. Meteorol. Soc. Jpn. Ser. II* **1997**, *75*, 181–189. [CrossRef]

35. Lorenc, A.C. Analysis methods for numerical weather prediction. *Q. J. Roy. Meteor. Soc.* **1986**, *112*, 1177–1194. [CrossRef]

36. Gao, J.; Xue, M.; Brewster, K.; Droegemeier, K.K. A Three-Dimensional Variational Data Analysis Method with Recursive Filter for Doppler Radars. *J. Atmos. Ocean. Technol.* **2004**, *21*, 457–469. [CrossRef]

37. Meng, Z.; Zhang, F. Tests of an Ensemble Kalman Filter for Mesoscale and Regional-Scale Data Assimilation. Part III: Comparison with 3DVAR in a Real-Data Case Study. *Mon. Weather Rev.* **2008**, *136*, 522–540. [CrossRef]

38. Yang, B.; Zhang, Y.; Qian, Y. Simulation of urban climate with high-resolution WRF model: A case study in Nanjing, China. *Asia Pac. J. Atmos. Sci.* **2012**, *48*, 227–241. [CrossRef]

39. Givati, A.; Lynn, B.; Liu, Y.; Rimmer, A. Using the WRF Model in an Operational Streamflow Forecast System for the Jordan River. *J. Appl. Meteorol. Climatol.* **2012**, *51*, 285–299. [CrossRef]

40. Aligo, E.A.; Gallus, W.A.; Segal, M. On the Impact of WRF Model Vertical Grid Resolution on Midwest Summer Rainfall Forecasts. *Weather Forecast.* **2009**, *24*, 575–594. [CrossRef]
41. Schwartz, C.S.; Kain, J.S.; Weiss, S.J.; Xue, M.; Bright, D.R.; Kong, F.; Thomas, K.W.; Levit, J.J.; Coniglio, M.C. Next-Day Convection-Allowing WRF Model Guidance: A Second Look at 2-km versus 4-km Grid Spacing. *Mon. Weather Rev.* **2009**, *137*, 3351–3372. [CrossRef]
42. Liu, J.; Bray, M.; Han, D. Exploring the effect of data assimilation by WRF-3DVar for numerical rainfall prediction with different types of storm events. *Hydrol. Process.* **2012**, *27*, 3627–3640. [CrossRef]
43. Liu, J.; Wang, J.; Pan, S.; Tang, K.; Li, C.; Han, D. A real-time flood forecasting system with dual updating of the NWP rainfall and the river flow. *Nat. Hazards* **2015**, *77*, 1161–1182. [CrossRef]
44. Tian, J.; Liu, J.; Wang, J.; Li, C.; Yu, F.; Chu, Z. A spatio-temporal evaluation of the WRF physical parameterisations for numerical rainfall simulation in semi-humid and semi-arid catchments of Northern China. *Atmos. Res.* **2017**, *191*, 141–155. [CrossRef]
45. Tian, J.; Liu, J.; Yan, D.; Li, C.; Yu, F. Numerical rainfall simulation with different spatial and temporal evenness by using a WRF multiphysics ensemble. *Nat. Hazard Earth Sys.* **2017**, *17*, 563–579. [CrossRef]
46. Hong, S.Y.; Lim, J.O.J. The WRF Single-Moment 6-Class Microphysics Scheme (WSM6). *Asia Pac. J. Atmos. Sci.* **2006**, *42*, 129–151.
47. Mlawer, E.J.; Taubman, S.J.; Brown, P.D.; Iacono, M.J.; Clough, S.A. Radiative transfer for inhomogeneous atmosphere: RRTM. *J. Geophys. Res. Atmos.* **1997**, *102*, 16663–16682. [CrossRef]
48. Dudhia; Jimy. Numerical Study of Convection Observed during the Winter Monsoon Experiment Using a Mesoscale Two-Dimensional Model. *J. Atmos. Sci.* **1989**, *46*, 3077–3107. [CrossRef]
49. Chen, F.; Dudhia, J. Coupling an Advanced Land Surface–Hydrology Model with the Penn State–NCAR MM5 Modeling System. Part I: Model Implementation and Sensitivity. *Mon. Weather Rev.* **2001**, *129*, 569–585. [CrossRef]
50. Janji, Z.I. *Nonsingular Implementation of the Mellor-Yamada Level 2.5 Scheme in the NCEP Meso Model*; NCEP Office Note; National Oceanic and Atmospheric Administration: Washington, DC, USA, 2001; Volume 43.
51. Kain, J.S.; Kain, J. The Kain—Fritsch convective parameterization: An update. *J. Appl. Meteorol.* **2004**, *43*, 170–181. [CrossRef]
52. Mazzarella, V.; Maiello, I.; Capozzi, V.; Budillon, G.; Ferretti, R. Comparison between 3D-Var and 4D-Var data assimilation methods for the simulation of a heavy rainfall case in central Italy. *Adv. Sci. Res.* **2017**, *14*, 271–278. [CrossRef]
53. Lea, B.; Loris, F.; Marco, G.; Ulrich, H. Satellite-Based Rainfall Retrieval: From Generalized Linear Models to Artificial Neural Networks. *Remote Sens.* **2018**, *10*, 939.
54. Wang, J.; Dong, X.; Xi, B. Investigation of Liquid Cloud Microphysical Properties of Deep Convective Systems: 2. Parameterization of Raindrop Size Distribution and its Application for Convective Rain Estimation. *J. Geophys. Res. Atmos.* **2018**, *123*, 637–651. [CrossRef]
55. Leutbecher, M.; Palmer, T.N. Ensemble forecasting. *J. Comput. Phys.* **2008**, *227*, 3515–3539. [CrossRef]
56. Droegemeier, K.K.; Smith, J.D.; Businger, S.; Doswell, C., III; Doyle, J.; Duffy, C.; Foufoula-Georgiou, E.; Graziano, T.; James, L.D.; Krajewski, V.; et al. Hydrological Aspects of Weather Prediction and Flood Warnings: Report of the Ninth Prospectus Development team of the U.S. Weather Research Program. *Bull. Am. Meteorol. Soc.* **2000**, *81*, 2665–2680. [CrossRef]

remote sensing

MDPI

Article

Estimating Urban Evapotranspiration at 10m Resolution Using Vegetation Information from Sentinel-2: A Case Study for the Beijing Sponge City

Xuanze Zhang * and Peilin Song

Key Laboratory of Water Cycle and Related Land Surface Processes, Institute of Geographic Sciences and Natural Resources Research, Chinese Academy of Sciences, Beijing 100101, China; songpl@igsnrr.ac.cn
* Correspondence: xuanzezhang@igsnrr.ac.cn

Abstract: Estimating accurately evapotranspiration (ET) in urban ecosystems is difficult due to the complex surface conditions and a lack of fine measurement of vegetation dynamics. To overcome such difficulties using recent developments of remote sensing technology, we estimate leaf area index (LAI) from Sentinel-2-based Normalized Difference Vegetation Index (NDVI) using the NDVI–LAI nonlinear relationship. By applying Sentinel-2-based LAI and land cover classification (LCC) to a carbon-water coupling model (PML-V2.1) with surface meteorological forcing data as input, we, for the first time, estimate monthly ET at 10m × 10m resolution for the Beijing Sponge City. Results show that for the whole sponge city during June 2018, the LAI, ET and gross primary productivity (GPP) are 0.83 m^2 m^{-2}, 1.6 mm d^{-1} and 2.8 gC m^{-2} d^{-1}, respectively. For different LCCs, lakes and rivers have the highest ET ($\geq$8 mm d^{-1}), followed by mixed forests and croplands (ET is 4–6 mm d^{-1} and LAI is 2–3 m^2 m^{-2}) with dominant contribution (>80%) from plant transpiration, while grasslands (2–4 mm d^{-1}) have 50–70% from transpiration due to smaller LAI (1~2 m^2 m^{-2}). The impervious surfaces occupying ~60% of the sponge city area, have the smallest ET (<2.0 mm d^{-1}) in which interception evaporation by impervious surface contributes 20–30%, and transpiration from greenbelts (0.5–1.0 m^2 m^{-2} of LAI) contributes 40–50%. These findings can provide a valuable scientific basis for policymaking and urban water use planning. This study proposes a Sentinel-2-based technology for estimating ET as a feasible framework to evaluate city-level hydrological dynamics in urban ecosystems.

Keywords: evaporation; evapotranspiration; LAI; NDVI; urban ecosystem; sponge city; PML-V2; Penman–Monteith equation; Sentinel-2

Citation: Zhang, X.; Song, P. Estimating Urban Evapotranspiration at 10m Resolution Using Vegetation Information from Sentinel-2: A Case Study for the Beijing Sponge City. *Remote Sens.* **2021**, *13*, 2048. https://doi.org/10.3390/rs13112048

Academic Editor: José A. Sobrino

Received: 14 May 2021
Accepted: 21 May 2021
Published: 22 May 2021

Publisher's Note: MDPI stays neutral with regard to jurisdictional claims in published maps and institutional affiliations.

1. Introduction

Owing to the high heterogeneity and complexity in urban ecosystems, it is rather difficult to monitor or predict the hydrological dynamics of urban surfaces [1]. Some megacities, e.g., Beijing—the capital city of China—have experienced strong urbanization, large population inflow, island effect and climate change during the past few decades [2]. These changes induce urban hydrological processes to be highly uncertain and make policymakers face tough challenges in water use planning and management. Therefore, there is an urgent need to accurately estimate urban hydrological processes.

Evapotranspiration (ET), as a key component of the urban hydrological processes and surface energy balance, plays an important role in regulating water resource supply and relieving the urban island effect (e.g., surface cooling) [3]. Different from natural ecosystems, the urban ecosystems include large proportions of artificial modifications in land cover, such as impervious surfaces including roofs, squares and cement or asphalt roads. These man-made reconstructions could contribute a large fraction of evaporation [4–6], but the quantification at city levels remains highly uncertain due to a lack of clearly distinguishing estimations of ET between impervious surfaces and vegetated or bare-soil lands. The

51

good news is that recent developments of fine resolution remote sensing for land use and land cover classification, vegetation dynamics and environmental monitoring provide new opportunities to estimate urban ET more accurately [7]. For example, the Sentinel-2, as part of the European Commission's Copernicus program with the launch of satellite Sentinel-2A on 23 June 2015, are monitoring variability in global land surface conditions at a 10–60m resolution and a 5–10-day revisit [8].

In this study, we take the sponge city project in Beijing city as a study case to estimate ET of the urban ecosystem at 10m resolution using the satellite-based land cover map and vegetation information derived from Sentinel-2 data. Beijing city has 5-year mean annual precipitation of 560 mm and mean annual temperature of 12 °C, with the potential evaporation of 550~600 mm year^{-1}. To reduce stress on water supply (e.g., ~30 m^3 per capita water use per year in Beijing) and urban environment, the Beijing Sponge City project was started on 4 December 2017, aiming at turning 20% of Beijing city into a sponge city covered area by 2020 [9]. Therefore, to evaluate the benefit of this project, it is essential to implement a city-level assessment of the project-induced ecohydrological changes at fine resolution.

2. Materials and Methods

2.1. Observational Forcing Datasets

2.1.1. Land Cover Map at 10m Resolution Derived from Sentinel-2

The land cover classification (LCC) global map at 10m resolution was obtained from FROM-GLC10 [7]. The FROM-GLC10 LCC data is developed based on Sentinel-2 data in 2017 with Google Earth Engine, and the overall accuracy of this LCC validated against the circa 2015 validation sample is 73% [7]. The LCC data includes 10 classes (i.e., cropland, forest, grassland, shrubland, wetland, tundra, impervious surface, bare land, and snow/ice). The most advances of the FROM-GLC10 LCC map compared to previous Landsat series-based LCC products are that it provides more spatial detail, better distinguish the forest from shrub or grassland classes, and better performance in coastal areas [7].

2.1.2. NDVI and LAI at 10m Resolution Derived from Sentinel-2

We calculate the Normalized Difference Vegetation Index (NDVI) from the Sentinel-2 reflected radiance by

$$NDVI = \frac{R_{nir} - R_{red}}{R_{nir} + R_{red}} \tag{1}$$

where R_{nir} and R_{red} are the spectral bands at near infrared (842 nm) and red (665 nm), respectively. The Sentinel-2 reflectance data are available at the USGS EROS Center (https://www.usgs.gov/centers/eros, accessed on 8 April 2021). The leaf area index (LAI) at a 10m resolution was derived from the retrieved Sentinel-2 NDVI using a nonlinear regression model between LAI and NDVI,

$$LAI = a * exp(b * NDVI) + c \tag{2}$$

where the parameters a, b and c are determined as 12.4, 6.4 and 0.6, respectively.

The determination process was based on MODIS-based NDVI and LAI products, which was described as: (i) The MODIS LAI (MOD15A2H) and surface reflectance (SR) products (MOD09A1) at 500m were collected over the study area (Beijing Sponge City) for the summer months (June, July, and August) from 2013 to 2019. The NDVI was then estimated using the 500m SR product (Equation (1)). (ii) MODIS LAI and NDVI values were collocated on the pixel basis. As the MODIS LAI product has a valid range between 0 and 6.9, with a precision of 0.1, we classified all NDVI values into 69 groups based on unique LAI values (eliminating the zero-LAI group). The probability density plots for each group are shown in Figure 1a. (iii) For each LAI-based value group, the probability density was fitted using the Gaussian distribution function. Then, the NDVI value corresponded by the maximum probability density was extracted and collocated with the specific LAI value.

The result shows a strong exponential relationship between NDVI and LAI, especially when the LAI value increased beyond 0.5–0.6 (Figure 1b). Therefore, the scatter values were fitted using the exponential model (Equation (2)), which resulted in an R^2 of 0.82, implying that such exponential model in Equation (2) is robust for the Beijing Sponge City.

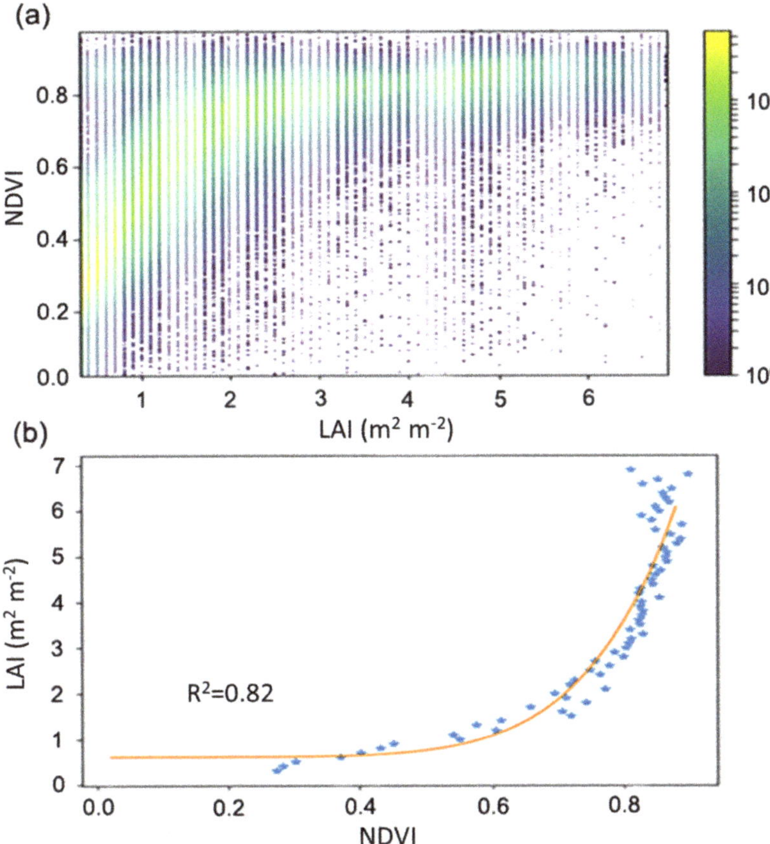

Figure 1. The nonlinear relationship between LAI and NDVI. (**a**) Frequency distribution of NDVI–LAI. Higher values of the scatter number in (**a**) indicate stronger relation between LAI and NDVI. (**b**) Regression between LAI and NDVI. The brown curve was fitted using Equation (2) from the NDVI–LAI values (blue stars) under the maximum probability density.

2.1.3. Surface Climate Driving Dataset

To estimate evapotranspiration of urban ecosystem at 10m resolution, a high-resolution surface climate forcing data including precipitation, surface air temperature, wind speed, surface pressure, specific humidity, downward shortwave and longwave radiations, etc., is needed to drive the terrestrial evapotranspiration model (PML-V2 model, see Section 2.2). In this study, we used the China Meteorological Forcing Dataset (CMFD) version 1 at $0.1° \times 0.1°$ and daily resolution for June 2018 as input for the PML-V2 model. The CMFD V1.0 dataset covered the period of 1979–2018 and was downscaled from station-based data, TRMM satellite-based precipitation, GEWEX-SRB shortwave radiation and the GLDAS forcing dataset [10]. The surface climate driving variables used for the Beijing Sponge City area were spatially bilinearly interpolated onto a 10m $\times$ 10m resolution. The monthly CO_2 concentration observed in June 2018 is set as 407 ppm.

2.2. PML-V2.1 Model

Version 2 of the Penman–Monteith–Leuning model (PML-V2) was developed by coupling the widely-used photosynthesis model [11] and a canopy stomatal conductance model [12] with the Penman–Monteith energy balance equation [13] to jointly estimate gross primary productivity (GPP), E_c and E_s [14–18]. The PML-V2 model also simulates the E_i based on a revised Gash-model scheme [19]. The PML-V2 model has been applied to successfully produce the MODIS LAI-based global GPP and ET products at a 500m and 8-day resolution from 2002 to present, which were noticeably better than most widely used GPP and ET products [16]. In this study, we incorporated modules of impervious surface evaporation (E_u) and open-water evaporation (E_w) into the PML model (PML-V2.1) to make it suitable for urban ecosystems. Key parameters used in the PML-V2.1 model are provided in Table 1. The following shows the detailed description for PML-V2.1.

Table 1. Key parameters used in the PML-V2.1 model.

Parameter	Definition	Unit	Land Cover Classification [a]							
			CRO	MIF	GRA	SHR	WET	WAT	IMP	BAR
α	Surface albedo for shortwave radiation	–	0.150	0.150	0.250	0.250	0.250	0.050	0.350	0.350
ε	Emissivity for longwave radiation	–	0.960	0.990	0.950	0.950	0.960	0.990	0.940	0.940
D_0	Reference vapor pressure deficit at stomatal conductance reduction	kPa	2.000	0.552	0.638	0.864	0.661	0.700	0.552	0.864
k_Q	Extinction coefficient of PAR	–	0.721	0.386	0.595	0.230	0.996	0.600	0.386	0.230
k_A	Extinction coefficient of available energy	–	0.899	0.899	0.900	0.888	0.888	0.700	0.899	0.888
S_{leaf}	Specific canopy rainfall storage capacity per unit leaf area	mm	0.010	0.198	0.227	0.014	0.022	0.000	0.198	0.014
F_{ER0}	Specific ratio of evaporation rate over rainfall intensity per unit vegetation cover	–	0.092	0.256	0.010	0.010	0.017	0.000	0.256	0.010
S_U	Specific canopy rainfall storage capacity per unit impervious surface area	mm	0.014	0.014	0.014	0.014	0.014	0.014	0.014	0.014
LAI_{ref}	Reference LAI	m	5.000	5.000	5.000	5.000	5.000	5.000	5.000	5.000
h	Canopy height	m	1.000	10.00	0.5000	10.000	0.500	0.500	10.00	0.500
$V_{m,25}$	Maximum catalytic capacity of Rubisco per unit leaf area at 25°C	$\mu\text{mol m}^{-2}\text{s}^{-1}$	22.560	28.450	29.560	18.770	24.440	0.000	28.450	18.770
β	Initial photochemical efficiency	–	0.029	0.029	0.029	0.029	0.029	0.000	0.029	0.029
η	Initial value of the slope of CO_2 response curve	$\text{mol m}^{-2}\text{s}^{-1}$	0.069	0.040	0.026	0.024	0.069	0.000	0.040	0.024
m	Ball-Berry coefficient	–	5.289	8.355	3.934	4.406	9.211	0.000	8.355	4.406
D_{min}	The threshold below which there is no vapor pressure constraint	kPa	1.499	0.711	0.650	1.493	0.664	1.000	0.711	1.493
D_{max}	The threshold above which there is no assimilation	kPa	6.500	3.500	5.199	5.797	5.188	6.500	3.500	5.797

[a] CRO: cropland, MIF: mixed forest, GRA: grassland, SHR: shrubland, WET: wetland, WAT: water body, IMP: impervious surface and BAR: bare land.

2.2.1. Energy Balance at Urban Land Surface

Based on the surface energy balance, the net radiation (R_n) can be balanced by the latent heat flux (LE), sensible heat flux (H) and ground heat flux (G). As for a biweekly or longer estimation, the G is often negligible ($G \ll H + LE$), then the H is given by

$$H = R_n - LE - G \approx R_n - LE \tag{3}$$

The net radiation at the surface is the sum of the net shortwave downward radiation and the net longwave downward radiation,

$$R_n = (1 - \alpha)SW + \left(LW - \epsilon\sigma T_a{}^4\right) \tag{4}$$

where the shortwave downward radiation SW (W m^{-2}), the longwave downward radiation LW (W m^{-2}) and the surface air temperature T_a (K) are from atmospheric forcing input data [10]. The shortwave albedo α (-) and the longwave emissivity ϵ (-) are from satellite-based estimations. σ is the Stefan–Boltzmann constant (5.67×10^{-8} W m^{-2} K^{-4}). The latent heat flux (LE, W m^{-2}) is calculated by $LE = \frac{1}{c}\lambda ET$ and

$$ET = E_c + E_s + E_i + E_u + E_w \tag{5}$$

where $\lambda = 2500 - 2.2(T_a - 273.15)$ is the latent heat of vaporization (kJ kg^{-1}) at T_a, and c (=86.4) is a conversion factor for units from (MJ m^{-2} d^{-1}) to (W m^{-2}). ET is the evapotranspiration (mm d^{-1}) summed from the canopy transpiration (E_c) and soil evaporation (E_s), interception evaporation (E_i), impervious surface evaporation (E_u) and open-water evaporation (E_w).

2.2.2. Canopy Transpiration (E_c) and Soil Evaporation (E_s)

The transpiration at canopy scales (E_c) is coupled with the photosynthesis process (A_{gs}) via the dynamical modulation of the canopy stomatal conductance (G_c), and the soil evaporation (E_s) depends on absorbed energy flux and soil water deficit,

$$E_c = \frac{\varepsilon A_c + (\rho c_p/\gamma)DG_a}{\varepsilon + 1 + G_a/G_c} \tag{6}$$

$$E_s = \frac{f\varepsilon A_s}{\varepsilon + 1} \tag{7}$$

where the surface available energy ($A = R_n - G$) is divided into canopy absorbed energy (A_c) and soil absorbed energy (A_s), $A_c = (1 - \tau)A$ and $A_s = \tau A$, $\tau = exp(-k_A LAI)$, $k_A = 0.6$. $\varepsilon = \frac{\Delta}{\gamma}$, and Δ is the slope of the curve relating saturation water vapor pressure to temperature (kPa °C^{-1}). ρ is the density of air (g m^{-3}); c_p is the specific heat of air at constant pressure (J g^{-1} °C^{-1}); D is vapor pressure deficit (kPa); G_a is the aerodynamic conductance (m s^{-1}); G_c (m s^{-1}) is the canopy conductance; f is a dimensionless variable that determines the water availability for soil evaporation.

The canopy stomatal conductance (G_c) is calculated by the photosynthesis process (A_{gs}) for each PFT with the constraint of D at surface.

$$G_c = \frac{mA_{gs}}{C_a(1 + D/D_0)} \tag{8}$$

$$A_{gs} = \frac{P_1 C_a}{k(P_2 + P_4)}\left\{kLAI + ln\frac{P_2 + P_3 + P_4}{P_2 + P_3\exp(kLAI) + P_4}\right\} \tag{9}$$

$$V_m = \frac{V_{m,25}\exp[a(T - 25)]}{1 + \exp[b(T - 41)]} \tag{10}$$

where $P_1 = A_m \beta I_0 \eta$; $P_2 = A_m \beta I_0$; $P_3 = A_m \eta C_a$; $P_4 = \beta I_0 \eta C_a$; $A_m = 0.5 Vm$. I_0 is the photosynthetically active radiation (PAR, in mol) from shortwave downward radiation. C_a is the atmospheric CO_2 concentration (in ppm or mol mol^{-1}). $V_{m,25}$, β, η, m, D_{min}, D_{max} and D_0 are the parameters (see Table 1) in the PML-V2.1 model.

Finally, the gross primary productivity (GPP) is calculated by

$$GPP = A_{gs} f_D \tag{11}$$

$$f_D = \begin{cases} 1, & D < D_{min} \\ \frac{D_{max} - D}{D_{max} - D_{min}}, & D_{min} < D < D_{max} \\ 0, & D > D_{max} \end{cases} \tag{12}$$

where f_D is the D constraint function; D_{min} and D_{max} are the parameters (see Table 1).

2.2.3. Interception Evaporation (E_i) by Canopy Vegetation

The rainfall interception evaporation (E_i) is calculated by the van Dijk model, which was developed by van Dijk and Bruijnzeel [19,20], who modified it from the original Gash model [21,22]. The modified Van Dijk model used in this study is described by

$$E_i = \begin{cases} f_V P & for \ P < P_{wet} \\ f_V P_{wet} + f_{ER}(P - P_{wet}) & for \ P \geq P_{wet} \end{cases} \tag{13}$$

with

$$f_V = 1 - exp\left(-\frac{LAI}{LAI_{ref}}\right) \tag{14}$$

$$P_{wet} = -ln\left(1 - \frac{f_{ER}}{f_V}\right)\frac{S_V}{f_{ER}} \tag{15}$$

$$f_{ER} = f_V F_{ER0} \tag{16}$$

$$S_V = S_{leaf} LAI \tag{17}$$

where P is rainfall rate (mm d^{-1}), and P_{wet} is reference threshold precipitation rate when the canopy is wet (mm d^{-1}). f_V describes the fraction area covered by intercepting leaves, which is determined by the leaf area index (LAI) and reference LAI (LAI_{ref}, see Table 1) for the specific vegetation types. f_{ER} is the ratio of the average evaporation rate over the average rainfall intensity during storms, and S_V is the canopy rainfall storage capacity (mm), which currently is parameterized as water storage in the leaf at the canopy level. The f_{ER0} and S_{leaf} are the parameters shown in Table 1.

2.2.4. Impervious Surface Evaporation (E_u)

Impervious surface evaporation (E_u) is calculated by the revised van Dijk model in this study,

$$E_u = \begin{cases} f_u P & for \ P < P_{wet} \\ f_u P_{wet} + f_{ER}(P - P_{wet}) & for \ P \geq P_{wet} \end{cases} \tag{18}$$

with

$$f_u = 1 - f_V - f_w \tag{19}$$

$$P_{wet} = -ln\left(1 - \frac{f_{ER}}{f_u}\right)\frac{S_u}{f_{ER}} \tag{20}$$

$$f_{ER} = f_u F_{ER0} \tag{21}$$

where f_u describes the fractional area covered by impervious surface in urban ecosystems, which is the rest fraction of vegetation coverage (f_V) and water body covered fraction (f_w). S_u is the impervious surface storage capacity (mm).

2.2.5. Open-Water Evaporation (E_w)

The open-water evaporation (E_w) in lakes, rivers and other water bodies for Northern China is parameterized on the basis of Dalton's Law [23],

$$E_w = 0.144(1 + 0.75U_{1.5})[D + \Delta(T_{1.5})(\alpha - 1)T_{1.5}] \tag{22}$$

where $U_{1.5}$ and $T_{1.5}$ is wind speed (m/s) and air temperature (°C) at 1.5 m height, respectively. $\Delta(T_{1.5})$ is the slope of the curve relating saturation water vapor pressure to temperature $T_{1.5}$. $\alpha - 1$ is a regulator coefficient for atmospheric stability.

3. Results

3.1. Validation of Estimated LAI and ET

The LAI estimated from the Sentinel-2-based NDVI was compared to the observed LAI for June 2018 in the Beijing Sponge City (Figure 2a). The observed LAI was measured within the region around 39.50–40.50° N, 115.40–116.10° E. The result shows that the Sentinel-2-based LAI has a high correlation with the observed values ($R^2 = 0.74$), indicating that the LAI at 10m resolution estimated from Sentinel-2 can be well applied to estimate ET for the Beijing Sponge City.

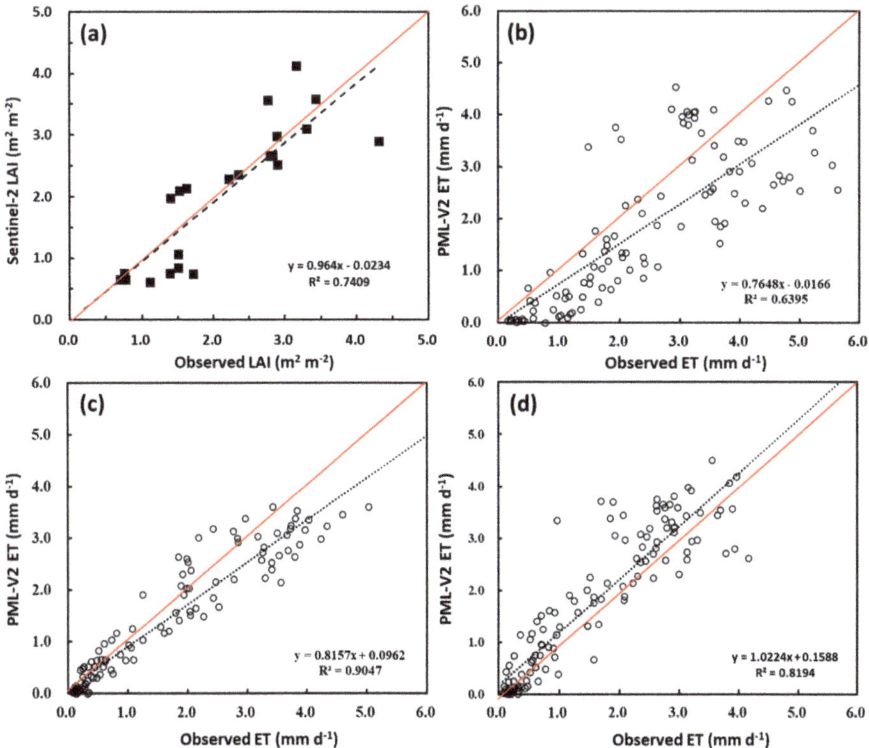

Figure 2. Validation of the Sentinel-2-retrieved LAI and the PML-V2-simulated ET. (**a**) The Sentinel-2-retrieved LAI compared to observed LAI for June 2018; (**b–d**) The PML-V2-simulated 8-day ET based on MODIS LAI compared to observed ET for the three flux tower sites at Daxing, Miyun and Guantao, respectively, in the Beijing area over 2008–2010.

We also validate the performance of PML-V2 in simulating ET in the Beijing region. Figure 2b–d shows the comparisons of PML-V2-estimated ET based on MODIS LAI with the observed ET over 2008–2010 from three field sites (i.e., Daxing, Miyun and Guantao) which are located within the Beijing region. The result indicates that PML-V2 has satisfied performance in simulating ET for Beijing Sponge City with (R^2 = 0.64–0.90). Therefore, based on the above good performance in the Sentinel-2-estimated LAI and the PML-V2-estimated ET, we further evaluate NDVI, LAI, and ET and related variables at 10m resolution for Beijing Sponge City.

3.2. Land Use and Vegetation Information in Beijing Sponge City

By analyzing the Sentinel-2-derived 10m resolution LCC map in 2017, we find that the Beijing Sponge City project (Figure 3) covers ~1265 km^2 over the central Beijing city, China, including impervious surface buildings (59.27%), grasslands (26.08%), mixed forest (7.34%), croplands (5.10%), and wetlands and water bodies (~2%). Figure 3 presents fine spatial details of the urban ecosystem, such as clear patterns of lakes, rivers and streets. Most grasslands are parks, and fixed forests are mainly concentrated in northwestern Beijing Sponge City, while the eastern parts are croplands.

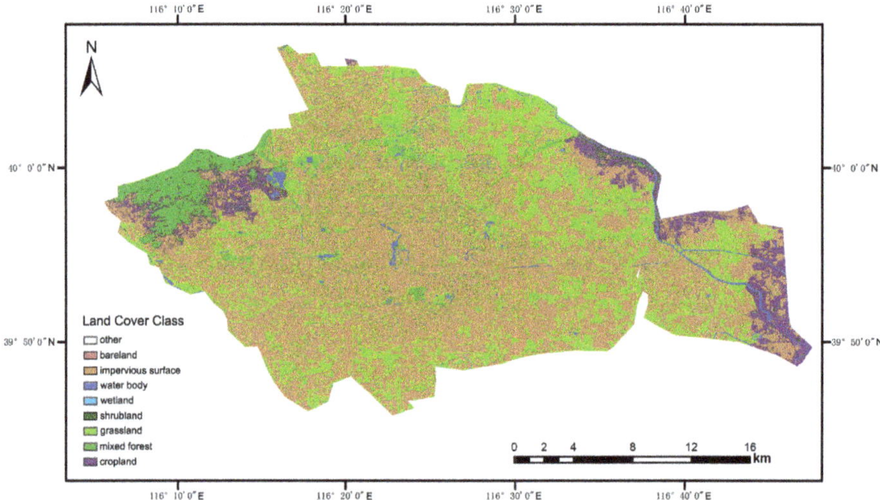

Figure 3. Land cover classification at 10m resolution for Beijing Sponge City. Land cover map for 2017 derived from the FROM-GLC10-based Sentinel-2.

We further analyze the NDVI and LAI for June 2018, which was composited from 10-day Sentinel-2 images in June 2018 in clear-sky conditions. We can see high spatial details in the NDVI from Figure 4a. The NDVI for lakes and rivers is ≤0.0, the impervious surfaces (e.g., large mansions and main streets) are 0.0–0.25, and grasslands and croplands are 0.1–0.5, while mixed forests and some parks with forest reserve have NDVI values of 0.5–0.7 (Figure 4a). A high NDVI indicates a high LAI in this study. Figure 4b shows the detailed pattern of LAI for different land cover classes for the sponge city area. As expected, the lakes and rivers have no LAI, and the impervious surfaces (e.g., large mansions and main streets) have only <1.0 m^2 m^{-2} of LAI, but 1~2 m^2 m^{-2} of LAI can be seen in many avenues with greenbelts. The mixed forests in northwestern Beijing Sponge City have LAI values ranging from 1 to 3 m^2 m^{-2}. Most grasslands and croplands have 1~2 m^2 m^{-2} of LAI, but some parks with forest reserves show the highest values (3–8 m^2 m^{-2}) of LAI (Figure 4b).

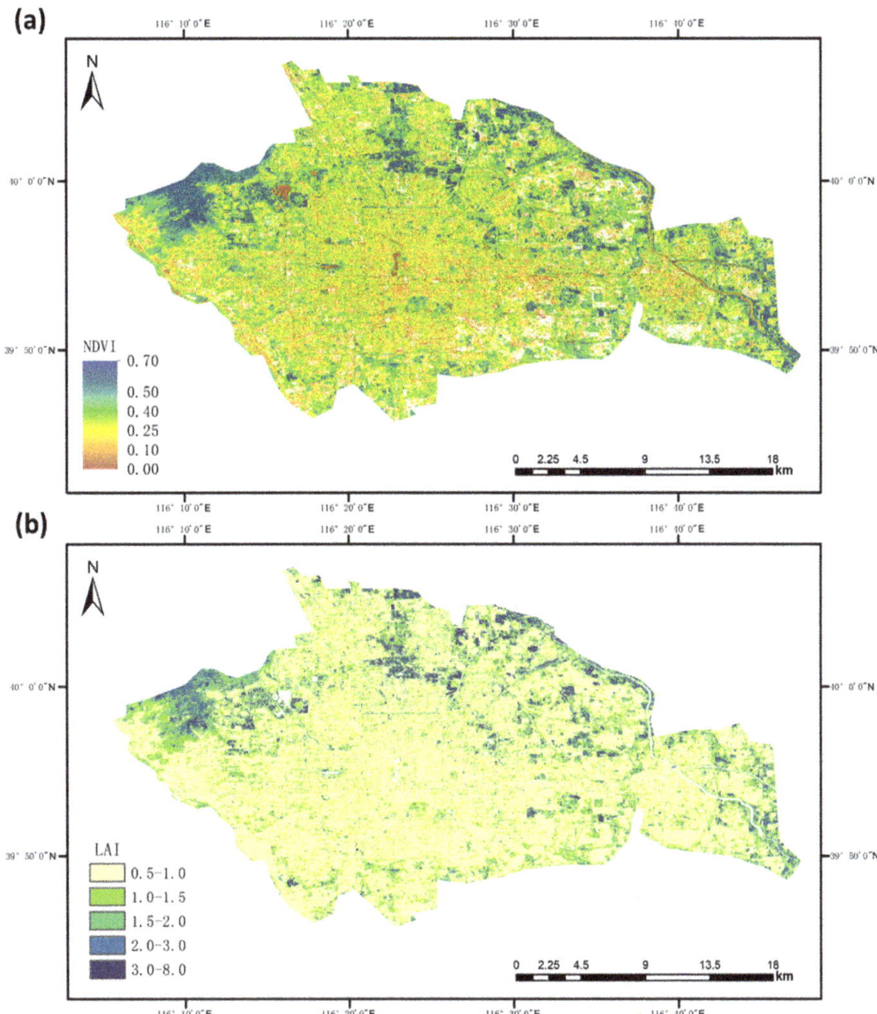

Figure 4. Vegetation index at 10m resolution for Beijing Sponge City. (**a**) NDVI and (**b**) LAI (m^2 m^{-2}) for June 2018 derived from the Sentinel-2.

3.3. ET and Related Variables in Beijing Sponge City

Based on the Sentinel-2-derived LCC and LAI data, we conducted daily simulations for June 2018 at 10m resolution using the PML-V2 model with daily climate forcing data from CMFD v1.0. Figure 5 presents the spatial patterns of simulated monthly ET and GPP averaged over the daily output for June 2018. Lakes and rivers have the highest ET ($\geq$8 mm d^{-1}) due to the full water supply for evaporation in Summer. There is no GPP in lakes and rivers as simulated by PML-V2. The vegetation production activities are strongest in mixed forests and croplands, with the GPP ranging from 8 to 16 gC m^{-2} d^{-1} (Figure 5b). In these mixed forests and croplands, the ET is also high (4–6 mm d^{-1}), where the plant transpiration (Ec) plays a dominant role with ratios of Ec to ET larger than 0.8 (Figure 6a). In addition, the impervious surfaces have very small ET (<2 mm d^{-1}) and GPP (<4 gC m^{-2} d^{-1}), indicating both Ec and soil evaporation (Es) are very small in these areas. The grasslands have 2–4 mm d^{-1} of ET and 4–8 gC m^{-2} d^{-1} of GPP in the sponge city (Figure 5), where the ratio Ec/ET are 0.5–0.7 (Figure 6a).

In summary, on average, for the whole sponge city, we find that the LAI in June 2018 is 0.83 m^2 m^{-2}; the ET is about 1.6 mm d^{-1}; the GPP is 2.8 gC m^{-2} d^{-1}. Table 2 further gives the evaluation for different districts in the sponge city. It shows that the central areas (i.e., Xicheng and Dongcheng districts) have the smallest LAI (0.66–0.7 m^2 m^{-2}), ET (~1.61 mm d^{-1}) and GPP (2.36–2.44 gC m^{-2} d^{-1}), while the western areas (i.e., Shijingshan and Haidian districts) have the highest LAI (0.93–1.05 m^2 m^{-2}), ET (~1.65 mm d^{-1}) and GPP (3.10–3.53 gC m^{-2} d^{-1}).

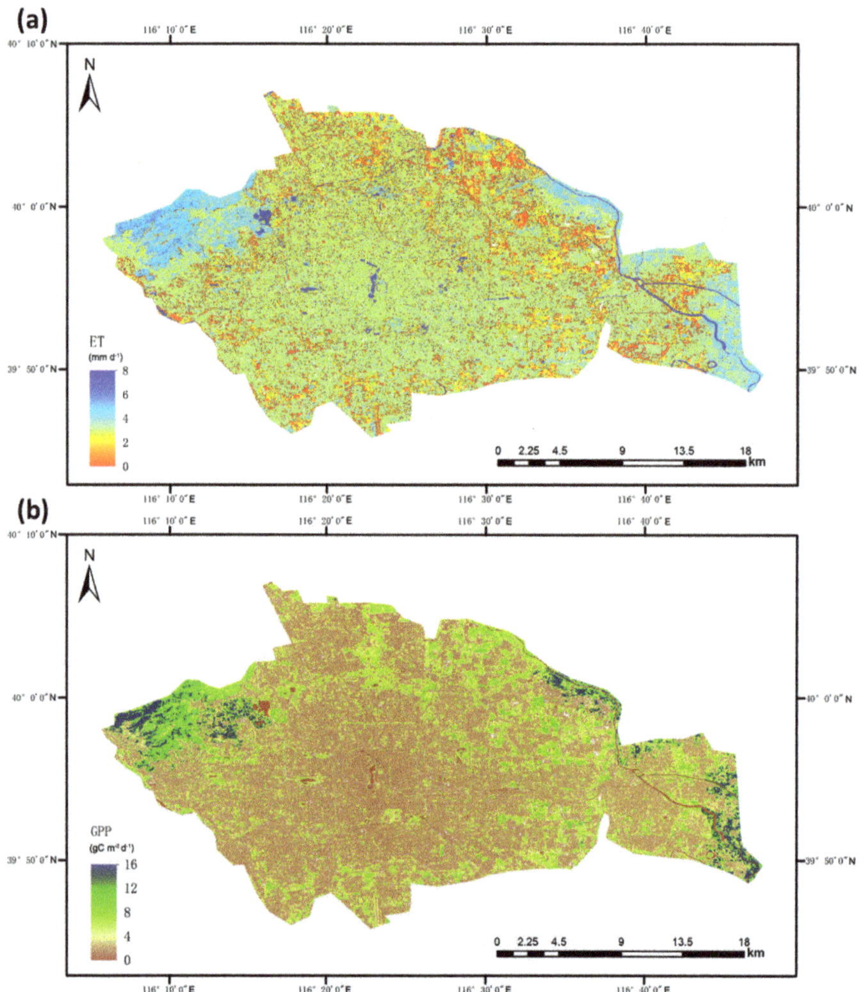

Figure 5. ET and GPP at 10m resolution for Beijing Sponge City estimated using the PML-V2 model and Sentinel-2 data. (**a**) Monthly ET (mm d^{-1}) in June 2018; (**b**) Monthly GPP (gC m^{-2} d^{-1}) in June 2018.

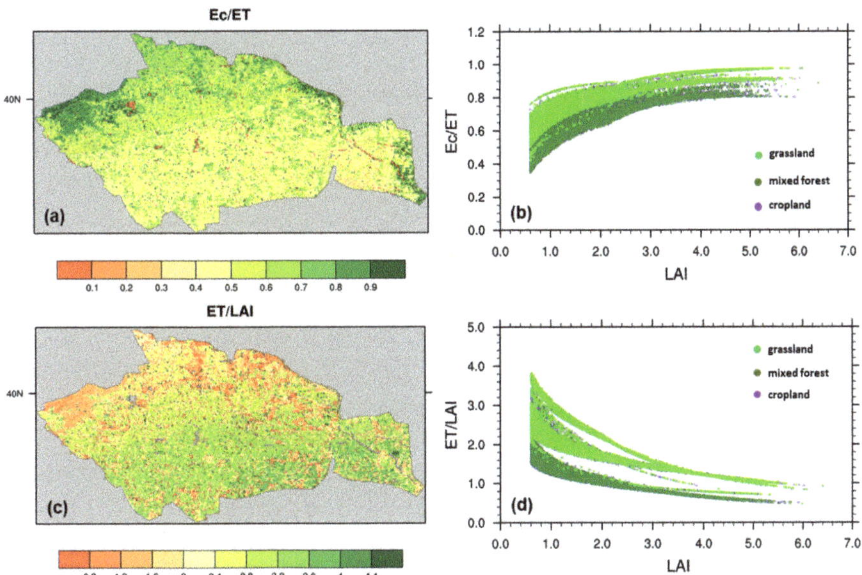

Figure 6. Spatial sensitivity of ET to LAI for the Beijing Sponge City. (**a**) Spatial pattern of the fraction of Ec to ET in June 2018; (**b**) Change in the fraction of Ec to ET with LAI; (**c**) Spatial pattern of the ratio of ET to LAI in June 2018; (**d**) Change in ratio of ET to LAI with LAI.

Table 2. The ecohydrological environment in different districts in Beijing Sponge City.

District in Beijing Sponge City	LAI ($m^2\ m^{-2}$)	ET ($mm\ d^{-1}$)	GPP ($gC\ m^{-2}\ d^{-1}$)
Xicheng	0.66	1.62	2.36
Dongcheng	0.70	1.60	2.44
Shijingshan	1.05	1.67	3.53
Haidian	0.93	1.64	3.10
Chaoyang	0.84	1.51	2.79
Fengtai	0.74	1.50	2.56
Tongzhou	0.86	1.66	3.01
Overall Mean	0.83	1.60	2.83

We further investigate how ET changes spatially with increasing LAI for June 2018. It is shown that the fraction Ec/ET increases with LAI for the three vegetation types (grassland, mixed forest and cropland) in the sponge city (Figure 6b). The Ec/ET for mixed forests increase from 0.4 (LAI = 0.5) to 0.8 (LAI > 3), while Ec/ET for grasslands and croplands show higher values, increasing from 0.6 (LAI = 0.5) to 0.9 (LAI > 3). The ratio ET/LAI represents the amount of water loss per unit LAI. In Figure 6c, we can find that ET/LAI for mixed forests and some grassland parks show the lowest ET/LAI (0.8–1.2), while impervious surfaces have the highest ET/LAI, with about 2-3 times larger values (2.4–3.6). The ET/LAI for the major vegetation types (grassland, mixed forest and cropland) decrease with LAI increase (Figure 6d); with LAI increasing from 0.5 to 5.0, the ET/LAI for mixed forests decrease from 2.0 to 0.6, and ET/LAI for grasslands and croplands from 3.0 to 1.0. This result indicates that grasslands and croplands have much higher water consumption per unit LAI than mixed forests.

Fractional contributions of other ET components to ET have been estimated (Figure 7). Soil evaporation (Es) contributes a relatively small fraction ($\leq$0.2) due to a very small fraction of bare lands and large vegetation coverage in mixed forests, grasslands and croplands in June 2018 in the Beijing Sponge City. The fraction of Ei/ET is small (≤ 0.1) in most LCC types but in grasslands is 0.1–0.2 (Figure 7b). According to a previous study, the city impervious surface could contribute less than 20% of ET [4]. Surprisingly, the impervious surface evaporation (Eu) for the Beijing Sponge City contributes 0.2–0.3 fractions to the ET in June 2018 in most impervious areas (Figure 7c). Finally, all ET from lakes and rivers are contributed by water body evaporation Ew (Figure 7d).

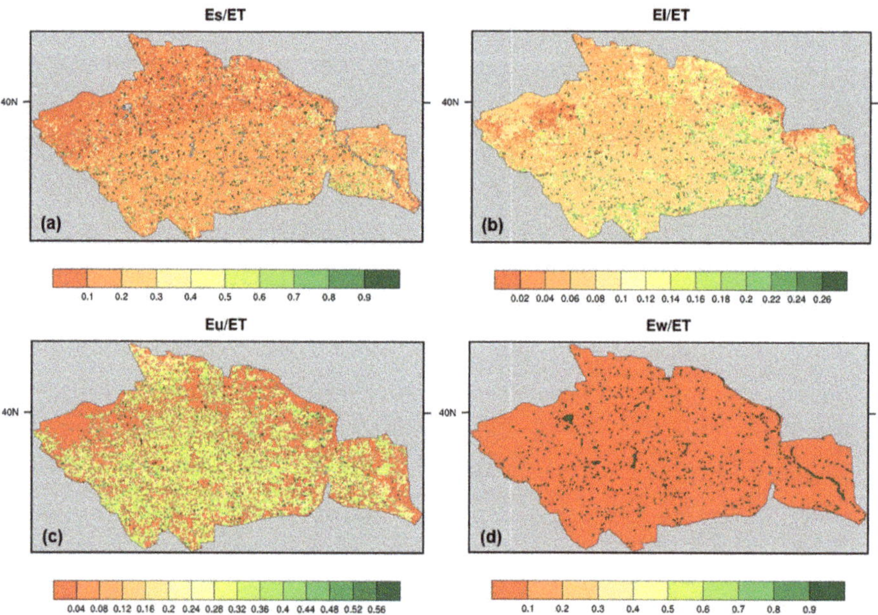

Figure 7. The fraction of Es, Ei, Eu and Ew to ET in June 2018 for the Beijing Sponge City. (**a**) Spatial pattern of fraction Es/ET; (**b**) Spatial pattern of fraction Ei/ET; (**c**) Spatial pattern of fraction Eu/ET; (**d**) Spatial pattern of fraction Ew/ET.

ET can be converted to latent heat flux (LE, in W m^{-2}) and plays an important role in regulating surface energy balance (Figure 8). For June 2018 in Beijing, the surface receives about 260–270 W m^{-2} of shortwave radiation, but about half is reflected in the atmosphere with the net radiation (Rn) for impervious surface less than 130 W m^{-2}. For mixed forests and grasslands, the Rn is about 140–150 W m^{-2}, and the croplands and water bodies have a higher Rn of 170–180 W m^{-2} (Figure 8b). Lakes and rivers have the highest LE (>250 W m^{-2}) but the smallest sensible heat flux (SH, <-60 W m^{-2}). The SH for mixed forests, forest parks and croplands are relatively small (-20 to 20 W m^{-2}), while both impervious surfaces and grasslands are high (40–60 W m^{-2}) (Figure 8d). This result indicates that the high surface air temperature (reflected by SH) in the summer of Beijing's central city is mostly contributed by impervious surfaces and grasslands.

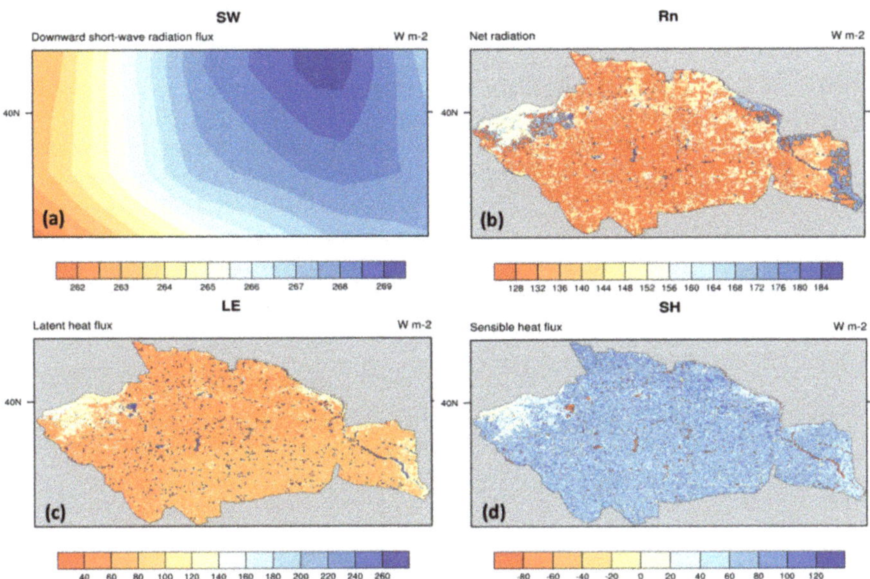

Figure 8. Energy fluxes in June 2018 for the Beijing Sponge City. (**a**) Downward shortwave radiation; (**b**) Surface net radiation; (**c**) Latent heat flux; (**d**) Sensible heat flux. Units are in W m^{-2}.

4. Discussion

In this study, we applied a water-carbon coupling model (PML-V2) with the use of Sentinel-2 LAI and land use and cover data and with surface meteorological forcing data as the input to estimate urban ET and its components at 10m resolution for the Beijing Sponge City. Our results indicate that the current vegetation coverage for the Beijing Sponge City is still at a low level (only with mean LAI < 1 m^2 m^{-2} in June 2018), and the city gains relatively limited benefits from urban ecosystem services.

Eddy covariance measurements to study evaporation from urban ecosystems [24,25] generally helped us to understand the combined evaporation from all land cover types, lacking the capability to divide individual contributions from such as impervious surfaces (roofs, roads and plazas, etc.) and vegetated areas (bare soil, forests, grasslands and croplands, etc.). For different land cover classes, by using an advanced water-carbon coupling ET model driven by Sentinel-2 LAI, we find that lakes and rivers have the highest ET ($\geq$8 mm d^{-1}), followed by mixed forests and croplands (ET is 4–6 mm d^{-1} and LAI is 2–3 m^2 m^{-2}), where the plant transpiration (Ec) plays the dominant role (>80%), then grasslands have 2–4 mm d^{-1} of ET, where the LAI is 1~2 m^2 m^{-2}, while impervious surfaces have smallest ET (<2.0 mm d^{-1}). In most impervious areas, the impervious surface evaporation (Eu) contributes 20-30% of ET, which is larger than the estimate (18%) from previous studies [4]. We have shown that another 40–50% of ET in impervious areas are contributed by plant transpiration (Ec) due to the small fractional area covered by greenbelts with trees and grassland (LAI is 0.5–1.0 m^2 m^{-2}). This study did not consider water vapor conversion from human water use activities, which also contributes to the impervious evaporation from building indoor water use [25].

5. Conclusions

First of all, we show the good performances of the nonlinear regression model (Equation (2)) for estimating Sentinel-2 LAI based on the strong exponential relationship between NDVI and LAI and the PML-V2.1 model of estimating ET and GPP at 10m resolution using Sentinel-2 LAI and land cover map. This Sentinel-2-based technology using the

PML-V2.1 model with surface meteorological forcing data as the input for estimating ET provides a new framework to evaluate city-level hydrological dynamics in urban ecosystems. Secondly, we find that plant transpiration from greenbelts with trees and grassland play an important role in most impervious areas for Beijing Sponge City. Thirdly, mixed forests, forest parks and croplands due to their high ET have much smaller surface heat contribution than the impervious and grasslands, providing better ecosystem services (e.g., cooling) for the sponge city.

Author Contributions: Conceptualization, X.Z.; methodology, X.Z.; software, X.Z. and P.S.; validation, X.Z. and P.S.; formal analysis, X.Z.; investigation, X.Z.; resources, X.Z.; data curation, X.Z.; writing—original draft preparation, X.Z.; writing—review and editing, X.Z. and P.S.; visualization, X.Z. and P.S.; supervision, X.Z.; project administration, X.Z.; funding acquisition, X.Z. All authors have read and agreed to the published version of the manuscript.

Funding: This research was funded by the Beijing Municipal Natural Science Foundation (grant number 8212017), the National Natural Science Foundation of China (grant number 42001019) and the Shanghai Sailing Program (grant number 19YF1413100). The APC was funded by the Beijing Municipal Natural Science Foundation (grant number 8212017).

Data Availability Statement: The resulting dataset in this study is available at https://doi.org/10.6 084/m9.figshare.14387630.v2 (accessed on 8 April 2021). More data and codes can be accessible from the corresponding author upon request.

Acknowledgments: We acknowledge the Beijing Water Science and Technology Institute (BWSTI) for providing technical and financial supports to this work. We thank the three anonymous reviewers and editors for their constructive suggestions and comments.

Conflicts of Interest: The authors declare no conflict of interest.

References

1. Teuling, A.J.; de Badts, E.A.G.; Jansen, F.A.; Fuchs, R.; Buitink, J.; Hoek van Dijke, A.J.; Sterling, S.M. Climate change, reforestation/afforestation, and urbanization impacts on evapotranspiration and streamflow in Europe. *Hydrol. Earth Syst. Sci.* **2019**, *23*, 3631–3652. [CrossRef]
2. Gong, P.; Liang, S.; Carlton, E.J.; Jiang, Q.; Wu, J.; Wang, L.; Remais, J.V. Urbanisation and health in China. *Lancet* **2012**, *379*, 843–852. [CrossRef]
3. Wang, X.; Liu, H.; Miao, S.; Wu, Q.; Zhang, N.; Qiao, F. Effectiveness of Urban Hydrological Processes in Mitigating Urban Heat Island and Human Thermal Stress during a Heat Wave Event in Nanjing, China. *J. Geophys. Res. Atmos.* **2020**, *125*. [CrossRef]
4. Ramamurthy, P.; Bou-Zeid, E. Contribution of impervious surfaces to urban evaporation. *Water Resour. Res.* **2014**, *50*, 2889–2902. [CrossRef]
5. Wang, L.; Huang, M.; Li, D. Where Are White Roofs More Effective in Cooling the Surface? *Geophys. Res. Lett.* **2020**, *47*. [CrossRef]
6. Zhang, Y.; Xia, J.; Yu, J.; Randall, M.; Zhang, Y.; Zhao, T.; Pan, X.; Zhai, X.; Shao, Q. Simulation and assessment of urbanization impacts on runoff metrics: Insights from landuse changes. *J. Hydrol.* **2018**, *560*, 247–258. [CrossRef]
7. Gong, P.; Liu, H.; Zhang, M.; Li, C.; Wang, J.; Huang, H.; Clinton, N.; Ji, L.; Li, W.; Bai, Y.; et al. Stable classification with limited sample: Transferring a 30-m resolution sample set collected in 2015 to mapping 10-m resolution global land cover in 2017. *Sci. Bull.* **2019**, *64*, 370–373. [CrossRef]
8. Sentinel-2. Overview of Sentinel-2 Mission. 2015. Available online: https://sentinel.esa.int/web/sentinel/missions/sentinel-2 (accessed on 8 April 2021).
9. The People's Government of Beijing Municipality. Announcement of Beijing Sponge City Project. 2017. Available online: http://www.beijing.gov.cn/zhengce/zhengcefagui/201905/t20190522_60725.html (accessed on 8 April 2021).
10. He, J.; Yang, K.; Tang, W.; Lu, H.; Qin, J.; Chen, Y.; Li, X. The first high-resolution meteorological forcing dataset for land process studies over China. *Sci. Data* **2020**, *7*, 25. [CrossRef]
11. Farquhar, G.D.; von Caemmerer, S.; Berry, J.A. A biochemical model of photosynthetic CO_2 assimilation in leaves of C 3 species. *Planta* **1980**, *149*, 78–90. [CrossRef]
12. Yu, Q.; Zhang, Y.; Liu, Y.; Shi, P. Simulation of the stomatal conductance of winter wheat in response to light, temperature and CO_2 changes. *Ann. Bot.* **2004**, *93*, 435–441. [CrossRef]
13. Monteith, J.L. Evaporation and environment. The state and movement of water in living organisms. In *Symposium of the Society of Experimental Biology*; Fogg, G.E., Ed.; Cambridge University Press: Cambridge, UK, 1965; Volume 19, pp. 205–234.
14. Leuning, R.; Zhang, Y.Q.; Rajaud, A.; Cleugh, H.; Tu, K. A simple surface conductance model to estimate regional evaporation using MODIS leaf area index and the Penman-Monteith equation. *Water Resour. Res.* **2008**, *44*. [CrossRef]

15. Zhang, Y.; Leuning, R.; Hutley, L.B.; Beringer, J.; McHugh, I.; Walker, J.P. Using long-term water balances to parameterize surface conductances and calculate evaporation at 0.05° spatial resolution. *Water Resour. Res.* **2010**, *46*. [CrossRef]

16. Zhang, Y.; Kong, D.; Gan, R.; Chiew, F.H.S.; McVicar, T.R.; Zhang, Q.; Yang, Y. Coupled estimation of 500 m and 8-day resolution global evapotranspiration and gross primary production in 2002–2017. *Remote Sens. Environ.* **2019**, *222*, 165–182. [CrossRef]

17. Zhang, Y.; Pena-Arancibia, J.L.; McVicar, T.R.; Chiew, F.H.; Vaze, J.; Liu, C.; Lu, X.; Zheng, H.; Wang, Y.; Liu, Y.Y.; et al. Multi-decadal trends in global terrestrial evapotranspiration and its components. *Sci. Rep.* **2016**, *6*, 19124. [CrossRef] [PubMed]

18. Gan, R.; Zhang, Y.; Shi, H.; Yang, Y.; Eamus, D.; Cheng, L.; Chiew, F.H.S.; Yu, Q. Use of satellite leaf area index estimating evapotranspiration and gross assimilation for Australian ecosystems. *Ecohydrology* **2018**, *11*. [CrossRef]

19. van Dijk, A.I.J.M.; Bruijnzeel, L.A. Modelling rainfall interception by vegetation of variable density using an adapted analytical model. Part 1. Model description. *J. Hydrol.* **2001**, *247*, 230–238. [CrossRef]

20. van Dijk, A.I.J.M.; Bruijnzeel, L.A. Modelling rainfall interception by vegetation of variable density using an adapted analytical model. Part 2. Model validation for a tropical upland mixed cropping system. *J. Hydrol.* **2001**, *247*, 239–262. [CrossRef]

21. Gash, J.H.C. An analytical model of rainfall interception by forests. *Q. J. R. Meteorol. Soc.* **1979**, *105*, 43–55. [CrossRef]

22. Gash, J.H.C.; Lloyd, C.R.; Lachaud, G. Estimating sparse forest rainfall interception with an analytical model. *J. Hydrol.* **1995**, *170*, 79–86. [CrossRef]

23. Hong, J.; Wang, S. Estimation and distribution characteristics of evaporation from water surfaces in Beijing, Tianjin and Tangshan areas. *Geogr. Res.* **1987**, *6*, 68–75. (In Chinese)

24. Hanna, S.; Marciotto, E.; Britter, R. Urban Energy Fluxes in Built-Up Downtown Areas and Variations across the Urban Area, for Use in Dispersion Models. *J. Appl. Meteorol. Climatol.* **2011**, *50*, 1341–1353. [CrossRef]

25. Zhou, J.; Liu, J.; Yan, D.; Wang, H.; Wang, Z.; Shao, W.; Luan, Y. Dissipation of water in urban area, mechanism and modelling with the consideration of anthropogenic impacts: A case study in Xiamen. *J. Hydrol.* **2019**, *570*, 356–365. [CrossRef]

remote sensing

Article

Impending Hydrological Regime of Lhasa River as Subjected to Hydraulic Interventions—A SWAT Model Manifestation

Muhammad Yasir, Tiesong Hu * and Samreen Abdul Hakeem

State Key Laboratory of Water Resources and Hydropower Engineering Science, Wuhan University, Wuhan 430072, China; muhammadyasir@whu.edu.cn (M.Y.); samreen@whu.edu.cn (S.A.H.)
* Correspondence: tshu@whu.edu.cn

Abstract: The damming of rivers has altered their hydrological regimes. The current study evaluated the impacts of major hydrological interventions of the Zhikong and Pangduo hydropower dams on the Lhasa River, which was exposed in the form of break and change points during the double-mass curve analysis. The coefficient of variability (CV) for the hydro-meteorological variables revealed an enhanced climate change phenomena in the Lhasa River Basin (LRB), where the Lhasa River (LR) discharge varied at a stupendous magnitude from 2000 to 2016. The Mann–Kendall trend and Sen's slope estimator supported aggravated hydro-meteorological changes in LRB, as the rainfall and LR discharge were found to have been significantly decreasing while temperature was increasing from 2000 to 2016. The Sen's slope had a largest decrease for LR discharge in relation to the rainfall and temperature, revealing that along with climatic phenomena, additional phenomena are controlling the hydrological regime of the LR. Reservoir functioning in the LR is altering the LR discharge. The Soil and Water Assessment Tool (SWAT) modeling of LR discharge under the reservoir's influence performed well in terms of coefficient of determination (R^2), Nash–Sutcliffe coefficient (NSE), and percent bias (PBIAS). Thus, simulation-based LR discharge could substitute observed LR discharge to help with hydrological data scarcity stress in the LRB. The simulated–observed approach was used to predict future LR discharge for the time span of 2017–2025 using a seasonal AutoRegressive Integrated Moving Average (ARIMA) model. The predicted simulation-based and observation-based discharge were closely correlated and found to decrease from 2017 to 2025. This calls for an efficient water resource planning and management policy for the area. The findings of this study can be applied in similar catchments.

Keywords: SWAT; double-mass analysis; coefficient of variability; seasonal ARIMA; MK-S trend analysis

Citation: Yasir, M.; Hu, T.; Abdul Hakeem, S. Impending Hydrological Regime of Lhasa River as Subjected to Hydraulic Interventions—A SWAT Model Manifestation. *Remote Sens.* **2021**, *13*, 1382. https://doi.org/10.3390/rs13071382

Academic Editors: Luca Brocca, Carl Legleiter and Yongqiang Zhang

Received: 4 February 2021
Accepted: 31 March 2021
Published: 3 April 2021

Publisher's Note: MDPI stays neutral with regard to jurisdictional claims in published maps and institutional affiliations.

1. Introduction

Dams are intended to offer substantial aid to humankind by ensuring an enhanced water availability for municipal, industrial, and agricultural uses, as well as increased capability of flood regulation and hydropower generation [1]. On the other hand, the construction of dams has considerably changed the natural flow regime of rivers worldwide. Above half of the 292 large river systems in the world have been affected by dams [2,3]. The influence of human activities in altering river discharge has profoundly increased in recent decades [4]. Over an intermediate time scale (e.g., decadal scale), human interferences in terms of water consumption, land-use change, dam construction, and sand mining, among others, are the powerful factors that escalate basin-scale hydrological changes. Therefore, a site-specific study is needed to disclose the governing effects of human disruptions on these hydrological changes [5–7].

To temper river floods, reduce water collection for irrigation, hydropower generation and facilitate navigation, dams have been created across big rivers around the world [8]. Dams have grown to one of the most perturbing human intrusions in river systems as the

67

number of dams and the total storage capacity of reservoirs rapidly increase [4]. Therefore, knowledge of dam construction and its regulating effects on river discharge is crucial for river and delta management and restoration. Highly regulated rivers in China are subject to large-scale ecosystem amendments made by hydrological alterations. Many of the earlier studies related to dam-induced hydrological alterations across river basins in China focused on the impacts of large dams that generally aim to control floods in large basins, such as the Lancang River [9], the Mekong River [10], the Pearl River [11], the Yangtze River [12], and the Yellow River [13,14]. In addition to large dams, the development of small dams has also been highlighted in national energy policies in China [15]. Therefore, small dam construction is intense in China, especially in South China, where hydropower resources are extensive. Thus, to fill in the knowledge void, the present study focused on the impact appraisal of reservoir functioning in the Lhasa River Basin, a Qinghai–Tibetan Plateau basin in South China (for more information, see Section 2.2). Several researchers have established a number of approaches with the objective of reckoning of the hydrologic modifications caused by human activity. However, hydrological modeling can be an effective alternative for hydrological analysis in different scenarios [16]. The SWAT (Soil and Water Assessment Tool) model developed by the authors of [17] has already been in widespread use for water resource management in many different rivers [18–23]. Additionally, there has been a general lack of applications of physically-based hydrological models to the Yarlung Tsangbo River Basin, especially the Lhasa River Basin [24]. The SWAT model was applied to the Lhasa River Basin in a recent study [25], where streamflow and sediment load were predicted for the Lhasa River in future. The SWAT model was applied to the Lhasa River Basin to simulate its streamflow variability under reservoir influence [26]. The SWAT model was utilized in [27] for hydrological drought propagation in the South China Dongjiang River Basin using the "simulated–observed approach". Their study estimated the effects of human regulations on hydrological drought from the perspective of the development and recovery processes using the SWAT model.

Streamflow forecasting is of great significance to water resource management and planning. Medium-to-long-term forecasting including weekly, monthly, seasonal, and even annual time scales is predominantly beneficial in reservoir operations and irrigation management, as well as the institutional and legal features of water resource management and planning. Due to their reputation, a large number of forecasting models have been developed for streamflow forecasting, including concept-based, process-driven models such as the low flow recession model, rainfall–runoff models, and statistics-based data-driven models such as regression models, time series models, artificial neural network models, fuzzy logic models, and the nearest neighbor model [28]. Of various streamflow forecasting methods, time series analysis has been most widely used in the previous decades because of its forecasting capability, inclusion of richer information, and more systematic way of building models in three modeling stages (identification, estimation, and diagnostic check), as standardized by Box and Jenkins (1976) [29]. The current study made use of "simulated–observed approach" after [27] for predicting the Lhasa River streamflow under reservoir operations in the Lhasa River Basin. SWAT-simulated and observed hydrological time series were used introduced to a stochastic AutoRegressive Integrated Moving Average (ARIMA) model. As a common data-driven method, the ARIMA model has been widely used in time series prediction due to its simplicity and effectiveness [30]. Adhikary et al. (2012) [31] used seasonal ARIMA (SARIMA) model to model a groundwater table. They took weekly time series and concluded that SARIMA stochastic models can be applied for ground water level variations. Valipour et al. (2013) [32] modeled the inflow of the Dez dam reservoir with SARIMA and ARMIA stochastic models. His research results showed that the SARIMA model yielded better results than the ARIMA model. Ahlert and Mehta (1981) [33] analyzed the stochastic process of flow data for the Delaware River by the ARIMA model. Yurekli et al. (2005) [34] applied SARIMA stochastic models to model the monthly streamflow data of the Kelkit River. Modarres and Ouarda (2013) [35] demonstrated the heteroscedasticity of streamflow time series with the ARIMA model in comparison to GARCH (Generalized Autoregressive Conditionally Heteroscedastic) models. Their results showed that ARIMA models performed better than GARCH models. Ahmad et al. (2001) [36] used the ARIMA model to analyze water quality data. Kurunç et al. (2005) [37] used the ARIMA and Thomas Fiering stochastic approach to forecast streamflow data of the Yesilurmah River. Tayyab et al. (2016) [38] used an auto regressive model in comparison to neural networks to predict streamflow.

The current study primarily aimed to (i) investigate the reservoir operations' impact on the Lhasa River discharge, (ii) apply the SWAT model to simulate Lhasa River streamflow under multiple reservoir functioning, and (iii) to predict Lhasa River streamflow under reservoir's influence using 'observed' and 'SWAT-simulated' hydrological data series as a step forward in overcoming the data scarcity problem of the area. The study was intended to benefit water resource managers and hydrological engineers in understanding the future hydrological regime in the Lhasa River Basin under reservoir functioning and aiding in developing better management practices and planning for hydrological resources in the area. The current study holds novelty in combining a physical-based hydrological model and a statistical time series forecasting model for the simulation and prediction, of the discharge of the Lhasa River respectively, one of the important rivers in the data-scarce Qinghai–Tibetan Plateau, which is under the influence of recent major hydraulic interventions in the form of the Zhikong and Pangduo hydropower reservoirs.

2. Materials and Methods

2.1. Study Area—Lhasa River Basin

The Lhasa River Basin (LRB), ranging from 29°19′ to 31°15′N and from 90°60′ to 93°20′E, is the economical and authoritative hub of the autonomous Qinghai–Tibetan plateau (QTP). The Lhasa River (LR) is the longest tributary of the Yarlung Tsangpo River; LRB covers a ≈32,321 km² basin area (ArcSWAT-estimated area by the digital elevation model used in the current study), comprising 13.5% of the total area of the Yarlung Tsangpo basin [39]. The LRB exhibits typical semi-arid monsoonal climate conditions, where the major proportion of received rainfall is concentrated in the summer season from June to September with the simultaneous generation of peak LR discharge during the same time. Peng et al. (2015) [24] showed that rainfall in summer is a governing feature in producing summer stream flow in the Lhasa River basin. Thus, the rainfall disproportion poses a direct influence on the rainfall-dependent runoff generation phenomena in the basin. The hydrometric and meteorological records for the LRB are maintained at the Pondo, Tanggya, and Lhasa hydrometric stations and the Damxung, Maizhokunggar, and Lhasa meteorological stations, respectively.

The LR stretches to a length of 551 km with a hydropower potential of 1.177 million kWh [39], and it is substantial in fulfilling the hydropower and agricultural requirements of the local community. The LR has been exposed to major hydraulic interventions in the form of reservoir development and confinement during the last and present decades. It is of vital importance to understand the hydrological phenomena of the LRB under the influence of hydraulic structures for a better understanding of the hydrological behavior of the study area to facilitate the understanding of future water resource availability and management in the area. The major hydraulic developments in the study area are the installation of Zhikong and Pangduo hydropower stations over the LR.

Zhikong and Pangduo Reservoirs Impoundment on Lhasa River

The Zhikong and Pangduo Dams were built in 2006 and 2013, respectively, on the LR. The Zhikong Dam is located 96 km upstream the urban Lhasa city in the middle and lower reaches of the LR, and it is 65 km downstream the Pangduo Dam, thus impounding the upper LR. To meet the substantially increasing power demand of the Tibet plateau, the Zhikong Dam was built with an installed power capacity of 100 MW and a reservoir water storage capacity of 0.225 billion m³. The other purposes of this installation include flood control in high rainfall months and irrigation water supply in low rainfall spells during the year. However, Wu et al. (2018) [39] showed that impoundment by the Zhikong Dam unswervingly altered the hydrological behavior of the downstream channels of the LR. The succeeding major hydrological intervention on the LR was the construction of the Pangduo Dam with 160 MW of installed capacity and 1.23 billion m^3 of reservoir water storage capacity. The development purposes of the Pangduo water conservancy project include irrigation water availability, power generation, and flood control. It is the pillar project and a leading reservoir developed for the enormous growth of the LRB.

2.2. Datasets Used

The current study utilized an arrangement of hydro-meteorological and geospatial information of the LRB in order to establish the desired results. The datasets included the following.

2.2.1. Geospatial Data of LRB

To extract the topographical features of the LRB, a 90 m resolution Advanced Thermal Emission and Reflection Radiometer, Global Digital Elevation Model (ASTER, GDEM) was developed for the study area (Figure 1). The land use features of the LRB were established using Landsat-8 Operational Land Imager (OLI) imagery with a 30 m resolution. The best cloud free images, with (137, 39) and (138, 39) path and row respectively, were used to delineate the land use map for the study area using maximum likelihood classification of the land features. The soil profile of the LRB was described using FAO-UNESCO (Food and Agricultural Organization-United Nations Educational, Scientific, and Cultural Organization) Harmonized World Soil Database version 1.2 (HWSD v1.2), a 30 arc-second raster database with over 15,000 different soil mapping units within the 1:5,000,000 scale FAO-UNESCO Soil Map of the World. All the datasets were projected to Universal Transverse Mercator 45N projection. All these datasets were mandatory input for SWAT model to simulate the LR streamflow.

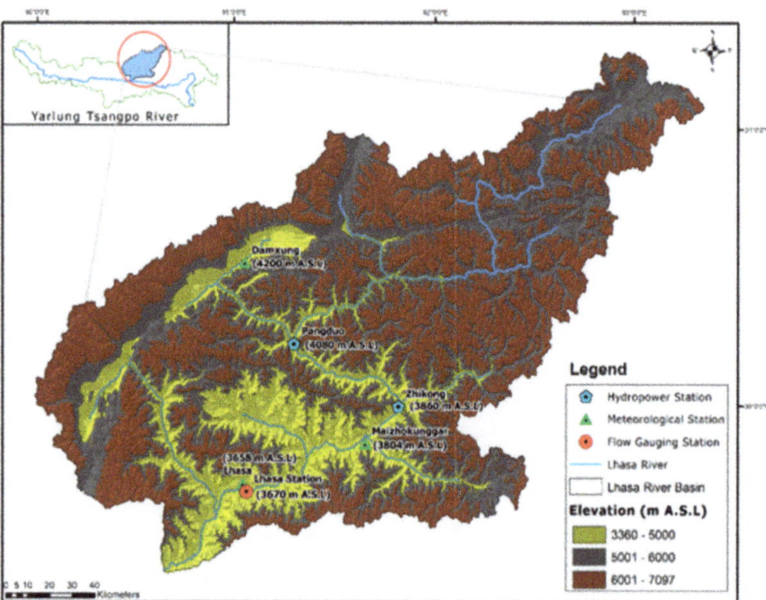

Figure 1. Location map of the Lhasa River Basin extracted from the Advanced Thermal Emission and Reflection Radiometer, Global Digital Elevation Model (ASTER GDEM) dataset showing hydrological and meteorological stations, hydropower plants, and some other features of the study area.

2.2.2. Hydro-Meteorological Data of LRB

The long-term continuous records for Lhasa River streamflow were obtained from the Lhasa hydrological station located in Lhasa city, 120 km below the Zhikong Dam near the basin outlet. The hydrological data records are maintained at three hydrometric stations, but the current study utilized the data records of the Lhasa station because they represented the total river discharge contributed from the entire catchment from 1956 to 2016. For data on the required climatic variables in the current study, the long-term data from three meteorological stations—Damxung, Maizhokunggar, and Lhasa—were used. The meteorological dataset includes records of daily precipitation, maximum and minimum temperature, relative humidity, wind speed, and sunshine hours. Data on these climate variables were fed into the SWAT model to simulate LR streamflow.

2.3. Method

2.3.1. Reservoir Impact Assessment on LR Discharge

The current study intended to estimate the impact of reservoir functioning by using the modest, graphical, and useful method of double-mass curve (DMC) analysis for the consistent and long-term trend examination of hydro-meteorological data. The concept of the double-mass curve is that a plot of the two cumulative quantities during the same period displays a straight line as long as the proportionality between the two remains unchanged, and the slope of the line represents the proportionality. The advantages of this method are that it can smooth a time series and eliminate random components in the series, thus showing the main trends of the time series. In last 30 years, Chinese researchers have explored the effects of soil and water conservation measures and land use/cover changes on runoff and sediment using the DMC method, resulting in some very good outcomes [40]. For the current study, the double-mass analysis of annual LR discharge and LRB precipitation for the long time span of 1956–2016 and the chosen study time period from 2000 to 2016 were done separately to ensure the accuracy and verification of the change points, if any, in the hydrological time series.

The coefficient of variation (CV) was used to determine the variability of climatic and hydrological changes in the LRB by using the long-term available hydro-meteorological time series. CV is defined as:

$$CV = \frac{\sigma}{\mu} * 100\% \tag{1}$$

where 'σ' is the standard deviation and 'μ' is the mean.

The construction and operation of water conservancy projects (dams, channel modifications, drainage works, etc.) have transformed the seasonal distribution of and caused abrupt changes in streamflow at the basin scales [41–43]. How to attribute the physical causes of hydrological variability and how to correctly identify the human-caused signals from natural hydrological variability are therefore important questions [44–46]. Understanding hydrological variability is initially needed to solve these queries, as well as for hydrological simulation and forecasting, water resource management, control of water disasters, and many other water activities [47]. However, the correct detection and attribution of complex variability in hydrological processes at multi-time scales are still challenging tasks [48,49], and the difficulty has not been resolved, although many methods are presently used [50,51]. There have been a large number of methods such as the moving T-test (Student's T-test) [52], moving F-test (also known Fisher-Snedecor distribution) [53], Mann–Kendall test [54], and Pettitt test [55]. The evaluation of a trend in time series of hydro-meteorological phenomena has been done using the non-parametric Mann–Kendall test (MK) [56–58] in the MS Excel software supported by XLSTAT 2014 macro. This test is extensively used and can deal with missing and distant data. The test has two parameters that are substantial for trend detection: a significance level (p) that represents the power of the test and a slope magnitude estimate (MK-S) that represents the direction and volume of the trend. The trends in time series were completed by a calculation of Kendall coefficient 'τ' [59–61]. In the current study, the MK trend at a significance level of 5% ($p < 0.05$) was applied to the hydro-meteorological time series. Specific attention was paid to the streamflow exposure of the LR to the impact of reservoir operations by analyzing the MK trend and changes in it with time.

2.3.2. SWAT Modelling of LR Streamflow

The SWAT model is among the most extensively applied open source, semi-distributed watershed-scale hydrologic models to simulate the water quantity, surface runoff, and quality of streamflow in river channels [62]. According to the working principle of the SWAT model, a watershed is initially divided into sub-basins, and each sub-basin is subdivided into hydrologic response units (HRUs) based on land use, topography, soil, and slope maps. The hydrologic cycle for each HRU is simulated based on the water balance, including precipitation, interception, surface runoff, evapotranspiration, percolation, lateral flow from the soil profile, and return flow from shallow aquifers. In this study, ArcSWAT2012 running on an ArcGIS 10.2 platform was used for watershed delineation and sub-basin discretization, resulting in 21 sub-basins that were further categorized to 149 HRUs. Srinivasan et al. (2010) [63] stated that since the accuracy of simulated streamflow may be reduced by not considering a reservoir or dam in a watershed, the calibration of the reservoir component is needed to improve the accuracy of simulated streamflow. The SWAT model provides four different reservoir outflow estimation methods: measured daily flow, measured monthly flow, average annual release rate for uncontrolled reservoir, and controlled outflow with target release. The selection of the method depends on available data regarding the reservoir. Therefore, the SWAT model was forced to simulate LR streamflow under the reservoir influence by using the default reservoir module of SWAT. The reservoir details including surface area, reservoir water volume, reservoir operational year, and monthly LR discharge data for the time span of 2000–2016 were poured into the model to simulate LR discharge under the hydraulic interventions of the Zhikong and Pangduo dams. The model was calibrated for the years 2005–2010 and validated for 2011–2016 with 500 simulations each using SWAT-CUP (SWAT Calibration and Uncertainty Procedures) algorithm. The global sensitivity method was used to rank the selected sensitive parameters.

The sensitivity and significance degree of each parameter were analyzed by the t-Stat and p-value, as well as the p-factor and r-factor; the higher the value of t-Stat, the greater its sensitivity is, and the lower the p-value, the greater the sensitivity is. The p-factor is the percentage of data that is enclosed by the 95PPU (95 Percent Prediction Uncertainty) band (ranging from 0 to 1, where 1 shows that all the prediction are within the 95PPU band), while the r-factor is the average width of the 95PPU band divided by the standard deviation of the measured variable (from 0 to ∞, with 0 showing perfect match). During the calibration and validation periods, the calculated monthly streamflow was compared with the observed data from the Lhasa hydrometric station using the Nash–Sutcliffe coefficient (NSE) [64], the coefficient of determination (R^2) [65], and the percent bias (PBIAS, %) [66]. Additionally, the observed and simulated discharge records were statistically tested for the correspondence between them using Pearson correlation [67], Spearman's correlation [67], and Kendall's rank correlation [68] for the credibility verification of SWAT modelling under the reservoir operations for LR streamflow simulation. Additionally, the forecasted river discharges using the observed and simulated hydrological time series were correlated using the same correlation tests.

2.3.3. ARIMA Forecasting of LR Discharge Time Series

An ARIMA time series model, which was pioneered by Box and Jenkins (1970), was employed in this study [28]. In the ARIMA (p, d, and q) model, where, p is autoregressive (AR), d is differencing, and q is moving average (MA). ARIMA models have two common forms: one is non-seasonal ARIMA (p, d, and q) and the other is seasonal ARIMA (p, d, and q) (P, D, and Q) S, where P, D, and Q represent seasonal parts and p, d, and q are non-seasonal parts of the model. SARIMA was applied in the current study using the observation-based LR discharge recorded at the Lhasa hydrometric station and the SWAT-simulation-based LR discharge to predict the future hydrological time series for the LRB. The SARIMA was used for predicting the LR discharge using both the observed and simulated hydrological time series in R. The SARIMA models can be used for stationary time series data, which was ensured through decomposition for non-stationarity and log transformation for de-seasonality of the SWAT-simulated and observed hydrological time series. The SARIMA model developed for the observation-based and simulation-based LR hydrological time series was trained for the time span of 2005–2012 and validated for the years 2013–2016 under the reservoir influence to minimize the ambiguity in the prediction of LR streamflow, and LR discharge forecasting was carried out from 2017 to 2025.

We constructed the model and depicted the autocorrelation function (ACF) and partial autocorrelation function (PACF) of model residuals to confirm autoregressive and moving average parameters. The final automatic ARIMA model selection was carried out in the R environment. The ACF and PACF of residuals were determined to evaluate the goodness of fit. The two most commonly used ARIMA model selection criteria, the Akaike's information criterion (AIC) and the Bayesian information criterion (BIC), were examined and compared for ARIMA model selection. The AIC was used for the purpose of selecting an optimal model fit to given data. The model that had the minimum AIC was selected as a parsimonious model [69–72]. The BIC was also utilized for the identification of the best fit model for LR flow prediction. The model with the least BIC was suitable for time series prediction [73].

AIC, in general case, is:

$$AIC = 2k = n\ln(SSE/n) \tag{2}$$

where k is the number of parameters in the statistical model, n is the number of observations, and SSE is square sum of error given by

$$SSE = \sum_{i=1}^{n} \varepsilon_i^2 \tag{3}$$

BIC, in general, is given by

$$BIC = k\ln(n) + n\ln(SSE/n) \tag{4}$$

R^2, root mean square error (RMSE), and mean absolute percentage error (MAPE) were selected to assess the ability of SARIMA in forecasting the LR discharge. A Ljung–Box test [74] at $p = 0.05$ was also performed to make sure best fit of the SARIMA model for both time series for LR discharge. Finally, we applied the best fitted models to forecast the monthly LR discharge using the observation-based and simulation-based hydrological time series. The schematic representation of study design is shown in Figure 2.

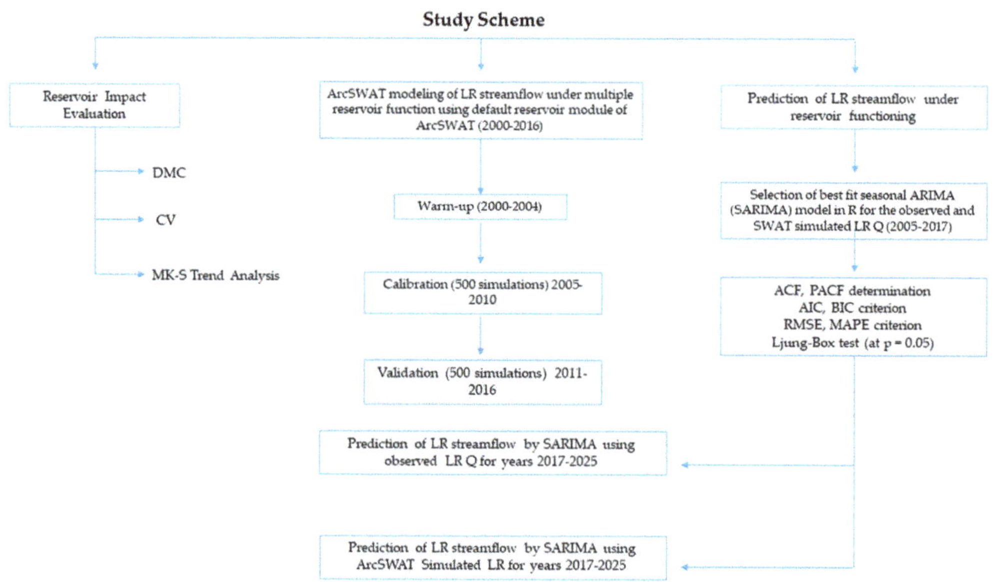

Figure 2. Schematic representation of reservoir impact evaluation, seasonal AutoRegressive Integrated Moving Average (SARIMA) and Soil and Water Assessment Tool (SWAT) model setup for the current study. LR: Lhasa River; DMC: double-mass curve; MK-S: non-parametric Mann–Kendall test; CV: coefficient of variability; PACF: partial autocorrelation function; AIC: Akaike's information criterion; BIC: Bayesian information criterion; RMSE: root mean square error; MAPE: mean absolute percentage error; Q: Discharge (m^3/s).

3. Results

3.1. Reservoir Impact Evaluation on Lhasa River Flow

3.1.1. Double-Mass Curve Analysis of Lhasa River Flow

To reckon LR discharge change under reservoir influence, DMC analysis, along with regression lines for two time spans, was carried out to better understand the hydrological phenomena and the likely change years in the time series. The double-mass curves for annually recorded rainfall and discharge, following the work of Searcy and Hardison (1960) [75], were individually applied for the time spans of 1956–2016 and 2000–2016 (Figures 3 and 4) respectively. The application of individual cumulative mass curves for two time spans was done with the aim of developing a more valid and reliable impact assessment in terms of change in the hydrological time series of the LR.

The DMC analysis of LR discharge from 1956 to 2016 revealed a nearly proportional behavior of the rainfall in correspondence to the measured LR discharge. However, we saw certain years of change along the time series that served as break points in the pattern of high and low flows in the LR discharge. The years for change are highlighted and indicated in Figure 3, where the pattern of streamflow breaks to differ from the preceding years. The result of the long-term DMC analysis was the identification of the change years, of which the years 2007 and 2013 were of particular significance for the current study. These two years marked the operation of the two major reservoirs (Zhikong and Pangduo, respectively) that were considered for the current study. The impact of chosen reservoir functioning on the hydrological behavior of the LR manifested itself in the long-term DMC analysis.

To further understand the phenomena of reservoir influence on LR discharge, double-mass analysis was applied to the time series from 2000 to 2016 (Figure 4), and we saw three identified change points in the time series during these years.

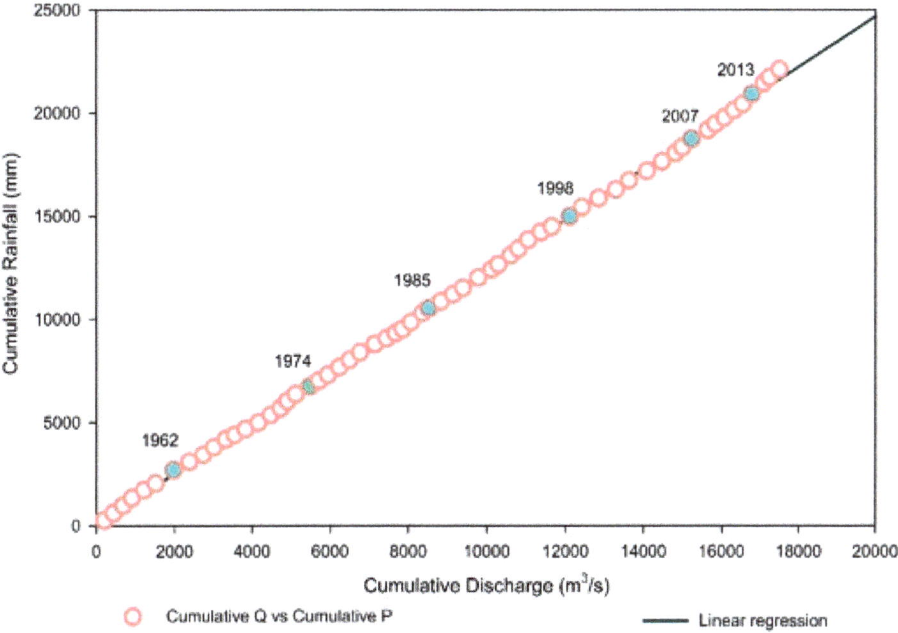

Figure 3. Double-mass curve for cumulative rainfall and cumulative discharge of the Lhasa River for the time span of 1956–2016. The years for change in hydrological time series are highlighted and supported by text.

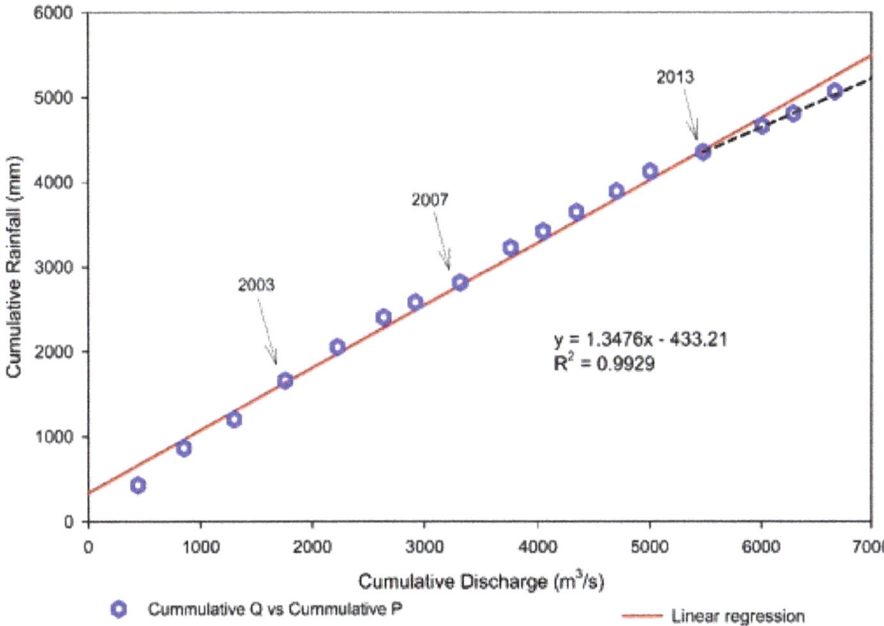

Figure 4. Double-mass curve for the cumulative rainfall and cumulative discharge of the Lhasa River for the time span of 2000–2016. The years for change in hydrological time series are supported by the text.

The year 2003 showed a change, as the maximum rainfall was recorded in this year and produced the simultaneously highest discharge during the year for the chosen study time period. The next identified change year was 2007, which deviated from the streak of data points along the regression line. This was the time when one of the selected reservoirs in the study was built on the LR. The Zhikong hydropower station was completed in 2006 and started functioning in September 2007. The most prominent break point in the hydrologic time series was identified in 2013 when the second major reservoir started operating on the LR, i.e., the Pangduo hydropower station, where the data points deviated from the regression line and indicated a peculiar hydrological behavior in the LRB. Yet again, the hydraulic interventions demonstrated themselves in the form of change points across the study time span of hydrologic time series.

3.1.2. Variation Assessment of Lhasa River Streamflow under Reservoir Operations

Figure 5 shows the inter-annual variation of the hydro-meteorological behavior of the LRB along the two time phases. The CV for the hydro-meteorological phenomena from 1956 to 1999 and from 2000 to 2016 exposed an aggravated variability in the latter time span compared to the previous time span. The CV of 24% for LRB precipitation during 1956–1999 was lowered to 20% in the second time span of 2000–2016; however, the CV values for both time spans were relatively closer, which means that the change in the pattern of precipitation advanced with a greater pace in the study time span compared to the former long time span of 1956–1999. Similarly, for the annual temperature, the CVs of 8% and 6% for the time spans of 1956–1999 and 2000–2016, respectively, were again closer values and revealed a faster temperature change in the LRB during 2000–2016.

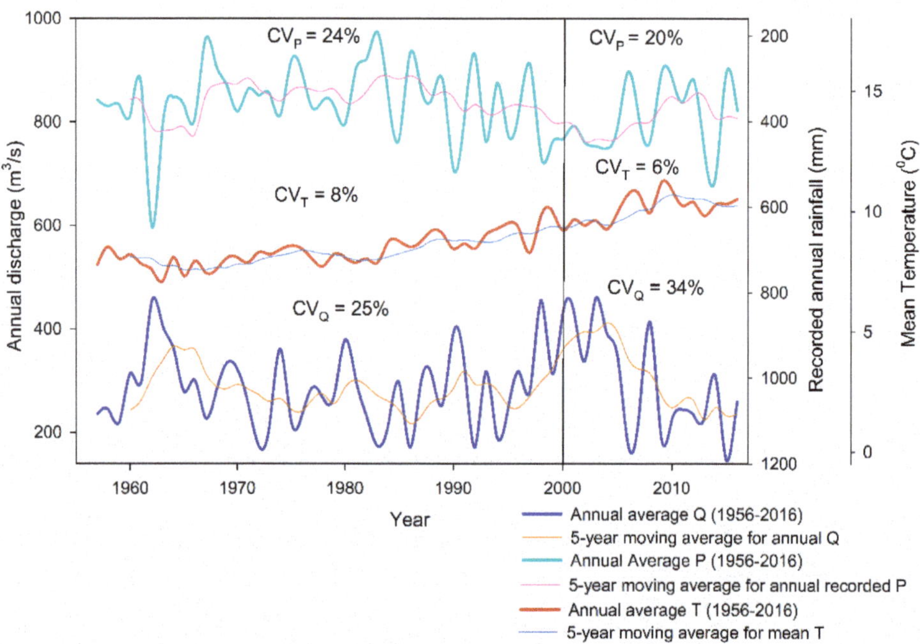

Figure 5. Changes of annual river discharge (at the Lhasa hydrometric station), annual mean temperature, and annual recorded precipitation for the Lhasa River Basin from 1956 to 1999 and from 2000 to 2016 (study time period with reservoirs functioning in the study area).

Since it is a typical QTP catchment, the LRB is prone to complex climate change phenomena [25]. These climate variables are closely associated with the hydrologic cycle. Particularly for the LRB, the LR discharge is furnished by the precipitation [25], and variability in rainfall poses a direct influence on the hydrological behavior of the LRB. Temperature is also an important feature in determining hydrological phenomena because it asserts its influence in the form of evapotranspiration, and, thus, variations in the temperature of the LRB may have a potential impact on the water resources in the area.

The CV values for LR discharge revealed an increased variability in 2000–2016 compared to 1956–1999. The CV of 25% for 1956–1999 copiously increased to 34% during the study time span of 2000–2016. With rainfall being the determining factor for discharge in the LRB, we saw that during the years of 1956–1999, the CV for rainfall and LR discharge were very close at 24% and 25%, respectively, thus indicating a close correspondence among them. Conversely, a large difference in the variability of rainfall and LR discharge was unveiled during the study time span of 2000–2016. This signified that during this time span, apart from the climatological justification, some other factor proclaimed its influence on the LR's hydrological dynamics.

The LR was subjected to some major hydraulic interventions during the years of 2000–2016, and the increased discharge alteration can be well-attributed to the reservoir operations in the LRB during this time period. The Zhikong and Pangduo reservoirs became operational in 2006 and 2013, respectively, establishing a clearly visible modification in the LR's hydrological regime, as presented in Figure 5. The LRB is experiencing an aggravated climatological variation accompanied with human interferences, resulting in a substantially altered hydrological phenomena in the area that warrants better planning and management practices in future.

3.2. MK-S Trend Analysis on Lhasa Streamflow under Reservoir Influence

Investigations of the trends in the time series of hydrological data were found to be an imperative means for the detection and understanding of changes in a rainfall–runoff process. Their results are exploitable in water management planning and flood-protection. Climatic changes, together with a different type and stage of human impact, are considered to be the main causes of rainfall–runoff changes [76]. In the current study, an MK-S test was applied to determine the direction and magnitude of the trend of hydro-meteorological phenomena of the LRB; the findings are presented as Table 1.

Table 1. Trend analysis on hydro-meteorological variables of the Lhasa River Basin. MK–τ represents Mann–Kendall's trend at $p = 0.05$ (bold values are significant at p-value), and S represents the Sen's slope estimator for change. The negative sign indicates a decrease.

	Rainfall (P)		Temperature (T)		Discharge (Q)	
	1956–1999	2000–2016	1956–1999	2000–2016	1956–1999	2000–2016
MK–τ	0.04	−0.13	**0.54**	**0.44**	0.03	−0.41
S	0.52 (mm yr^{-1})	−4.30 (mm yr^{-1})	0.03 (°C yr^{-1})	0.06 (°C yr^{-1})	0.30 (m^3 s^{-1} yr^{-1})	−14.02 (m^3 s^{-1} yr^{-1})

Dams influence variations in river discharge, particularly over seasonal time scales [77,78]. The seasonal variation in LR discharge is presented in Table 2, where maximum variation is shown to be have been experienced by the high flow months of the wet monsoonal season from June to October with a CV value of 62%, followed by the spring season from March to May with a CV value of 56%. The dry winter season from November to February was found to experience the minimum variability with a CV of 47%.

3.3. Lhasa River Streamflow Simulation and Prediction

3.3.1. SWAT Modeling of Lhasa River Flow under Reservoir Influence

In the current study, the SWAT model identified nine parameters sensitive to the runoff generation phenomena of the LRB. The sensitivity, ranges, and optimum values of the selected parameters for the study (as identified by SUFI-CUP) are presented in Table 3. The model ranked SOL_BD, EPCO, GW_REVAP, ESCO, and GW_DELAY as the most influential parameters in controlling the runoff phenomena in the LRB. This indicated that the LR discharge is predominantly controlled by the soil physical characteristics, evapotranspiration, and ground water processes in the LRB. This was supported by the previously discussed seasonal MK-S trend results for LR discharge, which also indicated a strong association of evapotranspiration phenomena and ground water movement in the LRB in the runoff generation process, particularly in the dry winter season.

Table 2. The change and trend on seasonal Lhasa River discharge for the time period of 2000–2016. CV stands for coefficient of variation, MK–τ represents Mann–Kendall's trend at $p = 0.05$ (bold values are significant at p), and S represents the Sen's slope estimator for change in LR discharge (m^3 s^{-1} month^{-1}). The negative sign indicates a decrease.

	Dry Winter Season (Nov-Feb)	Spring Season (Mar-May)	Wet summer Season (Jun-Oct)
CV	47%	56%	62%
MK–τ	**−0.28**	−0.14	**−0.27**
S	−0.6 (m^3 s^{-1} yr^{-1})	−0.4 (m^3 s^{-1} yr^{-1})	−5.6 (m^3 s^{-1} yr^{-1})

Table 3. Sensitivity of selected parameters in influencing Lhasa River flow.

No	Parameter	Parameter Description	Method chosen	Range Min-Max	Fitted Value	t-stat	p-Value	Rank
1.	r__SOL_BD	Soil bulk density (mg/m^3)	Relative	−0.5–0.5	0.17	2.287	0.045	1
2.	v__EPCO	Plant uptake compensation factor	Replace	−1–1	0.65	1.830	0.097	2
3.	v__GW_REVAP	Ground water "revap" coefficient	Replace	0.02–0.2	0.13	1.249	0.240	3
4.	v__ESCO	Soil evaporation compensation factor	Replace	0.01–1	0.85	−0.711	0.492	4
5.	v__GW_DELAY	Ground water delay (days)	Replace	0–500	12.50	0.630	0.542	5
6.	r__OV_N	Manning's "n" value for overland flow	Relative	−0.5–0.5	−0.02	0.397	0.699	6
7.	r__SOL_AWC	Available water capacity of soil layer (mm H$_2$O/mm soil)	Relative	−0.2–0.2	0.07	−0.204	0.842	7
8.	r__SOL_K	Saturated hydraulic conductivity (mm/h)	Relative	−0.5–0.5	0.47	0.182	0.858	8
9.	r__CN2	Initial SCS curve number for soil condition II	Relative	−0.2–0.0	−0.17	−0.022	0.982	9

"r" denotes the relative method, and "v" denotes the replace method.

The performance of the SWAT model in simulating LR discharge under the chosen reservoirs' influence for the time span of 2000–2016 is presented in Figure 6a. A comparison of observed and simulated LR discharge is shown in Figure 6b. The simulated hydrological time series corresponded appreciably well to the observed data series and regularly fluctuated with the precipitation pattern. The high peaks were very well captured by the SWAT model most of the time, particularly during the calibration years (2005–2010), with a few being under-estimated. For the validation years (2011–2016), the model again managed to capture the high peaks, but some peaks were under-estimated. The lower flow was consistently under-estimated by the model. A similar weakness of the SWAT model in capturing the low flows of the LR was reported in [25]. Overall, the model performed well in simulating the LR streamflow by conforming to the work of Moriasi et al. (2007) [79], where the modeling performance was acceptable if $R^2 > 0.5$, NSE > 0.5, and PBIAS < ±25%. The performance of the SWAT model during calibration and validation is presented in Table 4. However, the comparison between observed and simulated hydrological data series revealed an R^2 value of 0.75 (Figure 6b), and majority of the values were enclosed by the 95% prediction and confidence interval. Few of the high flow values were dispersed because they were under-estimated by the model. This confirmed the competency of the SWAT model in simulating the LR discharge under the reservoir operations selected for the current study.

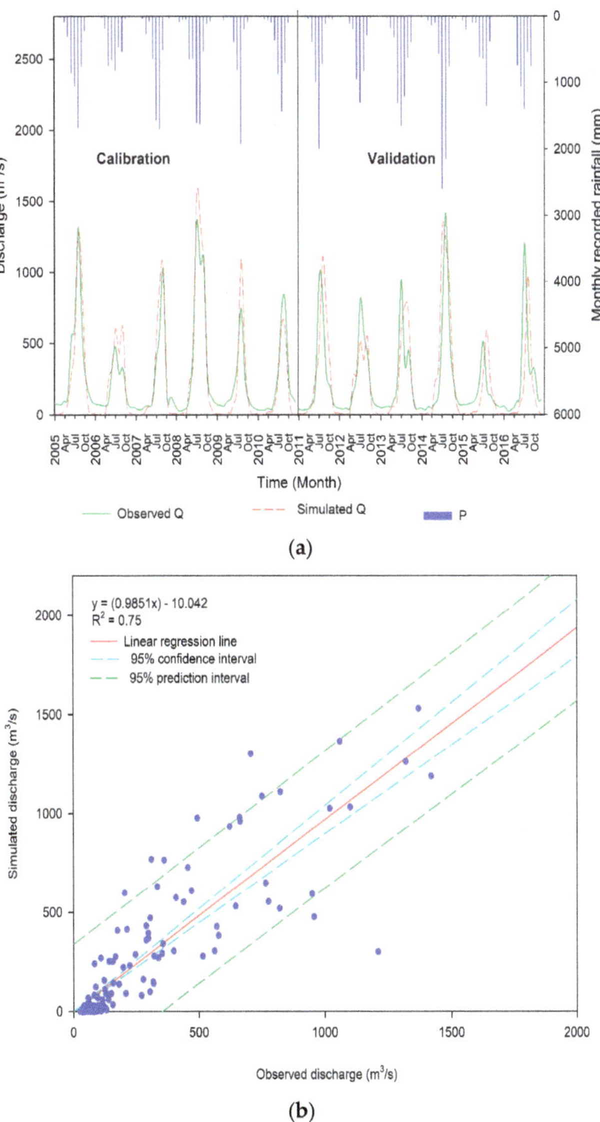

Figure 6. (**a**) SWAT simulation of Lhasa River discharge recorded at the Lhasa hydrometric station for time span of 2005–2016. (**b**) Comparison of observed and SWAT-simulated Lhasa River discharge from 2005 to 2016.

The association of observed and SWAT-simulated LR discharge was verified by the correlation tests presented in Table 5. All the correlation coefficients produced high values and thus proved that the SWAT-simulated results could be used to predict the future LR discharge from 2017–2025.

Table 4. Performance of the SWAT model in simulation of Lhasa River flow under reservoir operations. p-factor: percentage of data that is enclosed by the 95PPU band; r-factor: the average width of the 95PPU band divided by the standard deviation of the measured variable (from 0 to ∞, with 0 showing perfect match).

No.	Performance Criteria	2000–2016 (00–04 Warm-Up)	
		Calibration (05–10)	Validation (11–16)
1.	p-factor	0.96	0.50
2.	r-factor	1.09	0.47
3.	R^2	0.91	0.58
4.	NSE	0.86	0.50
5.	PBIAS	5.5	5.5

Table 5. Statistical correlation of observed and SWAT-simulated Lhasa River discharge.

No.	Correlation Coefficient	Value
1.	Pearson's correlation	**0.87**
2.	Spearman's correlation	**0.87**
3.	Kendall's rank correlation	**0.68**

Bold values are significant at $p = 0.05$.

3.3.2. Seasonal ARIMA Application for Predicting Hydrological Regime of Lhasa River Basin under Reservoir Operations

While making use of the observed LR hydrological time series to identify the future trend of LR streamflow under reservoir operations for the years 2017–2025, the SARIMA model $(1, 0, 0) (2, 1, 2)_{12}$ was found to be the optimum combination for forecasting of observed streamflow under the cumulative impact of reservoirs by justifying the performance evaluation criteria presented in Table 6 for attaining the lowest AIC and BIC values, a lower RMSE value of 0.29 m^3/s, and a MAPE value of only 4.02%—values which confirmed the validity of the model. The SARIMA model was validated for the years of 2013–2016. SARIMA produced closely corresponding predicted values for LR streamflow during the validation time span, with its correlation coefficient of $R^2 = 0.80$ revealing an efficient model that is capable of predicting the future discharge for the LR. The forecasted monthly LR discharge was seen to follow a decreasing trend during the time period of 2017–2025 under reservoir influence (Figure 7a).

Table 6. Performance of SARIMA model in predicting Lhasa River streamflow from 2017 to 2025 using observed and simulated hydrological time series.

Performance Criterion	Forecasted Q_{obs}	Forecasted Q_{sim}
AIC	76.7	199.4
BIC	91.28	209.12
RMSE	0.29 (m^3/s)	0.65 (m^3/s)
MAPE	4.02%	31.09%

The SARIMA model $(1, 0, 0) (2, 1, 0)_{12}$ was found to be the optimum combination for forecasting of SWAT-simulated streamflow under the cumulative impact of reservoirs. The SARIMA model produced correlation coefficient of $R^2 = 0.88$ for the validation years from 2013 to 2016 for SWAT-simulated and forecasted LR discharge with a relatively higher MAPE value of 31.09% (Table 6) for the simulation-based forecasted LR discharge. The predicted discharge using the SWAT-simulated hydrological time series likewise showed a decreasing discharge.

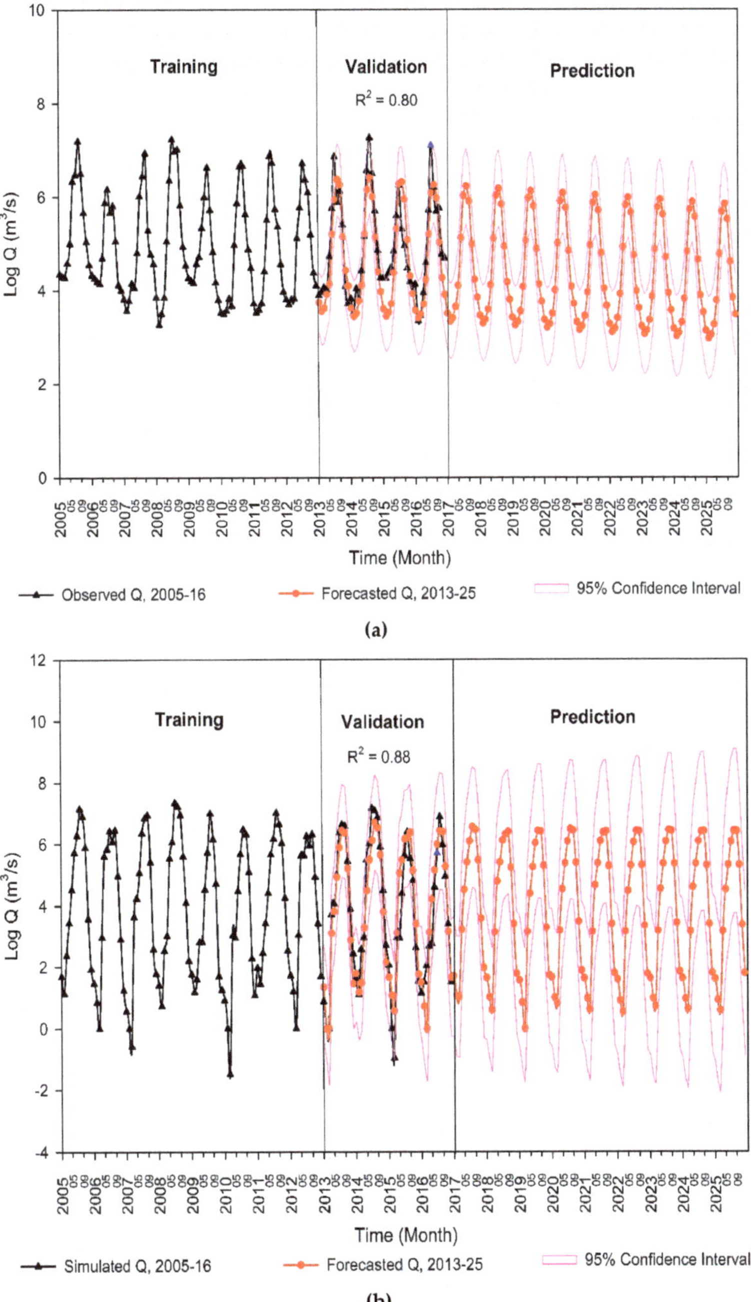

Figure 7. (**a**) Forecasted monthly Lhasa River streamflow for time span of 2013–2025 using the observed hydrological time series from 2005 to 2016. SARIMA model validation years from 2013 to 2016 are marked. (**b**) Forecasted monthly Lhasa River streamflow for time span of 2013–2025 using SWAT-simulated hydrological time series from 2005 to 2016. SARIMA model validation years from 2013 to 2016 are marked.

The comparison of observation-based and simulation-based LR discharge presented in Figure 8a showed a very close correspondence between both hydrological time series with an R^2 of 0.90. However, the simulation-based forecasted LR discharge was higher for high flow months in future. In advancing through the years from 2017 to 2025, the difference in the high peaks was seen to be increasing among the observation and simulation-based forecasted LR discharge, as presented in Figure 8b. However, both hydrological data series were shown to experience a decrease in the future years researched in the study.

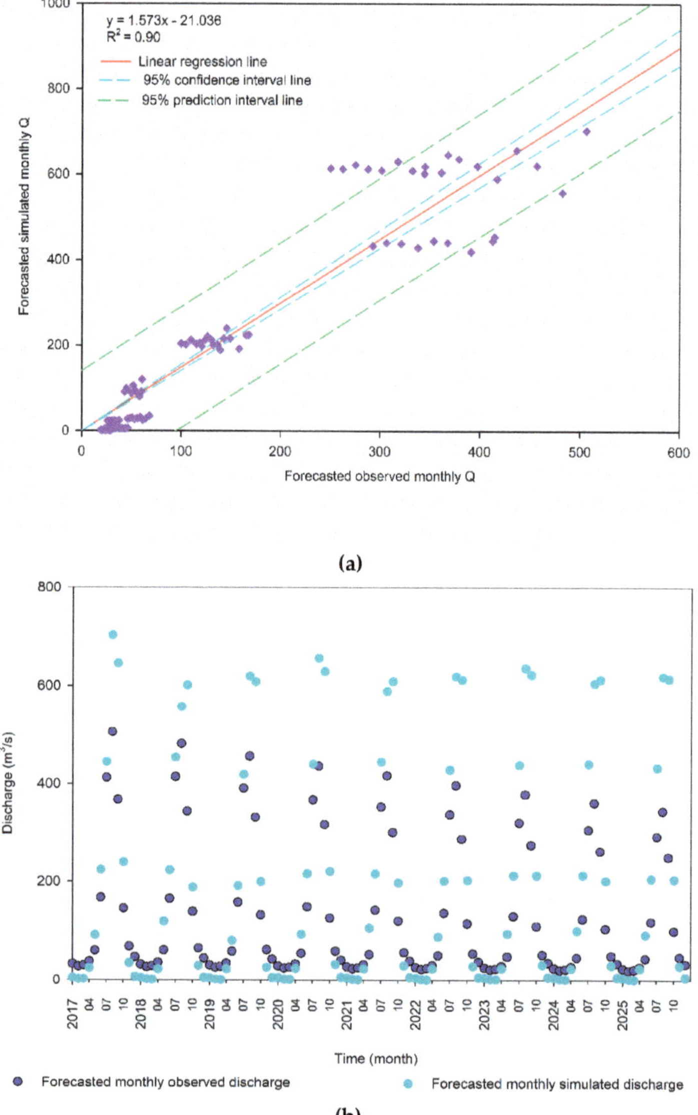

Figure 8. (**a**) Comparison between observation-based and SWAT-simulation-based Lhasa River forecasted flow from 2017 to 2025. (**b**) Scatter plot of observation-based and simulation-based forecasted monthly Lhasa River discharge for 2017–2025.

To corroborate the association of both forecasted hydrological time series, statistical correlational tests used in the study produced values of ≥0.80 and are presented in Table 7. This testifies to the credibility of the approach used in the current study and shows that simulation-based future LR discharge can be a replacement to observation-based discharge and be utilized for further analyses regarding water resource management, planning, distribution, hydropower generation, irrigation scheduling, and reservoir operational procedures in the LRB. This can prove to be an aid in overcoming the hydrological data scarcity issue because the LRB is a quintessential basin of the QTP with barely observed data [25].

Table 7. Statistical correlation between forecasted observation-based and simulation-based Lhasa River discharge, 2017–2025.

No.	Correlation Coefficient	Value
1.	Pearson's correlation	**0.95**
2.	Spearman's correlation	**0.95**
3.	Kendall's rank correlation	**0.80**

Bold values are significant at $p = 0.05$.

A flow–duration curve offers a practical approach for studying the flow characteristics of streams and for examining the association of one basin with another. A flow–duration curve is a cumulative frequency curve that shows the percent of time during which specified discharges were equaled or exceeded in a given period. A rather easier conception of the flow–duration curve is that it is a streamflow data demonstration combining the flow characteristics of a stream throughout the ranges of discharge in one curve [80]. The flow–duration curves for the observed LR discharge and forecasted observation-based and simulation-based LR discharge are presented in Figure 9.

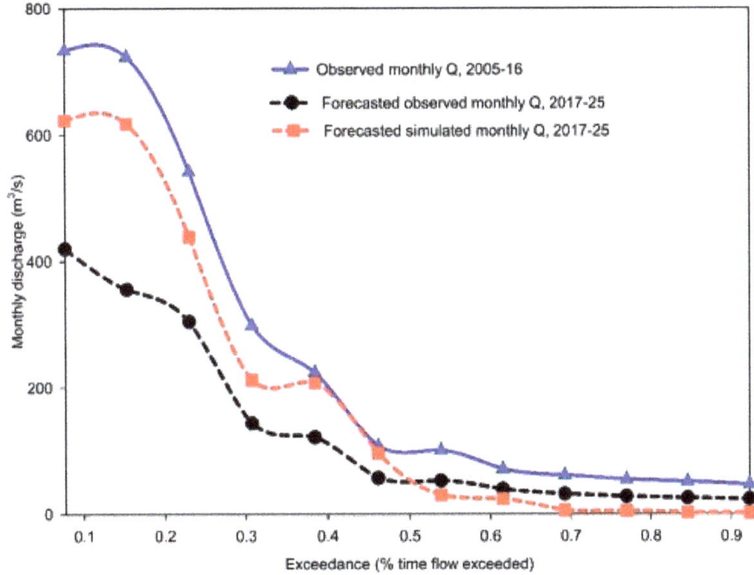

Figure 9. Flow–duration curves for the monthly observed, forecasted observation-based, and SWAT-simulation-based Lhasa River discharge.

We saw that the SWAT-simulation-based predicted LR discharge produced a steeper sloped curve following the similar high and low flow pattern of the observed LR discharge.

However, the simulation-based predicted low flows dropped drastically through the years taken for the study. On the contrary, the forecasted observation-based LR discharge revealed a flat sloped curve with a remarkably low peak events for the coming years. The authors of [25] also revealed a significantly decreased LR streamflow in future, though under the impact of climate change, and they attributed the decrease to the temperature change. Our study also revealed a considerable decrease in the future LR discharge under reservoir influence. The LRB is the largest inhabited QTP basin experiencing aggravated climate change and human interference impacts on its rainfall-dominated runoff generation mechanism. This suggests a call for better and strategic water resource management in the LRB. The findings from the current study can be used by water resource managers and hydropower engineers to develop flow–duration curves for the hydropower plants considered in the study by using their turbine capacity to estimate the required and available flow for producing power in future years. The study can also be replicated in basins with similar characteristic around the globe.

4. Discussion

We saw that the escalated variation in hydro-meteorological parameters revealed by the previously discussed CV values was reaffirmed by the MK-S test for the time period of 2000–2016 compared to the years 1956–1999. The trend of rainfall pattern in the LRB during the two time spans was shown to have undergone a decrease in the latter time span that was also of greater magnitude, as exposed by Sen's slope estimator. However, the decrease was not significant. Looking at the temperature change trend in the LRB during the two time spans, a significant increase in the temperature was revealed by the MK-S test, with a rapid rise in temperature in the study time span compared to the previous long-term time period. The authors of [81] also reported a significant increase in all the QTP basins including the LRB, with a range of 0.03–0.06 °C year^{-1}, which was similar to the findings of our study. The LR discharge trend revealed a non-significant increase in the former long-term time span. Conversely, for the study time span, we saw a significant decrease in the LR discharge with a larger magnitude, as estimated by the Sen's slope. We saw that the rainfall also decreased along the same time span, and the temperature was seen to have risen, which could have a serious impact on the LR discharge in the form of lower precipitation and aggravated evapotranspiration in the LRB. However, it can be seen that the decrease in the magnitude of LR discharge during 2000–2016 was far more, which clearly leads to the conclusion that in addition to the climatic phenomena, some other factors were causing the pronounced decrease in LR discharge during 2000–2016. During these years, major hydraulic interventions were witnessed in the LRB in the form of dam construction, including the Zhikong hydropower project (2006) and the Pangduo hydropower project (2013). These hydraulic structures impounded the water and caused the prominent decrease in LR discharge.

The MK-S test revealed a decreasing LR trend for all the seasonal flows, and the decrease reached significance in wet summer season and the dry winter season. Sen's slope estimator revealed that the greatest reduction in the LR discharge was experienced in the wet summer season, followed by the dry winter season. The spring season was shown to experience the lowest reduction in discharge. This phenomena of the LR discharge trend could be explained by the fact that the wet summer season was the peak flow time in the LRB, where the major portion of LR discharge was generated (~90%) during these months. This water was stored in the reservoirs built on the LR for the succeeding dry wet season with a minimum rainfall and was responsible for the significantly decreased LR discharge during peak flow season. The higher variability and non-significant decrease in the LR discharge during spring season could be attributed to the snow and glacier melting, which is a completely-climate driven phenomena in the LRB. Thus, the variation of LR discharge during spring season is highly prone to the snowfall received during the winter season. In LRB, from April to May is a melting season when air temperature is, on average, above zero [81]. This snowmelt contributes to LR discharge and hence stabilizes the effect of rising temperature in the form of evapotranspiration and may be a reason for smaller decrease in LR discharge during spring season. Increasing air temperatures lead to less snow accumulation in the winter and an earlier peak runoff in the spring, as well as reduced flows in summer and autumn [82–85]. The change in the streamflow regime results in a substantial impact on regional water resources and seasonal water supplies [86]. For the dry winter LR discharge, a significant decrease with a greater magnitude and lower variability compared to the spring season could have been a possible manifestation of the increase rate of temperature. The MK-S test showed a significant and intensified increase in temperature of the LRB for the years 2000–2016. Similar behavior was reported in [25] for the LRB, where the increase rate of the minimum temperature was found to be higher in spring and winter than in summer, whereas the maximum temperature showed the opposite trend. This increased minimum temperature is causing an overall warming of the LRB and is showing its effects in the form of evapotranspiration with minimal rainfall to balance it, thus leading to a decreased LR discharge during the dry winter season. Additionally, the water demand in the LRB during the dry winter months is met by ground water abstraction and the reservoir-stored water during the wet summer season. This is again a factor that affects the LRB's hydrological behavior in dry spells of a year.

While predicting the LRB streamflow, we saw that in both situations, the LR discharge was predicted to decrease under the reservoir influence. This could be attributed to the inertial characteristic of the ARIMA model forecasts. If the historical data rise rapidly right before the peak value, they cannot be foreseen by the ARIMA model and the peak value would therefore be underestimated; however, if the rise is slow and steady, the rising trend would be expected to continue after the peak by the ARIMA model [28]. Here in the case of observed LR discharge, we saw a decreasing development through the years 2000–2016, which was confirmed by the MK-S test results presented earlier. Thus, ARIMA predicted an obvious decrease in the observation-based forecasted LR discharge. For the SWAT-simulation based LR discharge, the ARIMA model again predicted a decreasing yet stable future streamflow for 2017–2025, following the same behavior as the SWAT-simulated flow (Figure 7b).

The current study was intended to overcome the data scarcity issue of the study area, which is a major concern of the QTP catchments. In the current study, data availability on reservoir operations was a major constraint. Additionally, the data quality and availability for hydro-meteorological parameters included in the study were a prime concern for the trans-boundary Lhasa River.

5. Conclusions

The current study was carried out to evaluate the reservoir construction's impact on LR discharge using double-mass curve analysis, estimating the CV for the closely related hydro-meteorological phenomena of LRB and identifying a trend on the hydro-meteorological behavior of the LRB using the well-known MK-S test. We found that:

1. The reservoir operation years showed themselves in the double-mass curve analysis for both the long-term (1956–2016) data series and the study time span of 2000–2016 as the break points in both curves.

2. CV values were individually calculated for two time spans of 1956–1999 and 2000–2016, and they showed that the variability in the hydro-meteorological phenomena for LRB was remarkably intensified in the latter time span compared to the years during the first time span. The variability in rainfall, temperature, and LR discharge escalated during the years 2000–2016. For the former time span, the LR discharge varied in accordance with the rainfall variation. For the latter time span, the variation in LR discharge was seen to be far greater than the rainfall variability. Additionally, the temperature change in the LRB was seen to be more rapid in the latter time period. However, the enormous variation in the LR discharge could not solely be attributed to the climatic factors, so some other factors are controlling the LR hydrological phenomena. This strong variation in LR discharge during the time span of 2000–

2016 was the outcome of the two dams built over the LR. The Zhikong hydropower plant and the Pangduo power plant—which began operations in 2006 and 2013, respectively—have been influencing the LR discharge and causing a substantial variation in the LR discharge.

3. The MK-S test revealed a non-significant increase in the rainfall and a subsequent increase in LR discharge for the time 1956–1999. However, for the time span of 2000–2016, the rainfall in LRB experienced a non-significant decrease, whereas the LR discharge significantly decreased with an amplified magnitude that could be well-attributed to the reservoir functioning in the LRB. The temperature in the LRB was found to significantly decrease for the time spans of 1956–1999 and 2000–2016. The increase in temperature was more in the latter time span and potentially affected the snowmelt, evapotranspiration, and (ultimately) the discharge of the LR on a seasonal scale.

While predicting the LR discharge from 2017 to 2025, SWAT-ARIMA coupling revealed:

4. The SWAT model was capable of simulating the LR discharge under reservoir influence, and simulation-based LR discharge can be a replacement to observed LR discharge. This could undisputedly aid in overcoming the hydrological data scarcity constraint in the LRB.

5. The best-fitted seasonal ARIMA has forecasted a closely corresponding decreasing LR discharge time series for the years 2017–2025 when using the observation-based and simulation-based LR hydrological data. The observation-based predicted discharge was seen to decline at a greater extent, but the simulation-based predicted discharge was seen to follow the similar behavior to the observed LR discharge but while decreasing in the years from 2017 to 2025.

6. This study revealed that a prominent climate change phenomena and human interferences are simultaneously affecting the hydrological regime of the LRB. This demands a more extensive study with a special influence on data availability for reservoir operations and procedures, water resource usage and allocation, ground water processes, etc. The study holds significance in assisting the water resource planning, management, availability, and requirement of water in hydropower generation, irrigation, domestic uses, etc., in the future for the LRB.

Author Contributions: This research is carried out in collaboration with all authors. Conceptualization, M.Y. and T.H.; methodology, M.Y.; software, M.Y. and S.A.H.; validation, M.Y. and S.A.H.; formal analysis, M.Y.; data curation, T.H.; writing—original draft preparation, M.Y.; writing—review and editing, T.H.; supervision, T.H.; project administration, T.H.; funding acquisition, T.H. All authors have read and agreed to the published version of the manuscript.

Funding: This study has been supported by the National Natural Science Foundation of China (Grant No. 91647204), and Wuhan Center of China Geological Survey (Grant No. 2020028; 2020023; YRSW-2020-359).

Acknowledgments: We thank three anonymous reviewers and editors for constructive suggestions and comments.

Conflicts of Interest: The authors declare no conflict of interest.

References

1. Wang, Y.; Zhang, N.; Wang, D.; Wua, J.; Zhang, X. Investigating the impacts of cascade hydropower development on the natural flow regime in the Yangtze River, China. *Sci. Total Environ.* **2018**, *624*, 1187–1194. [CrossRef] [PubMed]
2. Nilsson, C.; Reidy, C.A.; Dynesius, M.; Revenga, C. Fragmentation and Flow Regulation of the World's Large River Systems. *Science* **2005**, *308*, 405. [CrossRef] [PubMed]
3. Li, Q.; Yu, M.; Lu, G.; Cai, T.; Bai, X.; Xia, Z. Impacts of the Gezhouba and Three Gorges reservoirs on the sediment regime in the Yangtze River, China. *J. Hydrol.* **2011**, *403*, 224–233. [CrossRef]
4. Guo, C.; Jin, Z.; Guo, L.; Lu, J.; Ren, S.; Zhou, Y. On the cumulative dam impact in the upper Changjiang River: Streamflow and sediment load changes. *Catena* **2020**, *184*, 104250. [CrossRef]

5. Syvitski, J.P.M.; Vörösmarty, C.J.; Kettner, A.J.; Green, P. Impact of Humans on the Flux of Terrestrial Sediment to the Global Coastal Ocean. *Science* **2005**, *308*, 376. [CrossRef]
6. Wang, H.; Bi, N.; Saito, Y.; Wang, Y.; Sun, W.; Zhang, J.; Yang, Z. Recent changes in sediment delivery by the Huanghe (Yellow River) to the sea: Causes and environmental implications in its estuary. *J. Hydrol.* **2010**, *391*, 302–313. [CrossRef]
7. Dai, S.B.; Lu, X.X. Sediment load change in the Yangtze River (Changjiang): A review. *Geomorphology* **2014**, *215*, 60–73. [CrossRef]
8. *Dams and Development: A New Framework for Decision-Making: The Report of the World Commission on Dams*; World Commission on Dams: Cape Town, South Africa, 2000.
9. Zhao, Q.; Liu, S.; Deng, L.; Dong, S.; Yang, J.; Wang, C. The effects of dam construction and precipitation variability on hydrologic alteration in the Lancang River Basin of southwest China. *Stoch. Environ. Res. Risk Assess.* **2012**, *26*, 993–1011.
10. Li, D.; Long, D.; Zhao, J.; Lu, H.; Hong, Y. Observed changes in flow regimes in the Mekong River basin. *J. Hydrol.* **2017**, *551*. [CrossRef]
11. Zhang, Q.; Xiao, M.; Liu, C.L.; Singh, V.P. Reservoir-induced hydrological alterations and environmental flow variation in the East River, the Pearl River basin, China. *Stoch. Environ. Res. Risk Assess.* **2014**, *28*, 2119–2131. [CrossRef]
12. Gao, B.; Yang, D.; Zhao, T.; Yang, H. Changes in the eco-flow metrics of the Upper Yangtze River from 1961 to 2008. *J. Hydrol.* **2012**, *448–449*, 30–38. [CrossRef]
13. Yang, Z.; Yan, Y.; Liu, Q. Assessment of the flow regime alterations in the Lower Yellow River, China. *Ecol. Inform.* **2012**, *10*, 56–64. [CrossRef]
14. Zhang, Q.; Zhang, Z.; Shi, P.; Singh, V.P.; Gu, X. Evaluation of ecological instream flow considering hydrological alterations in the Yellow River basin, China. *Glob. Planet. Chang.* **2018**, *160*. [CrossRef]
15. Li, J.; Ma, L. *Background Paper: Chinese Renewables Status Report*; IAEA: Vienna, Austria, 2009.
16. Tahir, A.A.; Hakeem, S.A.; Hu, T.; Hayat, H.; Yasir, M. Simulation of snowmelt-runoff under climate change scenarios in a data-scarce mountain environment. *Int. J. Digit. Earth* **2019**, *12*, 910–930. [CrossRef]
17. Arnold, J.G.; Williams, J.R.; Maidment, D.R. Continuous-time water and sediment routing model for large basins. *J. Hydraul. Eng.* **1995**, *121*, 171–183. [CrossRef]
18. Shen, Z.Y.; Chen, L.; Chen, T. Analysis of parameter uncertainty in hydrological and sediment modeling using GLUE method: A case study of SWAT model applied to Three Gorges Reservoir Region, China. *Hydrol. Earth Syst. Sci.* **2012**, *16*, 121–132. [CrossRef]
19. Du, J.; Ru, H.; Zuo, T.; Li, Q.; Zheng, D.; Chen, A.; Xu, Y.; Xu, C.Y. Hydrological simulation by SWAT model with fixed and varied parameterization approaches under land use change. *Water Resour. Manag.* **2013**, *27*, 2823–2838. [CrossRef]
20. Zhou, F.; Xu, Y.; Chen, Y.; Xu, C.Y.; Gao, Y.; Du, J. Hydrological response to urbanization at different spatio-temporal scales simulated by coupling of CLUE-S and the SWAT model in the Yangtze River Delta region. *J. Hydrol.* **2013**, *485*, 113–125. [CrossRef]
21. Liechti, T.; Cohen, J.; Matos, P.; Boillat, J.L.; Schleiss, A.J. Influence of hydropower development on flow regime in the Zambezi River Basin for different scenarios of environmental flows. *Water Resour. Manag.* **2015**, *29*, 731–747. [CrossRef]
22. Bieger, K.; Hörmann, G.; Fohrer, N. Simulation of streamflow and sediment with the soil and water assessment tool in a data scarce catchment in the three Gorges region, China. *J. Environ. Qual.* **2014**, *43*, 37–45. [CrossRef] [PubMed]
23. Francesconi, W.; Srinivasan, R.; Pérez-Miñana, E.; Willcock, S.P.; Quintero, M. Using the soil and water assessment tool (swat) to model ecosystem services: A systematic review. *J. Hydrol.* **2016**, *535*, 625–636. [CrossRef]
24. Peng, D.; Chen, J.; Fang, J. Simulation of Summer Hourly Stream Flow by Applying TOPMODEL and Two Routing Algorithms to the Sparsely Gauged Lhasa River Basin in China. *Water* **2015**, *7*, 4041–4053. [CrossRef]
25. Tian, P.; Lu, H.; Feng, W.; Guan, Y.; Xue, Y. Large decrease in streamflow and sediment load of Qinghai–Tibetan Plateau driven by future climate change: A case study in Lhasa River Basin. *Catena* **2020**, *187*, 104340. [CrossRef]
26. Yasir, M.; Hu, T.; Samreen, A.H. Simulating reservoir induced Lhasa streamflow variability using ArcSWAT. *Water* **2020**, *12*, 1370. [CrossRef]
27. Wu, J.; Chen, X.; Yu, Z.; Yao, H.; Li, W.; Zhang, D. Assessing the impact of human regulations on hydrological drought development and recovery based on a 'simulated-observed' comparison of the SWAT model. *J. Hydrol.* **2019**, *577*, 123990. [CrossRef]
28. Wang, W. *Stochasticity, Nonlinearity and Forecasting of Streamflow Processes*; IOS Press: Amsterdam, The Netherlands, 2006.
29. Box, G.E.P.; Jenkins, G.M. *Time Series Analysis: Forecasting and Control*; Holden-Day: San Francisco, CA, USA, 1976.
30. Wang, Z.Y.; Qiu, J.; Li, F.F. Hybrid Models Combining EMD/EEMD and ARIMA for Long-Term Streamflow Forecasting. *Water* **2018**, *10*, 853. [CrossRef]
31. Adheikary, S.K.; Rahman, M.; Gupta, A.D. A stochastic modelling technique for predicting groundwater table fluctuations with time series analysis. *Int. J. Appl. Sci. Eng. Res.* **2012**, *1*, 238–249.
32. Valipour, M.; Banihabib, M.E.; Behbahani, S.M.R. Comparison of the ARMA and ARIMA and the autoregressive artificial neural network models in forecasting the monthly inflow of Dez dam reservoir. *J. Hydrol.* **2013**, *476*, 433–441. [CrossRef]
33. Ahlert, R.; Mehta, B. Stochastic analyses and transfer functions for flows of the upper Delaware River. *Ecol. Model.* **1981**, *14*, 59–78. [CrossRef]
34. Yurekli, K.; Kurunc, A.; Ozturk, F. Application of linear stochastic models to monthly flow data of Kelkit Stream. *Ecol. Model.* **2005**, *183*, 67–75. [CrossRef]
35. Modarres, R.; Ouarda, T. Modelling heteroscedasticty of streamflow time series. *Hydrol. Sci. J.* **2013**, *58*, 54–64. [CrossRef]

36. Ahmad, S.; Khan, I.H.; Parida, B. Performance of stochastic approaches for forecasting river water quality. *Water Res.* **2001**, *35*, 4261–4266. [CrossRef]

37. Kurunç, A.; Yürekli, K.; Çevik, O. Performance of two stochastic approaches for forecasting water quality and streamflow data from Yeşilırmak River, Turkey. *Environ. Model. Softw.* **2005**, *20*, 1195–1200. [CrossRef]

38. Tayyab, M.; Zhou, J.; Zeng, X.; Adnan, R. Discharge Forecasting By Applying Artificial Neural Networks at the Jinsha River Basin, China. *Eur. Sci. J.* **2016**, *12*, 108–127. [CrossRef]

39. Wu, X.; Li, Z.; Gao, P.; Huang, C.; Hu, T. Response of the Downstream Braided Channel to Zhikong Reservoir on Lhasa River. *Water* **2018**, *10*, 1144. [CrossRef]

40. Mu, X.M.; Zhang, X.Q.; Gao, P.; Wang, F. Theory of double mass curves and its applications in hydrology and meteorology. *J. China Hydrol.* **2010**, *30*, 47–51.

41. Vörösmarty, C.J.; Green, P.; Salisbury, J.; Lammers, R.B. Global water resources: Vulnerability from climate change and population growth. *Science* **2000**, *289*, 284–288. [CrossRef]

42. Poff, N.; Brown, L. Sustainable water management under future uncertainty with eco-engineering decision scaling. *Nat. Clim. Chang.* **2015**. [CrossRef]

43. Griffin, R.C. *Water Resource Economics: The Analysis of Scarcity, Policies, and Projects*; MIT Press: Cambridge, MA, USA, 2016.

44. McCabe, G.J.; Wolock, D.M. A step increase in streamflow in the conterminous United States. *Geophys. Res. Lett.* **2002**, *29*. [CrossRef]

45. Rosenzweig, C.; Neofotis, P. Detection and attribution of anthropogenic climate change impacts. *Wiley Interdiscip. Rev. Clim. Chang.* **2013**, *4*, 121–150. [CrossRef]

46. McCuen, R.H. *Modeling Hydrologic Change: Statistical Methods*; CRC Press: Boca Raton, FL, USA, 2016.

47. Varis, O.; Kajander, T.; Lemmelä, R. Climate and water: From climate models to water resources management and vice versa. *Clim. Chang.* **2004**, *66*, 321–344. [CrossRef]

48. Ganguli, P.; Ganguly, A.R. Space-time trends in US meteorological droughts. *J. Hydrol.* **2016**, *8*, 235–259.

49. Supratid, S.; Aribarg, T.; Supharatid, S. An integration of stationary wavelet transform and nonlinear autoregressive neural network with exogenous input for baseline and future forecasting of reservoir in flow. *Water Resour. Manag.* **2017**, 4023–4043. [CrossRef]

50. Zhang, Q.; Singh, V.P.; Li, K.; Li, J. Trend, periodicity and abrupt change in streamflow of the East River, the Pearl River basin. *Hydrol. Process.* **2014**, *28*, 305–314. [CrossRef]

51. Lloyd, C.E.M.; Freer, J.E.; Collins, A.L.; Johnes, P.J.; Jones, J.I. Methods for detecting change in hydrochemical time series in response to targeted pollutant mitigation in river catchments. *J. Hydrol.* **2014**, *514*, 297–312. [CrossRef]

52. Welch, B.L. The generalization of student's problem when several different population variances are involved. *Biometrika* **1947**, *34*, 28–35. [CrossRef]

53. Jackson, F.L.; Hannah, D.M.; Fryer, R.J.; Millar, C.P.; Malcolm, I.A. Development of spatial regression models for predicting summer river temperatures from landscape characteristics: Implications for land and fisheries management. *Hydrol. Process.* **2017**, *31*. [CrossRef]

54. Kisi, O.; Ay, M. Comparison of Mann-Kendall and innovative trend method for water quality parameters of the Kizilirmak River, Turkey. *J. Hydrol.* **2014**, *513*, 362–375. [CrossRef]

55. Pettitt, A.N. A non-parametric approach to the change-point problem. *Appl. Stat.* **1979**, *28*, 126–135. [CrossRef]

56. Helsel, D.R.; Frans, L.M. Regional Kendall test for trend. *Environ. Sci. Technol.* **2006**, *40*, 13. [CrossRef]

57. Libiseller, C. MULTMK/PARTMK. In *A Program for Computation of Multivariate and Partial MANN-Kendall Test*; LIU: Linköping, Sweden, 2004.

58. Kliment, Z.; Matouskova, M. Runoff changes in the Šumava Mountains (Bohemian Forest) and the foothill regions: Extent of influence by human impact and climate changes. *Water Resour. Manag.* **2009**, *23*, 1813–1834. [CrossRef]

59. Mann, H.B. Nonparametric test against trend. *Econometrica* **1945**, *13*, 245–259. [CrossRef]

60. Kendall, M.G. *Rank Correlation Methods*; Charles Griffin: London, UK, 1975.

61. Fu, G. Hydro-climatic trends of the Yellow River Basin for the last 50 years. *Clim. Chang.* **2004**, *65*, 149–178. [CrossRef]

62. Zhang, D.; Chen, X.; Yao, H.; Lin, B. Improved calibration scheme of SWAT by separating wet and dry seasons. *Ecol. Model.* **2015**, *301*, 54–61. [CrossRef]

63. Srinivasan, R.; Zhang, X.; Arnold, J.G. SWAT ungauged: Hydrological budget and crop yield predictions in the Upper Mississippi River Basin. *Trans. ASABE* **2010**, *53*, 1533–1546. [CrossRef]

64. Nash, J.E.; Sutcliffe, J.V. River flow forecasting through conceptual models: Part 1. A discussion of principles. *J. Hydrol.* **1970**, *10*, 282–290. [CrossRef]

65. Santhi, C.; Arnold, J.G.; Williams, J.R. Validation of the SWAT model on a large river basin with point and nonpoint sources. *J. Am. Water Resour. Assoc.* **2001**, *37*, 1169–1188. [CrossRef]

66. Legates, D.R.; McCabe, G.J. Evaluating the use of "goodness-of-fit" measures in hydrologic and hydroclimatic model validation. *Water Resour. Res.* **1999**, *35*, 233–241. [CrossRef]

67. De Winter, J.C.F.; Gosling, S.D.; Potter, J. Comparing the Pearson and Spearman correlation coefficients across distributions and sample sizes: A tutorial using simulations and empirical data. *Psychol. Methods* **2016**, *21*, 273–290. [CrossRef]

68. Kendall, M.; Gibbons, J.D. *Rank Correlation Methods. Charles Griffin Book Series*, 5th ed.; Oxford University Press: Oxford, UK, 1990; ISBN 978-0195208375.
69. Akaike, H. Information theory and an extension of the maximum likelihood principle. In *2nd International Symposium Information Theory*; Petrov, B.N., Csak, F., Eds.; Akademia Kiado: Budapest, Hungary, 1973; pp. 267–281.
70. Akaike, H. A new look at the statistical model identifiacation. *IEEE Trans. Autom. Control.* **1974**, *19*, 716–723. [CrossRef]
71. McQuarine, A.D.R.; Tsai, C.L. *Regression and Time Series Model Selection*; World Scientific: Singapore, 1998.
72. Yaya, O.; Fashae, A.O. Seasonal fractional integrated time series models for rainfall data in Nigeria. *Theor. Appl. Climatol.* **2014**, *120*, 99–108. [CrossRef]
73. Ghimire, B. Application of ARIMA Model for River Discharges Analysis. *J. Nepal Phys. Soc.* **2017**, *4*, 27–32. [CrossRef]
74. Ljung, G.M.; Box, G.E. On a measure of lack of fit in time series models. *Biometrika* **1978**, *65*, 297–303. [CrossRef]
75. Searcy, J.K.; Hardison, C.H. Double-mass curves. In *Manual of Hydrology: Part 1 General Surface-Water Techniques*; Geological Survey Water-Supply Paper 1541–B; U.S. Department of Interior: Washington, DC, USA, 1960.
76. Kliment, Z.; Matoušková, M.; Ledvinka, O.; Královec, V. Trend analysis of rainfall-runoff regimes in selected headwater areas of the Czech Republic. *J. Hydrol. Hydromech.* **2011**, *59*, 36–50. [CrossRef]
77. Chen, J.; Finlayson, B.L.; Wei, T.; Sun, Q.; Webber, M.; Li, M.; Chen, Z. Changes in monthly flows in the Yangtze River, China-with special reference to the Three Gorges Dam. *J. Hydrol.* **2016**, *536*, 293–301. [CrossRef]
78. Guo, L.; Su, N.; Zhu, C.; He, Q. How have the river discharges and sediment loads changed in the Changjiang river basin downstream of the three gorges dam? *J. Hydrol.* **2018**, *560*, 259–274. [CrossRef]
79. Moriasi, D.; Arnold, J.; VanLiew, M.; Bingner, R.; Harmel, R.; Veith, T. Model evaluation guidelines for systematic quantification of accuracy in watershed simulations. *ASABE* **2007**, *50*, 885–900. [CrossRef]
80. Searcy, J.K. *Flow-Duration Curves, Manual of Hydrology: Part 2, Low-Flow Techniques*; United States Geological Survey: Reston, VA, USA, 1959.
81. Cuo, L.; Li, N.; Liu, Z.; Ding, J.; Liang, L.; Zhang, Y.; Gong, T. Warming and human activities induced changes in the Yarlung Tsangpo basin of the Tibetan plateau and their influences on streamflow. *J. Hydrol. Reg. Stud.* **2019**, *25*, 100625. [CrossRef]
82. Mote, P.W.; Hamlet, A.F.; Clark, M.P.; Lettenmaier, D.P. Declining mountain snow pack in western North America. *Bull. Am. Metab. Soc.* **2005**, *86*, 39–49. [CrossRef]
83. Cayan, D.R.; Kammerdiener, S.A.; Dettinger, M.D.; Caprio, J.M.; Peterson, D.H. Changes in the onset of Spring in the Western United States. *Bull. Am. Metab Soc.* **2001**, *82*, 399–415. [CrossRef]
84. Stewart, I.; Cayan, D.C.; Dettinger, M.D. Changes in snowmelt runoff timing in Western North America under a 'business as usual' climate change scenario. *Clim. Chang.* **2004**, *62*, 217–232. [CrossRef]
85. IPCC. *Climate Change*; Synthesis report; IPCC: Geneva, Switzerland, 2007.
86. Li, L.; Xu, H.; Chen, X. Streamflow Forecast and Reservoir Operation Performance Assessment under Climate Change. *Water Resour. Manag.* **2010**, *24*, 83. [CrossRef]

remote sensing

MDPI

Article

Analyzing the Suitability of Remotely Sensed ET for Calibrating a Watershed Model of a Mediterranean Montane Forest

Steven M. Jepsen [1,*], Thomas C. Harmon [1] and Bin Guan [2,3]

[1] Department of Civil & Environmental Engineering and Environmental Systems Graduate Program, University of California Merced, 5200 N. Lake Road, Merced, CA 95343, USA; tharmon@ucmerced.edu

[2] Joint Institute for Regional Earth System Science and Engineering, University of California, Los Angeles, 607 Charles E. Young Drive East, Los Angeles, CA 90095, USA; bin.guan@jpl.nasa.gov

[3] Jet Propulsion Laboratory, California Institute of Technology, 4800 Oak Grove Drive, Pasadena, CA 91109, USA

* Correspondence: sjepsen@ucmerced.edu

Abstract: The ability to spatially characterize runoff generation and forest health depends partly on the accuracy and resolution of evapotranspiration (ET) simulated by numerical models. A possible strategy to increase the accuracy and resolution of numerically modeled ET is the use of remotely sensed ET products as an observational basis for parameter estimation (model calibration) of those numerical models. However, the extent to which that calibration strategy leads to a realistic representation of ET, relative to ground conditions, is not well understood. We examined this by comparing the spatiotemporal accuracy of ET from a remote sensing product, MODIS MOD16A2, to that from a watershed model (SWAT) calibrated to flow measured at an outlet streamgage. We examined this in the upper Kings River watershed (3999 km^2) of California's Sierra Nevada, a snow-influenced watershed in a Mediterranean climate. We assessed ET accuracies against observations from three eddy-covariance flux towers at elevations of 1160–2700 m. The accuracy of ET from the stream-calibrated watershed model surpassed that of MODIS in terms of Nash-Sutcliffe efficiency (+0.36 versus −0.43) and error in elevational trend (+7.7% versus +81%). These results indicate that for this particular experiment, an outlet streamgage would provide a more effective observational basis than remotely sensed ET product for watershed-model parameter estimation. Based on analysis of ET-weather relationships, the relatively large errors we found in MODIS ET may be related to weather-based corrections to water limitation not representative of the hydrology of this snow-influenced, Mediterranean-climate area.

Keywords: evapotranspiration; model; SWAT; calibration; regression; remote sensing; Sierra Nevada; flux tower; water limitation; vapor pressure deficit

Citation: Jepsen, S.M.; Harmon, T.C.; Guan, B. Analyzing the Suitability of Remotely Sensed ET for Calibrating a Watershed Model of a Mediterranean Montane Forest. *Remote Sens.* **2021**, *13*, 1258. https://doi.org/10.3390/rs13071258

Academic Editor: Yongqiang Zhang

Received: 24 February 2021
Accepted: 24 March 2021
Published: 26 March 2021

Publisher's Note: MDPI stays neutral with regard to jurisdictional claims in published maps and institutional affiliations.

1. Introduction

Accurate knowledge of evapotranspiration (ET) is needed for detailed mapping and characterization of water losses from terrestrial runoff [1,2] and impacts of drought and climate variability on forest health [3–5]. Advances in the ability to predict/characterize ET with ever increasing resolution (e.g., smaller spatial scale) will likely be made through the integration of remote sensing products with hydrologic modeling tools. Remote sensing products provide spatiotemporal estimates of ET based on satellite-measured light reflectance, meteorological data, and underlying mathematical models that are physical [6–9] or empirical in nature [10]. Two examples of remote sensing ET products, denoted ET$_{rs}$, are the MODerate Resolution Imaging Spectroradiometer (MODIS) ET product [11,12] and the Global Land Evaporation Amsterdam Model (GLEAM) ET product [13]. ET$_{rs}$ products such as these offer a potential source of observational data against which watershed numerical models can be calibrated, a process needed for the estimation of model parameters

91

that cannot be determined through direct observation [14,15]. ET_{rs} products are currently available at a much finer spatial resolution than the effective spatial resolution of most streamgages, the alternative and prevalent source of calibration data. The effective spatial resolution of a streamgage, which scales as the square root of the upstream contributing area unique to that gage, is typically orders of magnitude coarser than the spatial resolution of ET_{rs} products such as MODIS MOD16 or Landsat based provisional product [16,17]. This finer resolution of ET_{rs} affords the modeler the ability to estimate parameters at a correspondingly finer spatial resolution than is possible with stream-based calibration.

Previous studies have not examined whether ET_{rs} products are sufficiently accurate to substitute for stream discharge observations as an observational basis for watershed model calibration. For that substitution to make sense, the ET_{rs} product would need to be more accurate—relative to ground based observations—than the ET that is simulated by a watershed model calibrated to stream discharge. Otherwise, the calibration procedure would only move predictions from the watershed model further from reality in order to more closely match the remote sensing data. In order to evaluate the sufficiency of ET_{rs} accuracy, one would need to directly compare the accuracy of ET_{rs} to the accuracy of ET from watershed model calibrated to streamgage and not ET_{rs}. Such evaluation has not been carried out in previous studies calibrating watershed models to remotely sensed ET [18–27]. It is important to recognize that ET products from remote sensing and watershed models are both derived from models each having their own relative strengths and weaknesses. ET_{rs} products have been found to suffer in accuracy in certain types of environments such as nivean montane forest [8,28] and temperate grassland with dry surface conditions [9,29]. Meanwhile, watershed models are known to have especial difficulty during rainless periods [23] and periods of extreme runoff [30]. Inadequate attention has been given to determining when one type of model is accurate enough to serve as "observations" in calibration of another type of model.

The objective of this study was to determine if a specific remote-sensing ET product, MODIS MOD16A2, is accurate enough to be used to calibrate a watershed model of a snow-influenced, streamgage-equipped watershed in a Mediterranean climate. The guiding question was: Would calibrating the watershed model to ET_{rs} make the ET predictions from that watershed model more or less accurate than ET predictions from the same watershed model calibrated to observed streamflow? To address this, we compared the accuracy of ET from the following two models: (1) the model behind the MOD16A2 ET_{rs} product, and (2) a watershed model of ET calibrated to a streamgage. In addition, we analyzed ET-seasonality and ET-weather relationships from both models in order to identify environmental conditions conducive to model strengths and weaknesses. We applied this study to the upper Kings River watershed of California's Sierra Nevada, assessing all accuracies relative to ground based observations from eddy-covariance flux towers located along a 1160–2700 m elevational transect.

2. Study Area

The study area is the upper Kings River watershed in the southern Sierra Nevada mountain range of California, USA. The watershed collects runoff, largely as snowmelt, from a 3999 km^2 area ranging in elevation from 285 m in the western foothills to 4338 m along the Pacific Crest in the east (Figure 1, lower right). Different forks of the Kings River pass through a series of reservoirs operated for flood control and hydroelectric power (not shown) [31] to eventually empty into the Pine Flat Reservoir at the western outlet of the watershed (Figure 1, lower right). Runoff from the watershed provides water to over a million acreas of some of the world's most fertile and productive agricultural land [31].

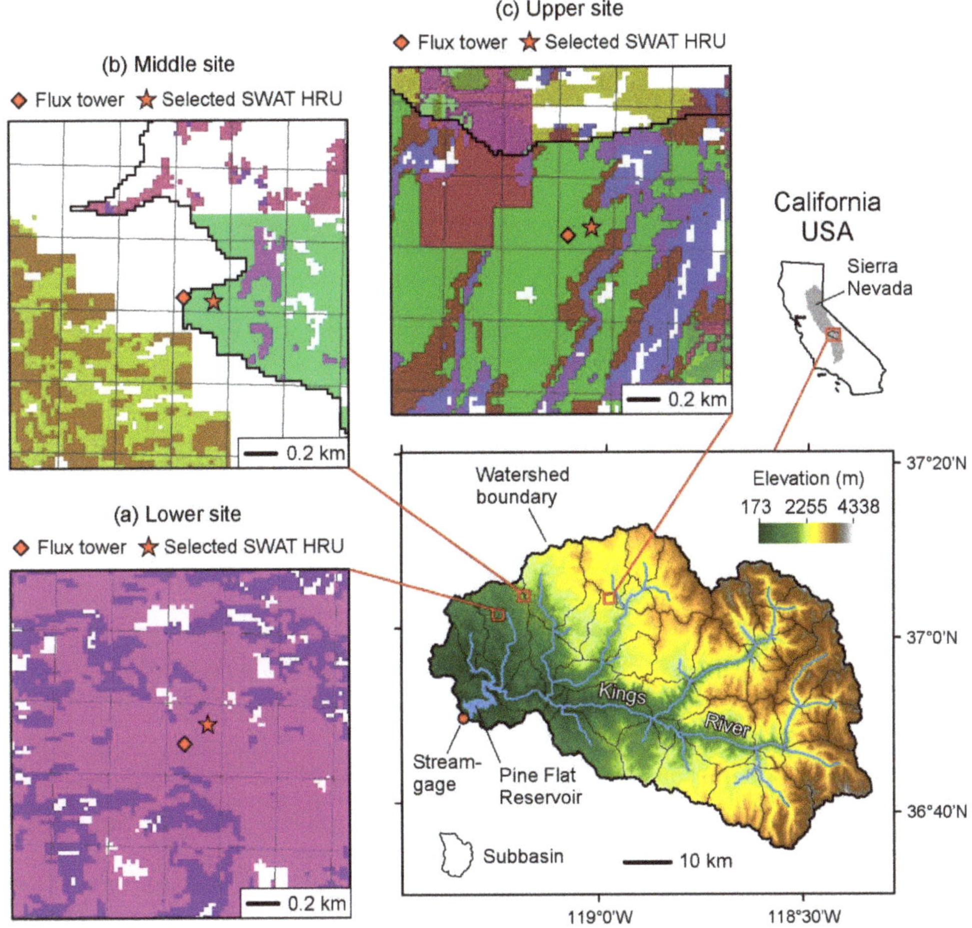

Figure 1. Location of (lower right) upper Kings River watershed in southern Sierra Nevada, California, and (**a**–**c**) flux towers within watershed. In (**a**–**c**), fishnets are grid cell boundaries of the MODIS MOD16A2 ET product, and HRUs are Hydrologic Response Units of SWAT watershed model (color-shaded areas) selected for analysis based on proximity to flux tower. Topography from USGS National Elevation Dataset [32], water bodies from USGS National Hydrography Dataset [33].

The climate of the area is Mediterranean, with cool moist winters and warm dry summers. Based on 1981–2010 climatic normals from the PRISM Norm81d product [34,35], the watershed receives approximately 999 mm of precipitation per year on average, most of which (85%) occurs during the wet six-month period of November–April. A little over half of all precipitation (54%) flows out of the watershed as the Kings River based on 1981–2010 full-natural streamflow below Pine Flat Reservoir [36]. Annual precipitation in the watershed increases with elevation, from approximately 530 mm in the lowermost areas to approximately 1200 mm in the uppermost areas (Supplement Figure S1). Average air temperature is 7.1 °C at the mean elevation of 2329 m, and decreases with elevation at a rate of −5.4 °C per km (Supplement Figure S2). The phase of precipitation shifts from rain to snow with increasing elevation, becoming mostly snow at elevations above approximately 2000 m [37,38]. A little over two-thirds of the watershed (68%) resides above this rain-snow transition elevation.

Soils are distributed over approximately 59% of the watershed up to an elevation of approximately 2700 m [39,40], with exposed bedrock dominating the higher elevations. The soils grade from thermic Alfisols at lowest elevations into frigid Entisols and Inceptisols at higher elevations. These soils are loamy to sandy, well- to excessively drained, with thicknesses ranging from 20 to 250 cm [39,40]. In addition to soil, the underlying regolith (weathered bedrock) is also known to be an important source of water to vegetation [41,42]. The dominant land cover types accounting for 99% of the watershed are evergreen forest (52%), shrub/scrub (30%), barren land (rock/sand/clay) (9.8%), grassland/herbaceous (6.0%), and open water (1.4%) [43].

The watershed contains the Southern Sierra Critical Zone Observatory (SSCZO) operated in cooperation with the Kings River Experimental Watersheds program (KREW) [37,44]. This observatory includes three eddy-covariance flux towers [28], maintained by the Goulden Lab at University of California, Irvine [45], which provided the ET observations used in this study (Section 3.3). For more information about the soils, vegetation, and climate of these sites, the readers are referred to Hunsaker et al. [37], Bales et al. [38], O'Geen et al. [42], Bales et al. [44], and Safeeq and Hunsaker [46].

3. Methods

3.1. Summary

Our methods were designed to address two objectives. The first objective was to determine which would be more accurate, ET predictions from a watershed model calibrated to streamgage (ET_{wm}) or ET predictions from a watershed model of the same area calibrated to ET_{rs} product (rather than streamgage). We did this by comparing the ET accuracy of two models: (1) MODIS MOD16, (2) SWAT calibrated to streamgage. Model accuracies were evaluated relative to ET observations at three eddy-covariance flux towers (Figure 1). If ET_{rs} is more accurate than ET_{wm}, then it follows that calibrating the watershed model to that remote-sensing ET product could potentially improve the watershed model's ET accuracy over that of a purely stream-based calibration approach. Conversely, if ET_{rs} is less accurate than ET_{wm}, then calibrating the watershed model to those ET_{rs} would only degrade the watershed model's ET accuracy relative to a stream-based calibration approach. ET observations and simulations were considered on a monthly basis during water years 2009–2018, with a water year defined to extend from October 1 through September 30. For metrics of ET model accuracy, we used the Nash-Sutcliffe efficiency (NSE) [47] which equals the fraction of variance in observations explained by a model, and model percent-bias in average ET (PBIAS) [48]. We also evaluated how well the two models producing ET_{rs} and ET_{wm} captured the relationship between average ET and elevation (1160–2700 m). Metrics of NSE and PBIAS were reported for the period of all water years during 2009–2018, water years with annual precipitation below the 2001–2019 median of 803 mm ("dry years") based on PRISM AN81d product [34,35], and water years with annual precipitation at or above that median.

The second objective of this study was to compare and contrast ET from flux towers and models in a way that identifies seasonal conditions associated with model strengths and weaknesses. We started this out with an examination of monthly time series plots. Next, we compared and contrasted ET "seasonality", defined as the set of monthly ET averages during the calendar year. Lastly, we examined relationships between monthly ET and weather variables of air temperature and vapor pressure deficit (Section 3.2), variables which influence atmospheric water demand and are also used to limit canopy conductance in the ET_{rs} model (Section 3.4).

3.2. Weather Data

Weather data for watershed modeling and examination of ET-weather relationships were obtained from daily 2.5-arcminute PRISM AN81d product [34,35,49] and GridMET product [50]. Air temperature and vapor pressure deficit were computed from PRISM values of minimum air temperature, maximum air temperature, and dewpoint temperature, along with the equation for saturation vapor pressure in Murray [51]. Downward solar radiation and wind speed were obtained from GridMET. Daily weather time series were assembled for each subbasin of the watershed (N = 47; Figure 1, lower right) by downsampling the gridded weather data to approximately 250-m resolution, masking the downsampled data to subbasin boundaries, and computing spatial averages (Supplement section S1). These weather time series were input to ArcSWAT to force spatial elements of the watershed model referred to as Hydrologic Response Units (HRUs) (Section 3.5). The weather data used for the examination of ET-weather relationships were taken from the model HRUs nearest to the flux towers as shown in Figure 1.

3.3. Flux Tower Observations of ET

The ET observations used in this study were collected at three flux towers (Figure 1) situated along an environmental gradient ranging from rain-dominated Ponderosa pine at 1160 m to snow-dominated Lodgepole pine at 2700 m [28] (Table 1). The observations were collected during the day at 30-minute intervals using the eddy covariance method [52], at a height of 5–10 m above the tallest trees [28]. We first obtained the 30-minute ET observations and "Flexible Filler" processing script (Matlab) from the Goulden Lab [45]. Using the Flexible Filler script, we filtered out 30-minute data during calm atmospheric conditions and filled the resulting gaps using the regression approach in Goulden et al. [28,52]. The filtered, gap-filled data were then aggregated to monthly values, requiring at least 50% data availability for each month. Dates of data availability slightly differed for each flux tower, as shown in Table 1.

Table 1. Site characteristics of eddy covariance flux towers providing ET observations for this study [28]. These sites are mapped in Figure 1. Water years of data used for this study are listed at right. Years with partial data availability, indicated with asterisk, provided less than 11 out of 12 months' worth of data.

Site Name (This Paper)	Local Site Name	Elevation (m)	Dominant Vegetation	Latitude (deg)	Longitude (deg)	Data Availability (Water Years)
Upper	Short Hair Creek	2700	Lodgepole pine	37.0671	−118.9871	2010–2011, 2012 *, 2015 *, 2016–2018
Middle	Providence 301	2015	White fir, pine, cedar	37.0673	−119.1948	2009–2018
Lower	Soaproot Saddle	1160	Ponderosa pine, oak	37.0311	−119.2563	2011–2018

3.4. Remote Sensing of ET

The ET_{rs} product used in this study was the 8-day, 0.25-arcminute global MOD16A2 product from MODIS sensors onboard the Terra and Aqua satellites [11,12,53]. The model behind this product uses a modifed Penman-Monteith approach adapted from Cleugh et al. [6] to predict evaporation from wet and dry soil, evaporation from wet canopy, and transpiration from dry canopy [12,54]. The model operates at a daily time step using meteorology from NASA's Global Modeling and Assimilation Office (GMAO) reanalysis dataset ($1.0 \times 1.25°$ resolution), 8-day composites of absorbed photosynthetically active radiation (FPAR) and leaf area index (LAI), and 16-day composites of albedo. Limitations on ET associated with water availability and stomatal physiology (e.g., Jarvis [55]) are modeled in the MOD16A2 product using canopy conductance correction factors consisting of linear functions of vapor pressure deficit and minimum air temperature [11,12,54]. These functions are parameterized by biome-specific look-up tables organized by land cover classification (Table 3.2 of Running et al. [54]).

There was some uncertainty as to how ET in individual MODIS cells, each approximately 500-m across, would correspond to flux tower observations which collect fluxes from areas (footprints) typically measuring 100–2000 m across [56]. For the comparisons to ET from SWAT and flux towers, and in consideration of plausible intra-footprint variability in MODIS data (see Supplement Figure S3), we decided to obtain the MODIS ET data from the single MODIS cells containing the flux tower locations (Figure 1). This decision followed from the results in Supplement Figure S3 showing that MODIS intra-footprint variability tends to be substantially less than same-cell differences between MODIS and flux tower. We filtered out any MOD16A2 data having QA/QC flags not of "good quality" and not free of "significant" cloud cover, then aggregated it to monthly values using time-weighted averaging and requiring at least 50% data availability for each month.

3.5. Watershed Modeling of ET

We modeled ET in the upper Kings River using the Soil & Water Assessment Tool (SWAT) version 2012, with ArcSWAT version 2012 for model construction. SWAT dates back to the late 1970s for research into land management impacts on water, sediment, nutrients, and pesticides [57–59]. It has been widely used to simulate hydrologic fluxes in snow-influenced watersheds such as this study area [60–65]. In SWAT, the water conservation equation is solved at a daily time step in HRUs, each of which is a numerical 1-D element ("bucket") of homogeneous land cover, soil, and slope. Each HRU exchanges water/energy with the atmosphere (including ET) and delivers runoff directly to the stream reach within enclosing subbasin (subbasins shown in Figure 1, lower right). We selected the Penman-Monteith option for potential ET; actual ET was computed based on potential ET and water availability in soil and underlying aquifer [66].

To construct the watershed model, we first built a "base model" in ArcSWAT using the parameters and input datasets for elevation, soil, land cover, meteorology, and snow cover detailed in Supplement section S1. We then initialized the base model with LAI and biomass from the end of a 30-year spin-up forced by dynamic steady-state weather (Supplement section S2), forming the "plant spin-up" model representing mature forest conditions at the start of simulations. Next, we applied a calibration procedure to the plant spin-up model as follows. First, we used a global sensitivity analysis (GSA) to identify the most influential parameters of the plant spin-up model to estimate via calibration. Using the Sobol method of GSA [67], we identified the influential parameters (N = 12) accounting for 99% of the variance in model streamflow error (Supplement section S3). Next, we calibrated the plant spin-up model by estimating values of influential parameters that minimized model errors relative to 2003–2010 monthly, full-natural streamflow at the watershed outlet [36], as described in Supplement section S4. We used for calibration the Sequential Uncertainty Fitting algorithm (SUFI-2) calibration and uncertainty tool of the SWAT-CUP software package [68–70]. Lastly, we validated the calibrated model to 2011–2019 monthly, full-natural streamflow using the parameter ranges obtained from calibration (Supplement Table S6). Simulations for calibration and validation were spun up to a total of 13 years of weather data, consisting of 10 years of dynamic steady-state weather (for aquifer equilibration) followed by three years of real weather (for soil and plant equilibration) (Supplement sections S2 and S4). In terms of model performance, the calibrated and validated models both showed "very good" results (Figure 2) based on criteria in Moriasi et al. [48] and Abbaspour et al. [68]. For the comparisons to flux tower and MODIS, we used monthly ET of the "best" simulation (Figure 2) taken from the HRUs nearest to the flux towers (Figure 1).

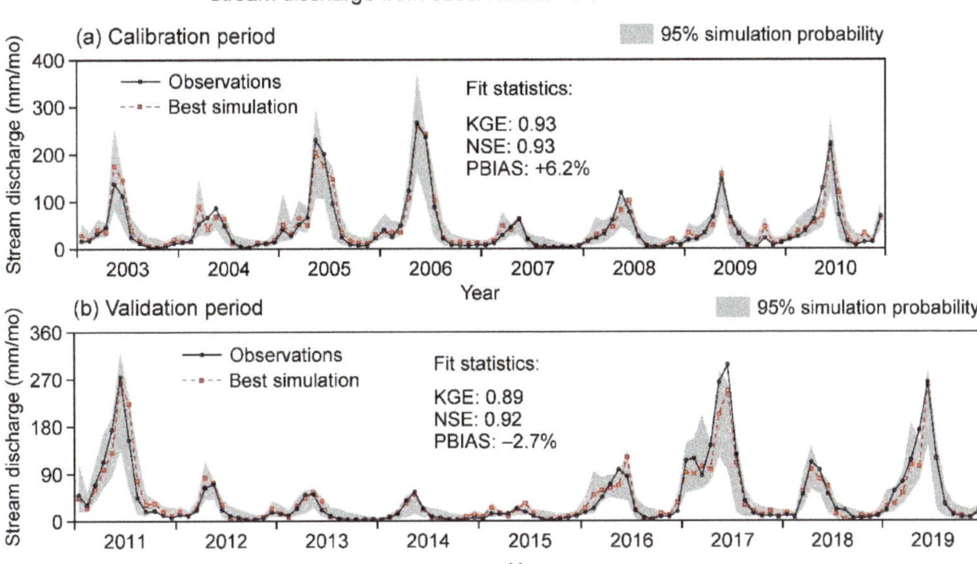

Figure 2. Monthly stream discharge at Pine Flat Dam from observations versus best simulation of SWAT watershed model. 95% simulation probability also shown, defined at the 2.5– to 97.5–percentile of monthly simulations. Fit statistics are for "best simulation" having the highest objective function, i.e., the Kling-Gupta efficiency (KGE) [71]. NSE = Nash-Sutcliffe efficiency [47], PBIAS = model percent bias [48].

4. Results

4.1. Long-term ET

Long-term ET observations were substantially greater than the MODIS model predictions and greater but more similar to the SWAT model predictions. Observed ET averaged across all months and flux towers was 54.6 mm/mo (Figure 3). This value was 13% greater than SWAT's prediction (48.5 mm/mo), 83% greater than MODIS's prediction (29.8 mm/mo) (Figure 3). In addition, observed long-term ET in the upper Kings River decreased with elevation at a markedly steeper rate than the MODIS model predicted. Based on the flux tower network, ET decreased with elevation at a rate of -0.013 mm mo^{-1} m^{-1} (Figure 3). This rate was fairly close (7.7%) to the SWAT model prediction of -0.012 mm mo^{-1} m^{-1} (Figure 3). In contrast, the MODIS model predicted an elevational trend of only -0.0025 mm mo^{-1} m^{-1}, one-fifth the observed value. These results showed that both magnitude and elevational trend in long-term ET were predicted much more accurately by SWAT than MODIS.

4.2. Monthly ET

ET observations from the flux towers followed a seasonal pattern that roughly tracked the length of daylight, with peak values occurring near the middle of calendar years and minimum values occurring near the end of calendar years (Figure 4). Outputs from the SWAT model followed a similar seasonal timing, although the shape of its annual waveforms was sometimes excessively narrowed and peaked, such as in year 2016 at the middle site (Figure 4, middle). The MODIS model produced annual waveforms in ET that were notably out of phase with observations and SWAT model predictions. ET-values from MODIS tended to be on a descending limb, or near their annual minima (within one month thereof), during summer when observed and SWAT-modeled ET were near their maxima. This out-of-phase characteristic of MODIS predictions was especially apparent during years 2016–2018 at the lower and middle sites (Figure 4). In addition, the annual ET waveforms from MODIS appeared less distinct than those from observations and SWAT.

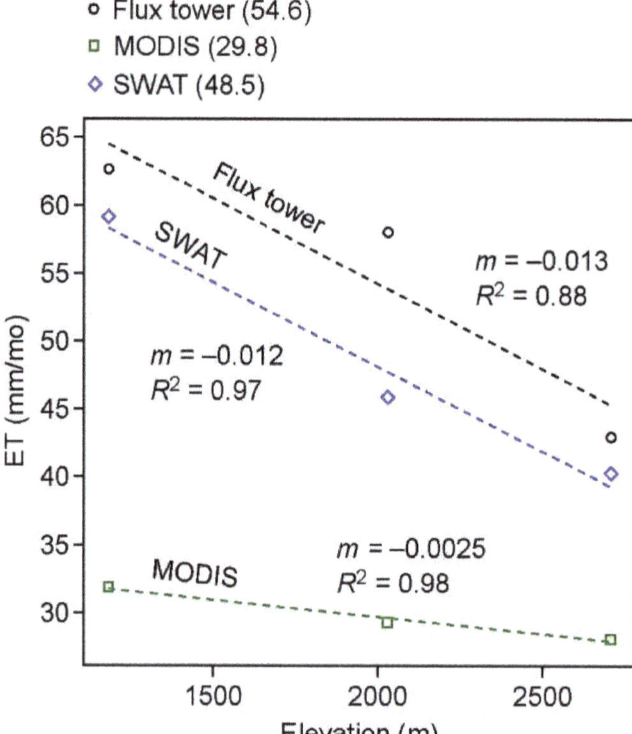

Figure 3. Average monthly ET versus elevation from flux towers, MODIS, and SWAT. Averages across all elevations are parenthesized in the legend. m = slope of linear-regression trendline, R^2 = coefficient of determination of linear-regression fit.

In terms of the NSE metric of model fitness, the SWAT model matched monthly ET observations better than the MODIS model. The NSE of SWAT across all flux tower sites was +0.36, ranging from +0.04 at the middle site to +0.68 at the upper site (Table 2). In comparison, the NSE of MODIS across all sites was −0.43, ranging from −0.73 at the middle site to −0.33 at the lower site. These negative NSE-values indicate a level of prediction efficiency lower than that provided by knowledge of long-term observed ET. Both SWAT and MODIS showed minimum prediction efficiency at the middle site (Table 2). In addition, both models showed slightly lower prediction efficiency (NSE lower by ~0.2) during wet years than dry years (Table 2).

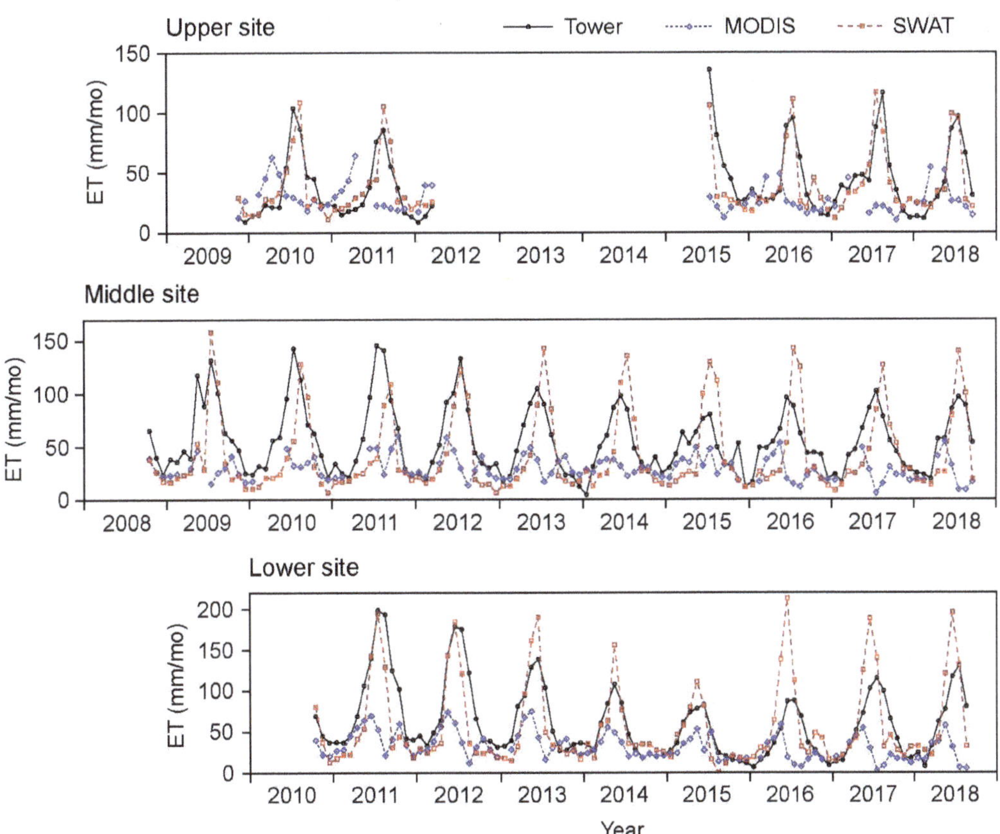

Figure 4. Monthly ET from flux towers, MODIS, and SWAT at three sites in upper Kings River watershed.

In terms of model bias, both SWAT and MODIS underestimated long-term ET as shown by negative PBIAS-values in Table 2. However, the underestimates from MODIS were substantially greater in magnitude than those from SWAT. The SWAT model underestimated average monthly ET across all sites by 13% (PBIAS = −13%). These underestimates ranged from 23% at the middle site to 5% at the other two sites. In comparison, the MODIS model underestimated average monthly ET across all sites by 47% (PBIAS = −47%), with underestimates ranging from 50% at the middle site to 35% at the upper site.

4.3. Seasonality in ET and Weather

Seasonal distributions in ET, potential ET (PET), and weather variables are plotted in Figure 5. Observed ET at the flux towers reached peak values in June or July, following a seasonal distribution ("seasonal curve") that was well aligned with PET from MODIS (Figure 5, left column). Air temperature and vapor pressure deficit reached peak values in July or August (Figure 5, middle column), following seasonal curves that closely tracked one another. This close tracking indicated a strong cross-correlation of these weather variables on a monthly time scale (R^2 = 0.89–0.94 for three sites, not shown). Based on these results, air temperature and vapor pressure deficit reached maximum seasonal values within approximately one month of observed ET and MODIS PET. This timing seemed reasonable given that these weather variables are known to strongly influence atmospheric water demand.

Table 2. Error statistics of monthly ET from MODIS, SWAT, and air temperature corrected MODIS, relative to flux tower observations. "All site" statistics (bottom) are for monthly ET data concatenated across the three sites. "All years" are water years 2009–2018, "wet years" are water years 2009–2011 and 2016–2017, and "dry years" are water years 2012–2015 and 2018. NSE = Nash-Sutcliffe efficiency [47], PBIAS = percent bias (positive = model overestimate) [48], N = number of months.

Site	Product	All years			Wet years			Dry years		
		NSE	PBIAS	N	NSE	PBIAS	N	NSE	PBIAS	N
Upper	MODIS	−0.56	−35	57	−0.66	−32	39	−0.42	−40	18
	SWAT	+0.68	−5.4	68	+0.67	−3.8	47	+0.70	−9.1	21
	Corrected MODIS	+0.77	+1.7	57	+0.78	+2.2	39	+0.75	+0.67	18
Middle	MODIS	−0.73	−50	104	−1.0	−57	46	−0.54	−43	58
	SWAT	+0.04	−23	120	−0.04	−31	60	+0.13	−15	60
	Corrected MODIS	+0.69	−1.5	104	+0.70	−12	46	+0.64	+9.3	58
Lower	MODIS	−0.33	−49	91	−0.45	−53	33	−0.23	−47	58
	SWAT	+0.41	−5.3	95	+0.27	+3.5	36	+0.53	−10.7	59
	Corrected MODIS	+0.60	−1.3	91	+0.63	−7.0	33	+0.57	+2.1	58
All sites	MODIS	−0.43	−47	252	−0.56	−49	118	−0.31	−44	134
	SWAT	+0.36	−13	283	+0.27	−14	143	+0.46	−12	140
	Corrected MODIS	+0.67	−0.89	252	+0.71	−7.1	118	+0.63	+4.9	134

A seasonal asynchronicity was observed between MODIS ET and MODIS PET. MODIS ET reached a seasonal *minimum* only 1–2 months after the seasonal *maximum* in MODIS PET (Figure 5, left). MODIS PET reached a maximum value in July at all three sites. In comparison, MODIS ET reached a minimum value only one month later at the lower and middle sites, and two months later at the upper site (Figure 5, left). Such occurrences of minimum ET near the time of maximum atmospheric water demand are indicative of pronounced limitations on ET in the MODIS model.

Seasonal curves of SWAT ET tended to be skewed toward times earlier in the year relative to SWAT PET (Figure 5, right). These ET and PET curves closely tracked one another between December and mid-summer. Then, in mid-summer, the ET curves began following descending limbs 1–2 months before descending limbs of PET. This difference in timing of descending limbs produced summer and fall deficits of ET relative to PET (Figure 5, right). Such skewness of ET relative to PET is indicative of a seasonal shift in limitation on ET in the SWAT model.

4.4. ET-weather Relationships

The SWAT- and MODIS-models produced markedly different relationships between monthly ET and weather. Monthly ET from SWAT, as well as flux towers, showed a significant positive relationship to air temperature (Figure 6, left and middle columns). At the flux towers, slopes of linear regression between ET and temperature ranged from 3.8 to 4.0 mm mo^{-1} (°C)$^{-1}$. These slopes for SWAT ranged from 3.0 to 5.2 mm mo^{-1} (°C)$^{-1}$. In contrast, monthly ET from MODIS showed either no significant relationship to temperature (p-value > 0.05) or a significant negative relationship (Figure 6, right column). At the lower and middle sites where no significant relationship to temperature occurred, there were distinct mid-temperature peaks in MODIS ET. At the lower site, MODIS ET reached a peak value at an intermediate air temperature of 16.1 °C (Figure 6i). At the middle site, this peak ET occurred at an intermediate temperature of 7.9 °C (Figure 6f). Mid-temperature

ET peaks such as these did not occur in the flux tower observations or SWAT predictions. Vapor pressure deficit, noted earlier to be strongly correlated to temperature, showed a significant positive relationship to ET from both flux towers and SWAT model (Supplement Figure S6). In contrast, vapor pressure deficit showed either no significant relationship (p-value > 0.05) or a significant negative relationship to ET from MODIS (Supplement Figure S6).

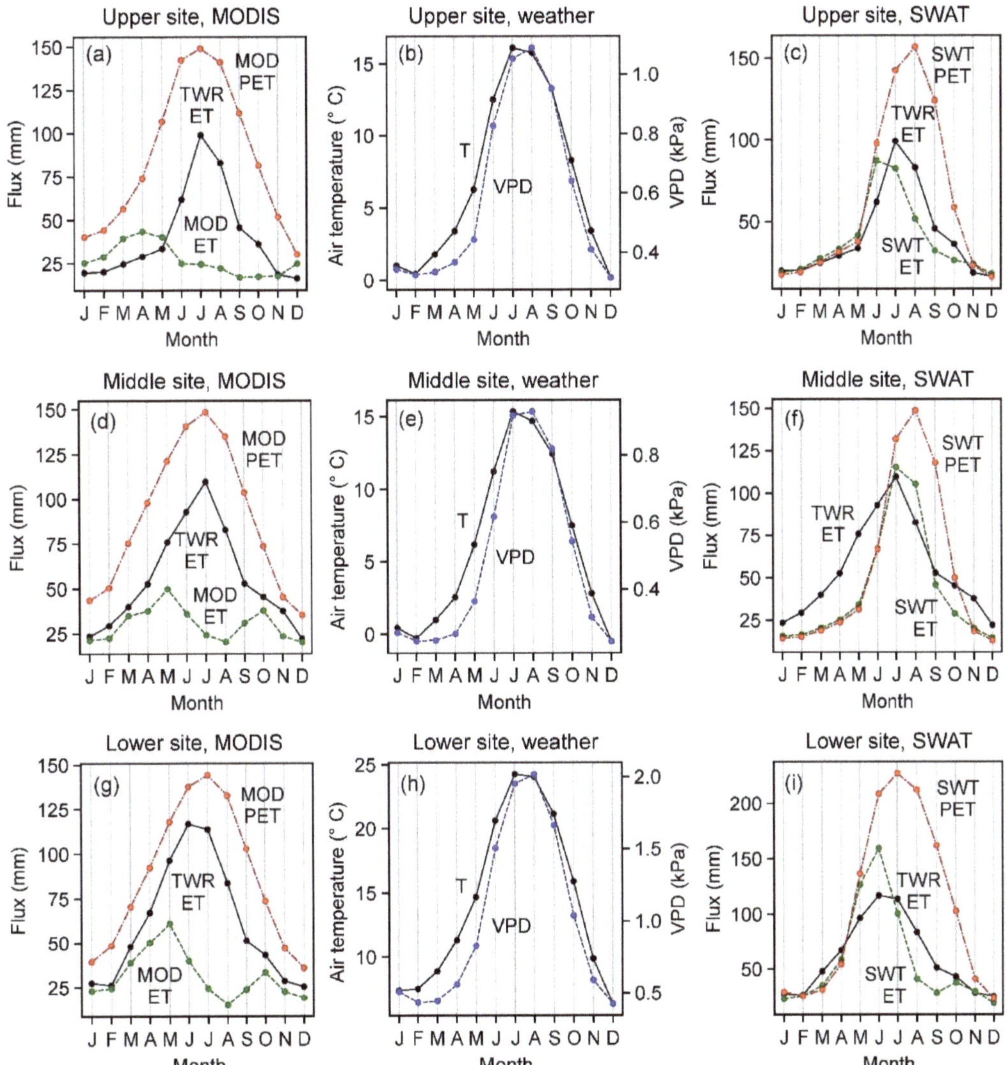

Figure 5. Monthly averages in air temperature (T), vapor pressure deficit (VPD), ET, and potential ET (PET) at the (**a**–**c**) upper site, (**d**–**f**) middle site, and (**g**–**i**) lower site. TWR = flux tower observation, MOD = MODIS value, SWT = SWAT value. MODIS PET (MOD PET) values have been multipled by 0.5.

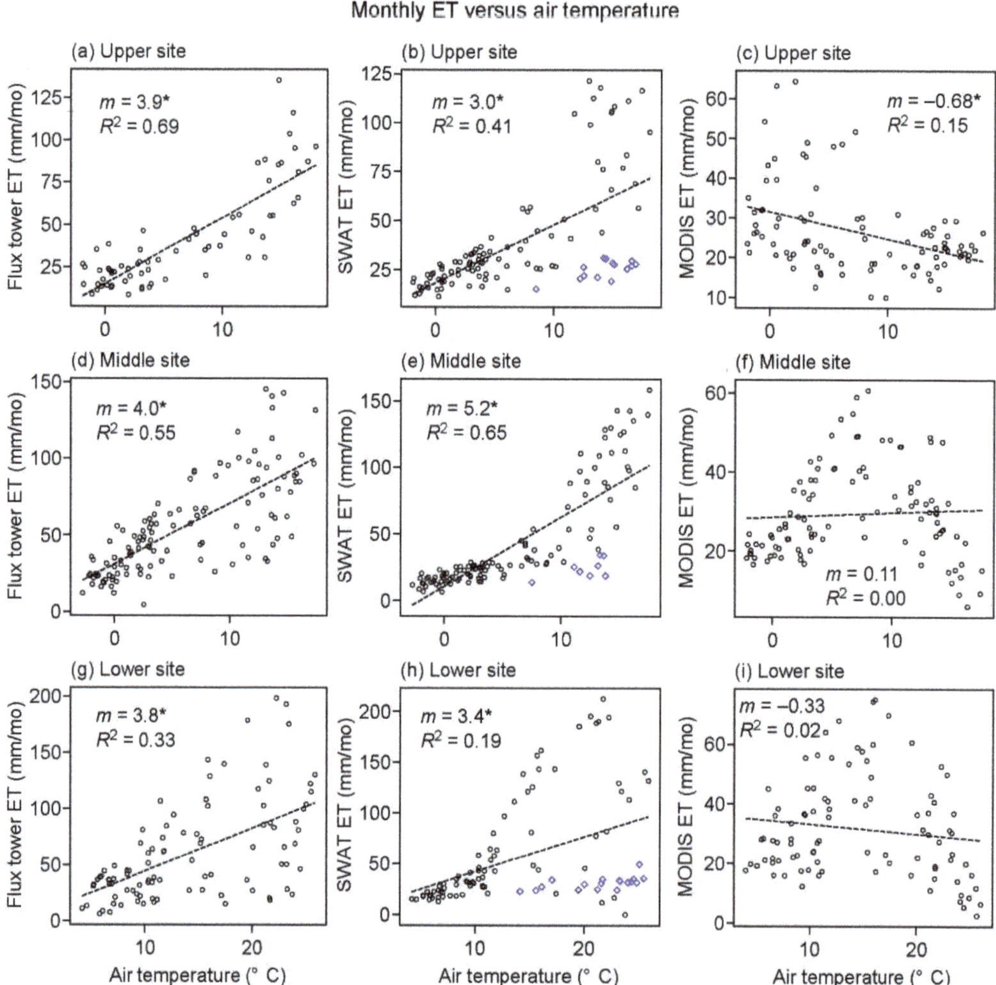

Figure 6. Monthly ET versus air temperature from different data sources at the (**a–c**) upper site, (**d–f**) middle site, and (**g–i**) lower site. Dashed lines are best fits from linear regression. m = slope of regression best-fit with asterisk where p-value < 0.05, R^2 = coefficient of determination of best-fit. In center column, SWAT modeled ET supplied primarily by aquifer water ("revap" flux) are plotted as blue-colored diamond symbols.

Scatterplots of monthly ET from SWAT exhibited kink-like features at the middle and upper sites, and a bifurcation pattern at the lower site (Figure 6, middle column). Data points within the elbows of these kinks and lower arms of bifurcation occurred during times when SWAT ET was supplied primarily by aquifer water (blue diamond symbols in Figure 6, middle column). At other times, SWAT ET was supplied primarily by water in soil overlying the aquifer. The transition from soil- to aquifer-sourced ET was observed (not shown) to coincide with the relatively abrupt declines in SWAT ET during late summer and fall (Figure 4), and may have contributed to the excessively narrow shapes of ET annual waveforms mentioned in Section 4.2.

5. Discussion

5.1. Need to Assess Remotely Sensed ET before Use in Watershed Model Calibration

Our results illustrate a case where spatiotemporal information about ET would be represented more accurately by a watershed model calibrated to streamflow than a watershed model calibrated to ET_{rs}. This assertion follows from our finding that the ET_{rs} (MODIS) data were less accurate than ET predictions from watershed model calibrated to streamflow. ET from the MODIS model had a PBIAS of -47% and NSE of -0.43 across all sites, compared to a PBIAS of -13% and NSE of $+0.36$ for ET from the stream-calibrated watershed model (Table 2). Moreover, the negative relationship between long-term ET and elevation (Figure 3) was underestimated by 81% in the MODIS model compared to only 8% in the stream-calibrated watershed model. In general, there would clearly be cases where a watershed model's representation of ET would likely be improved via calibration to ET_{rs}, one example being a watershed not instrumented with any streamgages. More study is needed, however, to understand the conditions (e.g., climate, modeling frameworks, density of observations) for which the use of ET_{rs} for watershed model calibration would produce superior ET accuracy over use of observed stream discharge.

5.2. Representation of Water limitation in Remote Sensing Products

Plants respond to water stress by regulating their stomata, which in turn modifies transpiration rate [8]. This regulation process is known to be a complex function of atmospheric conditions and plant water potential (including cell turgor pressure) [55,72]. The MODIS ET model accounts for this by numerically correcting the canopy conductance of water vapor using functions of minimum air temperature and vapor pressure deficit (VPD) [11,12,54]. The air temperature component of this conductance correction is meant to account for temperature limitation on plant growth while the VPD component is meant to account for water limitation. For weather at the lower site, the MODIS algorithm would predict an increase in canopy conductance with air temperature to approximately 15 °C, then a decrease in canopy conductance with further warming (Supplement Figure S7b). Below the transition temperature of 15 °C, the temperature-correction component dominates the overall correction to ET giving "temperature-limited" transpiration. Above the transition temperature, the VPD-correction component dominates giving "water-limited" transpiration. The transition temperature of 15 °C approximately coincides with the observed air temperature at which MODIS ET reaches a peak value at the lower site, 16 °C (Figure 6i). Based on this finding, the negative trend shown in Figure 6i between MODIS ET and air temperature for air temperatures > 16 °C, and absence of significant relationship between ET and temperature overall, can be explained by unrealistically high VPD-limitation on ET in the MODIS model.

This argument also seems to apply to the middle site. As a reminder, the air temperatures used in an ET-weather relationship were obtained from the watershed subbasin containing the selected site of interest (Section 3.2, Figure 1). Air temperatures at the middle site were 8 °C cooler on average than at the lower site (Figure 5e versus Figure 5h). This temperature difference exactly coincides with the -8 °C offset of maximum MODIS ET at the middle site relative to the lower site (Figure 6f versus Figure 6i). This can be explained as follows. The 8 °C difference in air temperature between the lower and middle sites occurred across a relatively short distance of approximately 7 km (Figure 1). A difference in weather across this short of distance would not be registered in the MODIS ET product because of its use of weather data at $1° \times 1.25°$ resolution [11,12]. A mismatch in resolution of weather forcings between the SWAT model ($0.042° \times 0.042°$) and MODIS model ($1° \times 1.25°$) would thus introduce an apparent ET offset of -8 °C at the middle site relative to the lower site. This argument does not seem to apply at the upper site because at that location, a clear transition from temperature-limited ET to water-limited ET with increasing air temperature did not occur (Figure 6c).

Based on this interpretation, the relatively large underestimates in warm-season ET from MODIS (Figures 4 and 5) stemmed from excessive VPD-limitation on canopy conductance in the MODIS model. In addition to atmospheric conditions, canopy conductance is known to be a function of plant water potential, which in turn depends on subsurface water availability and the ability of plants to access that water through their roots [72]. Weather, water availability, and plant roots are independent factors (at least to some degree), which is likely a reason for differing ET-VPD relationships across different geographic regions [73,74]. In the snow-influenced Mediterranean climate of the study area, snowmelt is known to be an important source of water to forest during the dry season. In such environments, VPD and actual water availability in the subsurface may be more loosely coupled than in the environments to which the MODIS model has been trained [75].

5.3. Regression-Based Correction to Remotely Sensed ET

ET from observations and the MODIS model showed markedly different relationships to weather variables (Section 4.4). Correlations between ET and air temperature were significantly positive at the flux towers (Figure 6, left) and either negative or not significant from MODIS (Figure 6, right). These contrasting ET-temperature relationships provided a possible basis for correcting MODIS ET to weather using linear regression. MODIS ET error expressed as a fraction of PET, defined as y_{regr} = (MODIS ET − flux tower ET)/(MODIS PET), was found to be well correlated to air temperature. Best fits from linear regression had slopes ranging from −0.010 to −0.024 $°C^{-1}$ and R^2-values of 0.25–0.66 (Figure 7a−c). The best of all fits was found at the upper site, where the R^2 was 0.66 (Figure 7c). We used this regression model to predict corrected values of MODIS ET, set equal to MODIS ET − [(MODIS PET) × y_{regr}]. The resulting predictions of corrected MODIS ET matched the flux tower observations better than both the original MODIS ET and the SWAT model (Figure 7d–f). The corrected MODIS ET had an NSE-value of +0.67 and a PBIAS of −0.9% across all sites, statistics considerably better than those of both the uncorrected MODIS model and the SWAT model (Table 2, bottom). In addition, most of the error in elevational trend in long-term MODIS ET was removed by the regression-based correction to weather (Figure 3 versus Figure 7f). The PBIAS of the corrected MODIS ET was noted to be 12% higher during dry years than wet years (Table 2, bottom), suggesting that regression models trained separately to dry and wet periods may provide further improvement to the correction method.

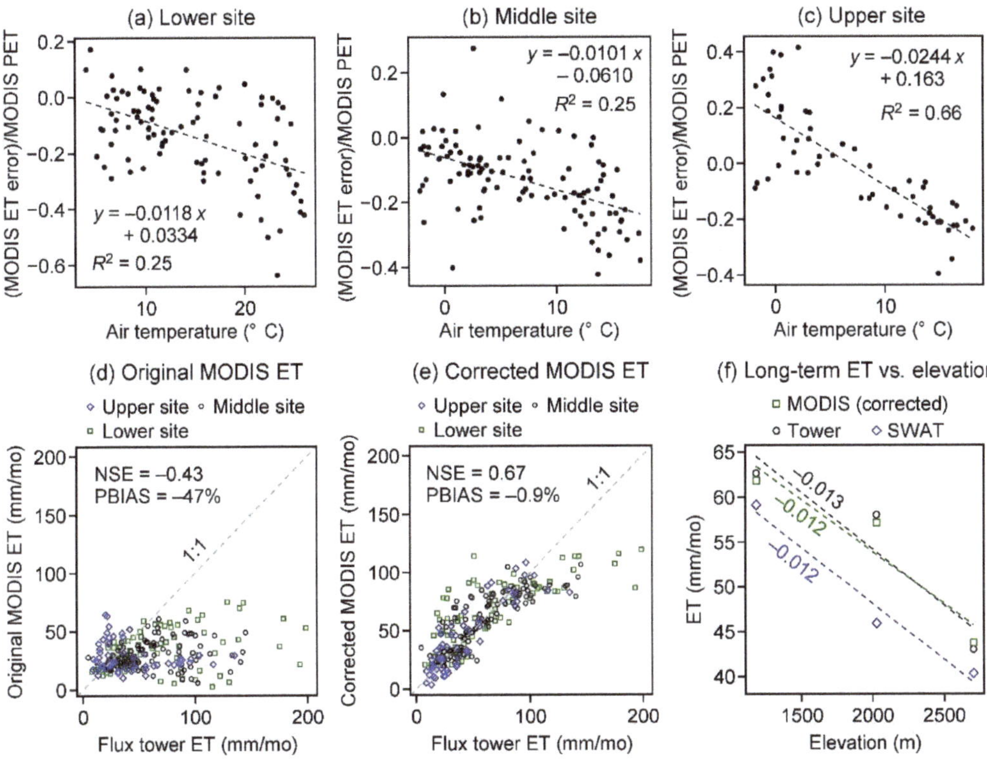

Figure 7. Weather correction to MODIS monthly ET using linear regression with air temperature as predictor variable. (**a–c**) Results of linear regression between MODIS ET error, as fraction of MODIS potential ET, and monthly air temperature for each site. MODIS ET error is defined as MODIS ET—flux tower ET. (**d**) Original (uncorrected) MODIS ET versus flux tower ET shown with 1:1 line. (**e**) Weather-corrected MODIS ET versus flux tower ET. (**f**) Average monthly ET from weather-corrected MODIS, flux tower, and SWAT versus elevation for comparison to Figure 3. Dashed lines are trendlines from linear regression, labeled with value of slope.

6. Conclusions

The concept of using remotely sensed ET products as "observations" for watershed model calibration offers great potential for resolving spatial heterogeneity in landscape properties. However, the extent to which that numerically resolved information corresponds to actual conditions on the ground has yet to be determined. That correspondence should depend on the relative accuracies of the two models involved: (1) the model behind the remote sensing product and (2) the watershed model not calibrated to the remote sensing product. We examined this by comparing the accuracy of ET from a remote sensing product, MODIS MOD16A2, to the accuracy of ET from a watershed model (SWAT) calibrated to streamflow. ET accuracies were evaluated relative to observations from three flux towers in a Mediterranean climate extending from rain-dominated Ponderosa pine at 1160-m elevation to snow-dominated Lodgepole pine at 2700-m elevation.

The accuracy of ET from the SWAT watershed model surpassed that from the MODIS model across time and space. SWAT explained 4–68% (36% overall) of the variance in monthly ET observations at the flux towers, while MODIS explained none of the observed variance as shown by negative values of Nash-Sutcliffe efficiency. Long-term ET observed across the towers decreased with elevation at a rate of -0.013 mm mo^{-1} m^{-1}. This elevational trend in long-term ET was slightly underestimated by SWAT, 7.7%, and largely underestimated by MODIS, 81%. These findings show that if the watershed model had been calibrated to remotely sensed ET rather than to stream discharge observations, the resulting accuracy of watershed model ET-predictions would have been substantially degraded.

The relatively large ET-errors from the MODIS model are interpreted to stem at least in part from an unrealistic dependence of canopy conductance on vapor pressure deficit (VPD). This interpretation is based on an erroneous reversal in slope of MODIS ET versus air temperature that approximately coincides with the transition from temperature- to VPD-controlled limitation on canopy conductance in the MODIS algorithm. This would explain the large underestimates in MODIS ET during the warmest times of the year when VPD reaches peak values. The empirical correction used in the MODIS algorithm to account for water limitation on ET may not represent the actual dynamics of water availability in the study area, which may be more loosely coupled to VPD than is assumed in the MODIS model.

Errors in monthly MODIS ET were found to be well correlated to air temperature. We showed that this could be used to "correct" ET-values from MODIS using linear regression with inputs of MODIS ET error, MODIS potential ET, and air temperature. This correction procedure removed much of the error in ET from the MODIS model, and produced ET predictions more accurate than those from the SWAT model. The regression-corrected MODIS ET may therefore serve as an improved source of "observations" for spatial calibration of a watershed model over the original MODIS ET data.

Supplementary Materials: The following are available online at https://www.mdpi.com/article/10.3390/rs13071258/s1: Section S1: Construction of base watershed model in ArcSWAT; Section S2: Watershed model initialization with LAI and biomass of mature forest; Section S3: Sensitivity analysis of influential watershed model parameters; Section S4: Watershed model calibration and validation; Table S1: Spatial datasets used in construction of watershed model in ArcSWAT 2012; Table S2: Parameter values of "base model" manually entered into tables of ArcSWAT project; Table S3: Modifications to plant database of SWAT model to more closely simulate biophysical parameters of mature Sierra Nevada forest; Table S4: SWAT model parameters varied in global sensitivity analysis using Sobol method; Table S5: Results of Sobol sensitivity analysis showing contribution of each SWAT parameter to total modeled variance in Kling-Gupta efficiency (KGE) of monthly streamflow; Table S6: Parameter ranges of calibrated SWAT model found using SWAT-CUP with the SUFI-2 (Sequential Uncertainty FItting Ver. 2) method; Figure S1: Annual precipitation versus elevation in upper Kings River watershed in 100-m elevation bins; Figure S2: Annual air temperature versus elevation in upper Kings River watershed in 100-m elevation bins; Figure S3: Range of MODIS 8-day ET-values within or touching a 500-m radius buffer around each flux tower location; Figure S4: SWAT model parameterization of snow areal depletion curve for upper Kings River watershed; Figure S5: Monthly leaf area index (LAI) and biomass from 30-year spin-up of SWAT base model to steady-state weather conditions; Figure S6: Monthly ET versus vapor pressure deficit (VPD) from different data sources at each of the three study sites; Figure S7: Weather correction to canopy conductance at the lower site based on MODIS model.

Author Contributions: Conceptualization, Methodology, Formal Analysis, Visualization, and Writing—Original Draft Preparation: S.M.J.; Investigation: S.M.J. and T.C.H.; Resources: S.M.J. and B.G.; Writing—Review and Editing: S.M.J., T.C.H. and B.G.; Supervision: T.C.H.; Project Administration: S.M.J. and T.C.H.; Funding Acquisition: T.C.H. All authors have read and agreed to the published version of the manuscript.

Funding: This work was supported by internal funding from the University of California, Merced.

Institutional Review Board Statement: Not applicable.

Informed Consent Statement: Not applicable.

Data Availability Statement: Flux tower ET observations and "Flexible Filler" processing script (Matlab) were publicly available from the Goulden Lab website, Department of Earth System Science, UC Irvine. These data could be found here: https://www.ess.uci.edu/~california/ (accessed on 24 March 2021). All other data sources were publicly available as described in the Supplementary Materials.

Acknowledgments: We thank the following individuals for their assistance with various technical matters: John T. Abatzoglou, Mike L. Goulden, Qin (Christine) Ma, Xiande Meng, Erin Mutch, and Amy Newsam. We also thank the reviewers for their helpful comments on the manuscript.

Conflicts of Interest: The authors declare no conflict of interest.

References

1. Abatzoglou, J.T.; Ficklin, D.L. Climatic and Physiographic Controls of Spatial Variability in Surface Water Balance over the Contiguous United States Using the Budyko Relationship. *Water Resour. Res.* **2017**, *53*, 7630–7643. [CrossRef]
2. Caracciolo, D.; Pumo, D.; Viola, F. Budyko's Based Method for Annual Runoff Characterization across Different Climatic Areas: An Application to United States. *Water Resour. Manag.* **2018**, *32*, 3189–3202. [CrossRef]
3. Anderegg, W.R.L.; Flint, A.; Huang, C.; Flint, L.; Berry, J.A.; Davis, F.W.; Sperry, J.S.; Field, C.B. Tree Mortality Predicted from Drought-Induced Vascular Damage. *Nat. Geosci.* **2015**, *8*, 367–371. [CrossRef]
4. Mildrexler, D.; Yang, Z.; Cohen, W.B.; Bell, D.M. A Forest Vulnerability Index Based on Drought and High Temperatures. *Remote Sens. Environ.* **2016**, *173*, 314–325. [CrossRef]
5. Young, D.J.N.; Stevens, J.T.; Earles, J.M.; Moore, J.; Ellis, A.; Jirka, A.L.; Latimer, A.M. Long-Term Climate and Competition Explain Forest Mortality Patterns under Extreme Drought. *Ecol. Lett.* **2017**, *20*, 78–86. [CrossRef]
6. Cleugh, H.A.; Leuning, R.; Mu, Q.; Running, S.W. Regional Evaporation Estimates from Flux Tower and MODIS Satellite Data. *Remote Sens. Environ.* **2007**, *106*, 285–304. [CrossRef]
7. Fisher, J.B.; Tu, K.P.; Baldocchi, D.D. Global Estimates of the Land–Atmosphere Water Flux Based on Monthly AVHRR and ISLSCP-II Data, Validated at 16 FLUXNET Sites. *Remote Sens. Environ.* **2008**, *112*, 901–919. [CrossRef]
8. Leuning, R.; Zhang, Y.Q.; Rajaud, A.; Cleugh, H.; Tu, K. A Simple Surface Conductance Model to Estimate Regional Evaporation Using MODIS Leaf Area Index and the Penman-Monteith Equation. *Water Resour. Res.* **2008**, *44*. [CrossRef]
9. Zhang, K.; Kimball, J.S.; Mu, Q.; Jones, L.A.; Goetz, S.J.; Running, S.W. Satellite Based Analysis of Northern ET Trends and Associated Changes in the Regional Water Balance from 1983 to 2005. *J. Hydrol.* **2009**, *379*, 92–110. [CrossRef]
10. Glenn, E.P.; Nagler, P.L.; Huete, A.R. Vegetation Index Methods for Estimating Evapotranspiration by Remote Sensing. *Surv. Geophys.* **2010**, *31*, 531–555. [CrossRef]
11. Mu, Q.; Heinsch, F.A.; Zhao, M.; Running, S.W. Development of a Global Evapotranspiration Algorithm Based on MODIS and Global Meteorology Data. *Remote Sens. Environ.* **2007**, *111*, 519–536. [CrossRef]
12. Mu, Q.; Zhao, M.; Running, S.W. Improvements to a MODIS Global Terrestrial Evapotranspiration Algorithm. *Remote Sens. Environ.* **2011**, *115*, 1781–1800. [CrossRef]
13. Miralles, D.G.; Holmes, T.R.H.; De Jeu, R.A.M.; Gash, J.H.; Meesters, A.G.C.A.; Dolman, A.J. Global Land-Surface Evaporation Estimated from Satellite-Based Observations. *Hydrol. Earth Syst. Sci.* **2011**, *15*, 453–469. [CrossRef]
14. Wagener, T.; Gupta, H.V. Model Identification for Hydrological Forecasting under Uncertainty. *Stoch. Environ. Res. Risk Assess.* **2005**, *19*, 378–387. [CrossRef]
15. Yilmaz, K.K.; Vrugt, J.A.; Gupta, H.V.; Sorooshian, S. Model Calibration in Watershed Hydrology. In *Advances in Data-Based Approaches for Hydrologic Modeling and Forecasting*; World Scientific: Singapore, 2010; pp. 53–105. [CrossRef]
16. Lin, P.; Pan, M.; Beck, H.E.; Yang, Y.; Yamazaki, D.; Frasson, R.; David, C.H.; Durand, M.; Pavelsky, T.M.; Allen, G.H.; et al. Global Reconstruction of Naturalized River Flows at 2.94 Million Reaches. *Water Resour. Res.* **2019**, *55*, 6499–6516. [CrossRef]
17. U.S. Geological Survey. Landsat Provisional Actual Evapotranspiration Science Product courtesy of the U.S. Geological Survey. Available online: https://www.usgs.gov/core-science-systems/nli/landsat/landsat-provisional-actual-evapotranspiration (accessed on 23 November 2020).
18. Zhang, Y.; Chiew, F.H.S.; Zhang, L.; Li, H. Use of Remotely Sensed Actual Evapotranspiration to Improve Rainfall–Runoff Modeling in Southeast Australia. *J. Hydrometeorol.* **2009**, *10*, 969–980. [CrossRef]
19. Zhang, Y.; Chiew, F.H.S.; Liu, C.; Tang, Q.; Xia, J.; Tian, J.; Kong, D.; Li, C. Can Remotely Sensed Actual Evapotranspiration Facilitate Hydrological Prediction in Ungauged Regions without Runoff Calibration? *Water Resour. Res.* **2020**, *56*. [CrossRef]
20. Zou, L.; Zhan, C.; Xia, J.; Wang, T.; Gippel, C.J. Implementation of Evapotranspiration Data Assimilation with Catchment Scale Distributed Hydrological Model via an Ensemble Kalman Filter. *J. Hydrol.* **2017**, *549*, 685–702. [CrossRef]
21. Wambura, F.J.; Dietrich, O.; Lischeid, G. Improving a Distributed Hydrological Model Using Evapotranspiration-Related Boundary Conditions as Additional Constraints in a Data-Scarce River Basin. *Hydrol. Process.* **2018**, *32*, 759–775. [CrossRef]
22. Becker, R.; Koppa, A.; Schulz, S.; Usman, M.; aus der Beek, T.; Schüth, C. Spatially Distributed Model Calibration of a Highly Managed Hydrological System Using Remote Sensing-Derived ET Data. *J. Hydrol.* **2019**, *577*, 123944. [CrossRef]
23. Gui, Z.; Liu, P.; Cheng, L.; Guo, S.; Wang, H.; Zhang, L. Improving Runoff Prediction Using Remotely Sensed Actual Evapotranspiration during Rainless Periods. *J. Hydrol. Eng.* **2019**, *24*, 04019050. [CrossRef]
24. Herman, M.R.; Hernandez-Suarez, J.S.; Nejadhashemi, A.P.; Kropp, I.; Sadeghi, A.M. Evaluation of Multi- and Many-Objective Optimization Techniques to Improve the Performance of a Hydrologic Model Using Evapotranspiration Remote-Sensing Data. *J. Hydrol. Eng.* **2020**, *25*, 04020006. [CrossRef]
25. Jiang, L.; Wu, H.; Tao, J.; Kimball, J.S.; Alfieri, L.; Chen, X. Satellite-Based Evapotranspiration in Hydrological Model Calibration. *Remote Sens.* **2020**, *12*, 428. [CrossRef]

26. Jin, X.; Jin, Y. Calibration of a Distributed Hydrological Model in a Data-Scarce Basin Based on GLEAM Datasets. *Water* **2020**, *12*, 897. [CrossRef]

27. Nesru, M.; Shetty, A.; Nagaraj, M.K. Multi-Variable Calibration of Hydrological Model in the Upper Omo-Gibe Basin, Ethiopia. *Acta Geophys.* **2020**, *68*, 537–551. [CrossRef]

28. Goulden, M.L.; Anderson, R.G.; Bales, R.C.; Kelly, A.E.; Meadows, M.; Winston, G.C. Evapotranspiration along an Elevation Gradient in California's Sierra Nevada. *J. Geophys. Res. Biogeosci.* **2012**, *117*, G03028. [CrossRef]

29. Mu, Q.; Jones, L.A.; Kimball, J.S.; McDonald, K.C.; Running, S.W. Satellite Assessment of Land Surface Evapotranspiration for the Pan-Arctic Domain. *Water Resour. Res.* **2009**, *45*. [CrossRef]

30. Martinec, J.; Rango, A. Merits of Statistical Criteria for the Performance of Hydrological Models. *J. Am. Water Resour. Assoc.* **1989**, *25*, 421–432. [CrossRef]

31. McFarland, J.R.; Tufenkjian, C.L. *The Kings River Handbook*, 5th ed.; Kings River Conservation District and Kings River Water Association: Fresno, CA, USA, 2009; Available online: http://www.krcd.org/_pdf/Kings_River_Handbook_2009.pdf (accessed on 8 August 2018).

32. U.S. Geological Survey. National Elevation Dataset (NED) 1 arc-second 2013 1 x 1 degree ArcGrid. Reston, VA. Available online: https://nationalmap.gov/ (accessed on 12 November 2014).

33. U.S. Geological Survey. National Hydrography Dataset (NHD) Medium Resolution for California 20140718 State or Territory Shapefile. Reston, VA. Available online: https://nationalmap.gov/ (accessed on 19 February 2016).

34. Daly, C.; Halbleib, M.; Smith, J.I.; Gibson, W.P.; Doggett, M.K.; Taylor, G.H.; Curtis, J.; Pasteris, P.P. Physiographically Sensitive Mapping of Climatological Temperature and Precipitation across the Conterminous United States. *Int. J. Climatol.* **2008**, *28*, 2031–2064. [CrossRef]

35. PRISM Climate Data, Northwest Alliance for Computational Science and Engineering, Oregon State University, Corvallis. Available online: http://prism.oregonstate.edu (accessed on 9 October 2020).

36. California Data Exchange Center (CDEC) ; California Department of Water Resources. Monthly Full Natural Streamflow of Kings River at Pine Flat Dam, Station ID KGF. Available online: http://cdec.water.ca.gov/dynamicapp/wsSensorData (accessed on 15 October 2020).

37. Hunsaker, C.T.; Whitaker, T.W.; Bales, R.C. Snowmelt Runoff and Water Yield along Elevation and Temperature Gradients in California's Southern Sierra Nevada. *J. Am. Water Resour. Assoc.* **2012**, *48*, 667–678. [CrossRef]

38. Bales, R.; Stacy, E.; Safeeq, M.; Meng, X.; Meadows, M.; Oroza, C.; Conklin, M.; Glaser, S.; Wagenbrenner, J. Spatially Distributed Water-Balance and Meteorological Data from the Rain–Snow Transition, Southern Sierra Nevada, California. *Earth Syst. Sci. Data* **2018**, *10*, 1795–1805. [CrossRef]

39. Natural Resources Conservation Service; U.S. Department of Agriculture. State Soil Geographic (STATSGO) Data Base: Data Use Information; Miscellaneous Publication Number 1492; Fort Worth, TX, USA. 1994. Available online: http://www.fsl.orst.edu/pnwerc/wrb/metadata/soils/statsgo.pdf (accessed on 11 March 2021).

40. Web Site for Official Soil Series Descriptions and Series Classification, Natural Resources Conservation Service, U.S. Department of Agriculture. Available online: https://soilseries.sc.egov.usda.gov (accessed on 8 October 2020).

41. Klos, P.Z.; Goulden, M.L.; Riebe, C.S.; Tague, C.L.; O'Geen, A.T.; Flinchum, B.A.; Safeeq, M.; Conklin, M.H.; Hart, S.C.; Berhe, A.A.; et al. Subsurface Plant-Accessible Water in Mountain Ecosystems with a Mediterranean Climate. *WIREs Water* **2018**, *5*, e1277. [CrossRef]

42. O'Geen, A.; Safeeq, M.; Wagenbrenner, J.; Stacy, E.; Hartsough, P.; Devine, S.; Tian, Z.; Ferrell, R.; Goulden, M.; Hopmans, J.W.; et al. Southern Sierra Critical Zone Observatory and Kings River Experimental Watersheds: A Synthesis of Measurements, New Insights, and Future Directions. *Vadose Zone J.* **2018**, *17*, 180081. [CrossRef]

43. U.S. Geological Survey. NLCD 2011 Land Cover Conterminous United States. Sioux Falls, SD. Available online: https://www.mrlc.gov/ (accessed on 7 April 2020).

44. Bales, R.C.; Hopmans, J.W.; O'Geen, A.T.; Meadows, M.; Hartsough, P.C.; Kirchner, P.; Hunsaker, C.T.; Beaudette, D. Soil Moisture Response to Snowmelt and Rainfall in a Sierra Nevada Mixed-Conifer Forest. *Vadose Zone J.* **2011**, *10*, 786–799. [CrossRef]

45. Data Access Page: Measurement of Energy, Carbon and Water Exchange Along California Climate Gradients, Goulden Lab, Department of Earth System Science, University of California, Irvine. Available online: https://www.ess.uci.edu/~{}california/ (accessed on 12 December 2019).

46. Safeeq, M.; Hunsaker, C.T. Characterizing Runoff and Water Yield for Headwater Catchments in the Southern Sierra Nevada. *J. Am. Water Resour. Assoc.* **2016**, *52*, 1327–1346. [CrossRef]

47. Nash, J.E.; Sutcliffe, J.V. River flow forecasting through conceptual models: Part 1. A discussion of principles. *J. Hydrol.* **1970**, *10*, 282–290. [CrossRef]

48. Moriasi, D.N.; Arnold, J.G.; Van Liew, M.W.; Bingner, R.L.; Harmel, R.D.; Veith, T.L. Model Evaluation Guidelines for Systematic Quantification of Accuracy in Watershed Simulations. *Trans. ASABE* **2007**, *50*, 885–900. [CrossRef]

49. Daly, C.; Smith, J.I.; Olson, K.V. Mapping Atmospheric Moisture Climatologies across the Conterminous United States. *PLoS ONE* **2015**, *10*, e0141140. [CrossRef]

50. Abatzoglou, J.T. Development of Gridded Surface Meteorological Data for Ecological Applications and Modelling. *Int. J. Climatol.* **2013**, *33*, 121–131. [CrossRef]

51. Murray, F.W. On the Computation of Saturation Vapor Pressure. *J. Appl. Meteorol.* **1967**, *6*, 203–204. [CrossRef]

52. Goulden, M.L.; Munger, J.W.; Fan, S.-M.; Daube, B.C.; Wofsy, S.C. Measurements of Carbon Sequestration by Long-Term Eddy Covariance: Methods and a Critical Evaluation of Accuracy. *Glob. Change Biol.* **1996**, *2*, 169–182. [CrossRef]

53. Running, S.; Mu, Q.; Zhao, M. MOD16A2 MODIS/Terra Net Evapotranspiration 8-Day L4 Global 500m SIN Grid V006. Distributed by NASA EOSDIS Land Processes DAAC. 2017. Available online: https://doi.org/10.5067/MODIS/MOD16A2.006 (accessed on 26 August 2019).

54. Running, S.W.; Mu, Q.; Zhao, M.; Moreno, A. *User's Guide: MODIS Global Terrestrial Evapotranspiration (ET) Product (MOD16A2/A3 and Year-End Gap-Filled MOD16A2GF/A3GF) NASA Earth Observing System MODIS Land Algorithm (For Collection 6)*; User's Guide Version 2.1; Land Processes Distributed Active Archive Center (LP DAAC), 2019. Available online: https://lpdaac.usgs.gov/documents/378/MOD16_User_Guide_V6.pdf (accessed on 15 August 2019).

55. Jarvis, J.G. The Interpretation of the Variations in Leaf Water Potential and Stomatal Conductance Found in Canopies in the Field. *Philos. Trans. R. Soc. Lond. B Biol. Sci.* **1976**, *273*, 593–610. [CrossRef]

56. Baldocchi, D.; Falge, E.; Gu, L.; Olson, R.; Hollinger, D.; Running, S.; Anthoni, P.; Bernhofer, C.; Davis, K.; Evans, R.; et al. FLUXNET: A New Tool to Study the Temporal and Spatial Variability of Ecosystem–Scale Carbon Dioxide, Water Vapor, and Energy Flux Densities. *Bull. Am. Meteorol. Soc.* **2001**, *82*, 2415–2434. [CrossRef]

57. Srinivasan, R.; Arnold, J.G. Integration of a Basin-Scale Water Quality Model with GIS. *J. Am. Water Resour. Assoc.* **1994**, *30*, 453–462. [CrossRef]

58. Arnold, J.G.; Srinivasan, R.; Muttiah, R.S.; Williams, J.R. Large Area Hydrologic Modeling and Assessment Part I: Model Development. *J. Am. Water Resour. Assoc.* **1998**, *34*, 73–89. [CrossRef]

59. Gassman, P.W.; Reyes, M.R.; Green, C.H.; Arnold, J.G. The Soil and Water Assessment Tool: Historical Development, Applications, and Future Research Directions. *Trans. ASABE* **2007**, *50*, 1211–1250. [CrossRef]

60. Fontaine, T.A.; Cruickshank, T.S.; Arnold, J.G.; Hotchkiss, R.H. Development of a Snowfall–Snowmelt Routine for Mountainous Terrain for the Soil Water Assessment Tool (SWAT). *J. Hydrol.* **2002**, *262*, 209–223. [CrossRef]

61. Ahl, R.S.; Woods, S.W.; Zuuring, H.R. Hydrologic Calibration and Validation of SWAT in a Snow-Dominated Rocky Mountain Watershed, Montana, U.S.A. *J. Am. Water Resour. Assoc.* **2008**, *44*, 1411–1430. [CrossRef]

62. Zhang, X.; Srinivasan, R.; Debele, B.; Hao, F. Runoff Simulation of the Headwaters of the Yellow River Using the SWAT Model with Three Snowmelt Algorithms. *J. Am. Water Resour. Assoc.* **2008**, *44*, 48–61. [CrossRef]

63. Ficklin, D.L.; Stewart, I.T.; Maurer, E.P. Climate Change Impacts on Streamflow and Subbasin-Scale Hydrology in the Upper Colorado River Basin. *PLoS ONE* **2013**, *8*, e71297. [CrossRef]

64. Watson, B.M.; Putz, G. Comparison of Temperature-Index Snowmelt Models for Use within an Operational Water Quality Model. *J. Environ. Qual.* **2014**, *43*, 199–207. [CrossRef] [PubMed]

65. Grusson, Y.; Sun, X.; Gascoin, S.; Sauvage, S.; Raghavan, S.; Anctil, F.; Sáchez-Pérez, J.-M. Assessing the Capability of the SWAT Model to Simulate Snow, Snow Melt and Streamflow Dynamics over an Alpine Watershed. *J. Hydrol.* **2015**, *531*, 574–588. [CrossRef]

66. Neitsch, S.L.; Arnold, J.G.; Kiniry, J.R.; Williams, J.R. *Soil and Water Assessment Tool Theoretical Documentation Version 2009*; Technical Report 406; Texas Water Resources Institute: College Station, TX, USA, 2011; p. 618. Available online: https://swat.tamu.edu/media/99192/swat2009-theory.pdf (accessed on 11 April 2019).

67. Chan, K.; Tarantola, S.; Saltelli, A.; Sobol', I.M. Variance-Based Methods. In *Sensitivity Analysis*; Wiley: Chichester, NY, USA, 2000; pp. 167–197.

68. Abbaspour, K.C.; Yang, J.; Maximov, I.; Siber, R.; Bogner, K.; Mieleitner, J.; Zobrist, J.; Srinivasan, R. Modelling Hydrology and Water Quality in the Pre-Alpine/Alpine Thur Watershed Using SWAT. *J. Hydrol.* **2007**, *333*, 413–430. [CrossRef]

69. Yang, J.; Reichert, P.; Abbaspour, K.C.; Xia, J.; Yang, H. Comparing Uncertainty Analysis Techniques for a SWAT Application to the Chaohe Basin in China. *J. Hydrol.* **2008**, *358*, 1–23. [CrossRef]

70. Arnold, J.G.; Moriasi, D.N.; Gassman, P.W.; Abbaspour, K.C.; White, M.J.; Srinivasan, R.; Santhi, C.; Harmel, R.D.; van Griensven, A.; Van Liew, M.W.; et al. SWAT: Model Use, Calibration, and Validation. *Trans. ASABE* **2012**, *55*, 1491–1508. [CrossRef]

71. Gupta, H.V.; Kling, H.; Yilmaz, K.K.; Martinez, G.F. Decomposition of the Mean Squared Error and NSE Performance Criteria: Implications for Improving Hydrological Modelling. *J. Hydrol.* **2009**, *377*, 80–91. [CrossRef]

72. Porporato, A.; Laio, F.; Ridolfi, L.; Rodriguez-Iturbe, I. Plants in Water-Controlled Ecosystems: Active Role in Hydrologic Processes and Response to Water Stress: III. Vegetation Water Stress. *Adv. Water Resour.* **2001**, *24*, 725–744. [CrossRef]

73. Novick, K.A.; Ficklin, D.L.; Stoy, P.C.; Williams, C.A.; Bohrer, G.; Oishi, A.C.; Papuga, S.A.; Blanken, P.D.; Noormets, A.; Sulman, B.N.; et al. The Increasing Importance of Atmospheric Demand for Ecosystem Water and Carbon Fluxes. *Nat. Clim. Change* **2016**, *6*, 1023–1027. [CrossRef]

74. Massmann, A.; Gentine, P.; Lin, C.J. When Does Vapor Pressure Deficit Drive or Reduce Evapotranspiration? *J. Adv. Model. Earth Syst.* **2019**, *11*, 3305–3320. [CrossRef] [PubMed]

75. He, M.; Kimball, J.S.; Yi, Y.; Running, S.W.; Guan, K.; Moreno, A.; Wu, X.; Maneta, M. Satellite Data-Driven Modeling of Field Scale Evapotranspiration in Croplands Using the MOD16 Algorithm Framework. *Remote Sens. Environ.* **2019**, *230*, 111201. [CrossRef]

remote sensing

MDPI

Article

Data Assimilation for Rainfall-Runoff Prediction Based on Coupled Atmospheric-Hydrologic Systems with Variable Complexity

Wei Wang [1,2], Jia Liu [1,*], Chuanzhe Li [1], Yuchen Liu [1] and Fuliang Yu [1]

1   State Key Laboratory of Simulation and Regulation of Water Cycle in River Basin, China Institute of Water Resources and Hydropower Research, Beijing 100038, China; wangwei_hydro@163.com (W.W.); lichuanzhe@iwhr.com (C.L.); melodieyu@163.com (Y.L.); yufl@iwhr.com (F.Y.)
2   College of Hydrology and Water Resources, Hohai University, Nanjing 210098, China
*   Correspondence: jia.liu@iwhr.com; Tel.: +86-150-1044-3860

Abstract: The data assimilation technique is an effective method for reducing initial condition errors in numerical weather prediction (NWP) models. This paper evaluated the potential of the weather research and forecasting (WRF) model and its three-dimensional data assimilation (3DVar) module in improving the accuracy of rainfall-runoff prediction through coupled atmospheric-hydrologic systems. The WRF model with the assimilation of radar reflectivity and conventional surface and upper-air observations provided the improved initial and boundary conditions for the hydrological process; subsequently, three atmospheric-hydrological systems with variable complexity were established by coupling WRF with a lumped, a grid-based Hebei model, and the WRF-Hydro modeling system. Four storm events with different spatial and temporal rainfall distribution from mountainous catchments of northern China were chosen as the study objects. The assimilation results showed a general improvement in the accuracy of rainfall accumulation, with low root mean square error and high correlation coefficients compared to the results without assimilation. The coupled atmospheric-hydrologic systems also provide more accurate flood forecasts, which depend upon the complexity of the coupled hydrological models. The grid-based Hebei system provided the most stable forecasts regardless of whether homogeneous or inhomogeneous rainfall was considered. Flood peaks before assimilation were underestimated more in the lumped Hebei model relative to the other coupling systems considered, and the model seems more applicable for homogeneous temporal and spatial events. WRF-Hydro did not exhibit desirable predictions of rapid flood process recession. This may reflect increasing infiltration due to the interaction of atmospheric and land surface hydrology at each integration, resulting in mismatched solutions for local runoff generation and confluence.

Keywords: WRF-3DVar data assimilation; coupled atmospheric-hydrologic system; rainfall-runoff prediction; lumped Hebei model; grid-based Hebei model; WRF-Hydro modeling system

Citation: Wang, W.; Liu, J.; Li, C.; Liu, Y.; Yu, F. Data Assimilation for Rainfall-Runoff Prediction Based on Coupled Atmospheric-Hydrologic Systems with Variable Complexity. *Remote Sens.* **2021**, *13*, 595. https://doi.org/10.3390/rs13040595

Academic Editors: Yongqiang Zhang, Dongryeol Ryu and Donghai Zheng
Received: 31 December 2020
Accepted: 3 February 2021
Published: 7 February 2021

1. Introduction

The last decades have witnessed significant changes in climate and hydrological conditions. The increased frequency of extreme storm floods has led to major risks of damage due to weather-related hazards. Forecasting of such high-intensity floods on a shorter time scale has immense benefits such as saving lives, protecting economic assets, and improving quality of life [1–3]. For mesoscale mountain areas along the Daqing River of northern China, steep slopes, combined with high intensity and short duration convective rainfall, substantially shorten hydrological lead times. In addition, due to the lack of high-resolution and dense observations, the "throughfall" observed by rain gauges cannot reflect the realistic rainfall distribution in space and time, thus the accuracy of forecasting is limited by the layout of the rain gauge network. For processes of runoff and routing, different dependent processes are added and derived within models including soil infiltration, overland transport, and channel routing, which result in the complexities and uncertainties in deducing the generation mechanisms of flash floods. In this study, we selected two upstream mountainous catchments along the Daqing River in which there is an urgent need for accurate flood prediction to prevent and reduce risks facing the construction of the downstream including Xiong'an New Area.

Although recent advances have improved rainfall forecasting [4], several challenges remain. One such challenge is the reproduction of the magnitude and the disturbance patterns of rainfall that can assimilate suitable observations into numerical weather prediction (NWP) models [3]. Rainfall is among the variables generated with the greater errors in NWP models, while it plays an important role in forecasting the atmospheric-hydrological processes for its influence on the time and scale of floods [5,6]. There are three main sources of error in rainfall prediction: the initial conditions, the lateral boundary conditions, and the physical approximations in the model equations. Data assimilation allows atmospheric information to be extracted from multiple data sources, thereby improving the reliability of coarse resolution data and the complexity of atmospheric motion, reducing the initial and lateral boundary errors [7,8]. Routray et al. [9] found that weather research forecasting (WRF) can be used to assimilate observations from different sources and contribute to a better understanding of mesoscale rainfall convective activity within the Indian monsoon region. Kumar and Varma [10] further explored a short duration intense rainfall event in India, demonstrating the potential of WRF to adapt to rainfall forecast accuracy. Fierro et al. [11] conducted a data assimilation study in the eastern part of the USA that showed that WRF, in conjunction with data assimilation, could significantly improve models of local short-term rainfall processes. Although data assimilation can help NWP models to more accurately capture rainfall and enable rainfall-runoff conversion by constructing an atmospheric-hydrological model system, its potential to further improve flood forecasting has not been fully investigated.

A reliable atmospheric-hydrological model system is required to improve rainfall predictions and hydrological forecasts for early flood hazard mitigation [12]. A promising method is the coupling of hydrological models to a regional model such as NWP, in order to rapidly obtain high-resolution rainfall and flood forecasting data. In [13–15], Lin et al. and Lu et al. discussed the implementation and improvement of the Canadian regional mesoscale compressible community mode (MC2) rainfall forecasting in the Huaihe Basin of southern China, concentrating on the Huaihe sub-basin coupled to the Xinanjiang hydrological model. Wu et al. [16] further explored MC2 and the multiple linear regression integrated forecast and found that high-resolution rainfall distributions were problematic at finer temporal and spatial scales, requiring data assimilation or sub-grid-scale parameterization. Yucel et al. [4] and Moser et al. [17] tested WRF data assimilation as the input for flood forecasting in the Black Sea and Iowa, respectively, both finding an enhancement in the accuracy of flood warnings.

With regard to the selection of coupled models, it is subject to a diversity of laws and non-universality, which makes it difficult to accurately express physical movement processes. The hydrological model has a more comprehensive physical foundation including lumped, grid-based, and fully distributed setups [18,19]. Not only the above physically-based models are used, but machine learning models are also widely applied in rainfall-runoff forecasting (i.e., artificial neural networks (ANNs) [20], support vector machines (SVM) [21], and the recent emergence of theory-guided data science (TGDS) [22,23]. For flood forecasting, which is affected by the discretization construction method, different construction expressions determine variations between heterogeneity analysis and model calculation structure, and further influence the accuracy of physical expressions in the prediction processes of the hydrological model [24–26]. To analyze the scale of hydrological processes, large-scale studies are still the mainstay. During the last ten years, a focus has been made on downscaling and modeling through appropriate discrete methods [27]. The development of models with high prediction accuracy and computational efficiency is a key issue for basin-scale flood forecasting. Liu et al. [28] conducted coupled lumped hydrological modeling and WRF flood forecasting on a 135.2 km^2 catchment with a 10 km resolution; Li et al. [29] used the rainfall of a 20 km WRF output to drive the distributed Luxihe model, extending the forecast period of flood forecast in the Liujiang catchment (58,270 km^2). Rogelis [30] compared the flow results of different resolution data (minimum resolution 1.67 km) driven by WRF on a 380 km^2 catchment driving different lumped hydrological models. Previous studies have mostly focused on humid regions; consequently, runoff methods are mostly based on saturation excess, and limited discussion of the appropriate construction of atmospheric-hydrological model systems have been conducted for semi-humid and semi-arid areas.

The coupling system can also integrate land surface models (LSM) with hydrological models. Most LSM and hydrological models incorporate the same descriptions of water balance, albeit with different aims [31,32]. LSM evolves from land-atmosphere coupling models with the purpose of solving the surface energy balance equation and providing the necessary lower boundary conditions for the atmosphere [31,33]. Inversely, hydrological models focus less on radiation and more on hydrological changes (i.e., the lateral route of water along land surfaces). Such models are the most complicated among the current coupling systems due to their complex structure and the sensitive parameters to be determined in the relevant physical processes as well as hard parameters (fixed parameters written directly into the source code during the compilation of the model).

Limited research has been undertaken to model atmospheric-hydrological processes in semi-humid and semi-arid regions of northern China. Consequently, there is a lack of effective atmospheric-hydrological coupling forecasting systems for this region. Herein, we used WRF models, three-dimensional variational (3DVar) data assimilation modules coupled to three model sets of varying complexity to construct the required model systems. To test the influence of various levels of complexity, three types models were selected, namely the lumped Hebei model, the grid-based Hebei model, and the WRF-Hydro model. These models were both standalone and coupled with the WRF model and three-dimensional variational (3DVar) data assimilation module. Four typical storm flood events with different spatial and temporal rainfall distributions, all of which occur in the upper catchment of the Daqing catchment, were explored before and after data assimilation. The purpose of this study was to investigate the impact of data assimilation on forecasting different types of rainfall-runoff events after coupling with variable hydrological structures. It should be noted that the atmospheric-hydrologic coupling in this study refers to "one-way" coupling of the three standalone hydrological model structures with the WRF and 3DVar data assimilation module, which means that the hydrological models are driven by the WRF and 3DVar outputs without feedback to the atmospheric modeling processes. The results obtained in this way can simply reflect the direct effects of data assimilation on rainfall as well as runoff forecasts.

There were four basic questions we aimed to explore:

- What are the differences in the improvement of rainfall before and after data assimilation for the storm events with different spatial and temporal distributions in semi-humid areas of northern China?
- What are the corresponding runoff effects of different coupling systems on the improved rainfall from WRF and its assimilation mode?
- What differences exist between the runoff processes modeled by the coupled systems of different complexity before and after assimilation?
- How does the complexity of the hydrological structure affect the transmission of rainfall improvement from data assimilation to runoff?

2. Study Area and Events

2.1. Study Area

We studied two mountainous catchments of the Daqing River (i.e., Fuping (2210 km^2) and Zijingguan (1760 km^2)). Two catchments lie along the upper reach of the Shazhi River on the south branch and the upper reach of the Juma River on the north branch, respectively (Figure 1). Low vegetation coverage, bare hills, mountains, and thin soil layers cause uneven infiltration capacity distribution across the study areas, and the surface often features extra-infiltration flow, which means surface flow occurs when the intensity of infiltration exceeds the intensity of precipitation. Rainfall occurs primarily during the flood season from late May to early September. The combined effects of the western Pacific subtropical high, the westerly cold vortex, and the western Taihang Mountain uplift influence the heavy convective rainfall that prevails in the region. Additionally, due to steep terrain, it is easy to form high intensity, short duration floods with a large peak flow. These floods cause damage in the region. According to statistics of torrential rain and flood data from 1958 to 2015 in both catchments, floods that occurred more than once a decade occurred five times in Fuping and six times in Zijingguan, with increasing frequency in extreme events in recent years.

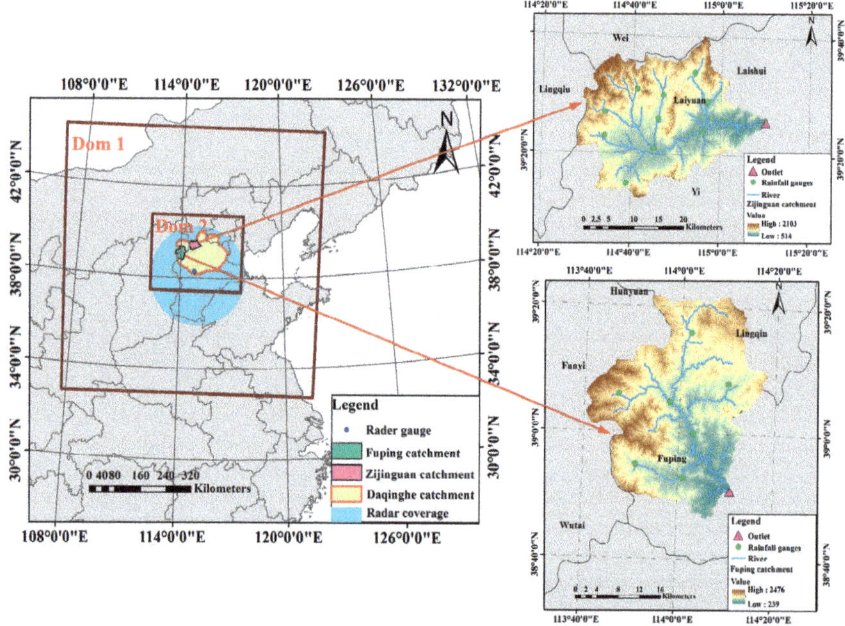

Figure 1. Locations of the Fuping and Zijingguan catchments covered by the weather research forecasting (WRF) nested domains and the weather radar.

2.2. Storm Events

Four storm events in the Fuping and Zijingguan catchments were screened. The events were divided according to their spatial and temporal distribution. In comparison to the southern part of China, it is difficult to find absolute homogeneous rainfall events in northern China, either in spatial or temporal dimensions. Herein, we calculated the storm events' coefficient of variance (*Cv*) values to indicate the relative response to the spatio-temporal homogeneousness of the flood events [34]. *Cv* values were calculated as follows:

$$Cv = \sqrt{\frac{\sum_{i=1}^{N}\left(\frac{x_i}{\bar{x}} - 1\right)^2}{N}} \tag{1}$$

where $\bar{x}$ is the average of x_i. When calculating *Cv* in the temporal dimension, x_i is the catchment areal rainfall at the *i*th hour and *N* is the total number of hours of the storm event. In this case, time *Cv* represents the temporal deviation of the catchment areal rainfall at each time step. When calculating Cv in the spatial dimension, x_i is the 24 h rainfall accumulation at the ith gauge and N is the number of rain gauges. In this case, the Cv value reflects the spatial deviation of the rainfall accumulation at each rain gauge.

The results are shown in Table 1, where larger Cv values represent a more uneven spatio-temporal distribution. Based on this criterion, the spatio-temporal consistency of the rainfall-runoff events listed in Table 1 were classified. Events 1 and 2 had relatively uniform spatial and temporal distributions, whereas Events 3 and 4 featured rainfall with uneven spatial and temporal distributions. As shown in Figure 2, the rainfall-runoff events had a flood recession time with different lengths, so we used a 72-h time window for Event 1 and Event 2, a 60-h time window for Event 3, while for Event 4, a 36-h time window was adopted. Different time windows were used to calculate the statistics when evaluating the forecasting results. Event 4 occurred on 21 July 2012, bringing the largest flood disaster in 60 years to Beijing and the Daqing River. The rainfall at the largest monitoring point corresponded to an event that occurred once in more than 500 years. In the Zijingguan catchment, its 24 h maximum single station cumulative rainfall was 355 mm, and the maximum flow at the outlet reached 2580.0 m^3/s. Figure 2 and Table 1 show other rainfall and flood characteristics.

Table 1. Storm events and spatio-temporal rainfall evenness characterized by Cv.

Storm Event	Catchment	Storm Duration	24 h Rainfall Accumulation (mm)	Peak Flow (m^3/s)	Temporal Cv	Spatial Cv
Even 1	Fuping	29 July 2007 20:00 to 30 July 2007 20:00	63.4	29.7	0.6011	0.3975
Event 2	Fuping	30 July 2012 10:00 to 31 July 2012 10:00	50.5	70.7	1.0823	0.1927
Event 3	Fuping	11 August 2013 07:00 to 12 August 2013 07:00	30.9	46.6	2.3925	0.7400
Event 4	Zijingguan	21 July 2012 04:00 to 22 July 2012 04:00	172.2	2580.0	1.8865	0.6098

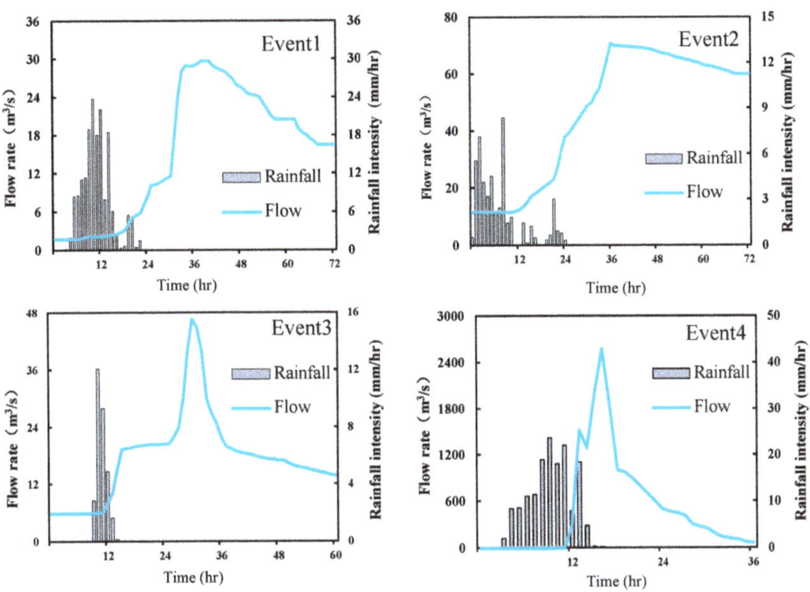

Figure 2. Observed rainfall-runoff processes for the four selected storm events.

3. Three Atmospheric–Hydrologic Coupled Systems

3.1. The Numerical Weather Prediction (NWP) Model

The mesoscale NWP model has limited skill in convective precipitation forecast, even in the WRF model. One of the considerable reasons for the poor performance is due to its nonlinear and chaotic nature, since solving quasistationary meso-β and meso-γ processes is complicated [35,36], and the power spectrum of the turbulence in convective motion has a resolution of a few kilometers [37,38]. 3DVar data assimilation could merge the fine rainfall information into the modeling system to obtain accurate high-precision rainfall data. The following are detailed descriptions of the WRF model configurations and the 3DVar data assimilation techniques used in this study.

3.1.1. Weather Research Forecasting (WRF) Model Configurations

As a next-generation mesoscale forecasting model and data assimilation system, the simulated scale used in the WRF spans tens of meters to thousands of kilometers, and is mainly used to enhance understanding and forecasting of mesoscale weather, assist operational forecasting, and improve atmospheric research. Detailed descriptions of specific WRF model settings are shown in Table 2 and Figure 1. The inner research area includes the Taihang Mountains, Bohai Sea, and major cities in Beijing, Tianjin, and Hebei. A warm-up time period is necessary for both WRF and WRF-Hydro. A seven-day length warm-up time period was set to generate the restart file, which serves mainly as the starting condition for WRF, and was also used to provide the initial soil moisture condition for WRF-Hydro to generate more realistic runoff. The WRF has an output time step of 1 h and an integration step of 6 s. Global Forecast System (GFS) data in each 6 h window was to provide the initial side boundary and atmospheric background conditions.

Table 2. Detailed configurations of the weather research forecasting (WRF).

Subject	Chosen Option	Subject	Chosen Option
Driving data	GFS each 6 h	WRF output interval	1 h
Integration time step	Dom1: 18 s Dom2: 6 s	domain center Vertical discretization	39°04′15″ N, 113°59′26″ E 40 layers
Horizontal resolution	Dom1: 9 km Dom2: 3 km	Pressure Projection resolution	50 hPa Lambert
Horizontal grid number	Dom1: 140 × 140 Dom2: 150 × 120	Longwave radiation Shortwave radiation	RRTM Dudhia

The parameterization scheme chosen can affect the mode of operation of the WRF; different parameterization schemes have different emphases on physical processes. The diversity of parameterization schemes and the corresponding differences in rainfall events in specific regions result in difficulties inherent to the accurate simulation of spatial and temporal rainfall distributions and cumulative rainfall. For the four rainfall processes in this study, we chose the optimal physical parameterization based on the relevant experimental research on the selection of sensitive parameterization schemes for ensemble rainfall forecasting (more details can be found in Tian et al. [39,40]). The microphysical parameterization schemes chosen for our forecasts included Purdue-Lin (Lin) [41] and WRF Single-Moment 6 (WSM6) [42]; the cumulus convection schemes included the Kain-Fritsch (KF) [43] and Grell-Devenyi (GD) [44] ensembles; and the planetary boundary layer schemes included the Yonsei University (YSU) [42] and Mellor-Yamada-Janjic (MYJ) [45] schemes. The specific parameterization details of the four studied storms are shown in Table 3. It should be noted that the cumulus parameterization was not used in the innermost domain, where the convective rainfall generation was assumed to be explicitly resolved.

Table 3. Optimal physical parameterization for the studied storm events.

Ensemble Scenarios		Event 1	Event 2	Event 3	Event 4
Microphysics scheme	Lin	✔			
	WSM6		✔	✔	✔
Cumulus convection	KF	✔	✔		
	GD			✔	✔
Planetary boundary layer	YSU	✔	✔	✔	
	MYJ				✔

3.1.2. Data Assimilation with WRF-3DVar

3D variational assimilation has numerous advantages, for example, that the objective function contains physical processes and utilizes the model itself as a dynamic constraint (i.e., it can efficiently represent complex nonlinear constraint relationships). The objective function of WRF-3DVar assimilation can be expressed as follows:

$$J(X) = \frac{1}{2}(X - X_b)^T B^{-1}(X - X_b) + \frac{1}{2}(H(X) - Y_0)^T R^{-1}(H(X) - Y_0) \tag{2}$$

where X is a parameter reflecting the atmosphere and surface conditions; X_b is the background field at the time of change; B is the corresponding background field error covariance matrix; and H is the observation function containing X variables. H can change the variables in the atmospheric model from model space projected into the observation space. Y_0 represents the assimilated observation vector and R is the error covariance matrix from observation.

Global Telecommunications System (GTS) data and Doppler radar data were used as the assimilation data in this study. In previous studies, we have demonstrated that the assimilation of radar reflectivity can improve local rainfall processes, especially for data <500 m, and can thus improve the reliability of assimilation information for forecasting systems [46,47]. Taking into account the spatial resolution of data from different sources and maintaining the stability of the atmospheric motion, we chose to assimilate <500 m radar reflectivity data in the inner domain (Dom2) and GTS data in the outer domain (Dom1). This assimilation method was supported by multi-source data assimilation experiments carried out in the same study area [47].

GTS data are a collection of traditional ground and upper air meteorological data including barometric pressure, wind direction and velocity, temperature, and humidity. GTS data have wide coverage in both the vertical and horizontal directions. These data are updated at 6 h intervals and are therefore widely used in large-scale atmospheric studies. GTS data provided by the National Center for Atmospheric Research (NCAR) were assimilated for the outer domain, sourced from sounders, ground-based weather stations, pilot balloons, and aircraft observations. An observation preprocessor with defined observation error covariance was employed to ensure quality control of the GTS data prior to assimilation, and a default U.S. Air Force (AFWA) OBS error file was also used for processing the GTS data and identifying instrumental and sensor errors.

The forecasting radar site of the Shijiazhuang s-band Doppler weather radar and specific coverage is shown in Figure 1. Its reflectance spatial resolution was 1 km × 1° with a scan radius of 250 km (i.e., it covered both the Fuping and Zijingguan catchments). The radar completed a cycle every 6 min at nine different scan elevation angles (0.5°, 1.5°, 2.4°, 3.4°, 4.3°, 6.0°, 9.9°, 14.6°, and 19.5°). Raw data for the radar were obtained from the National Meteorological Administration of China and, following quality control, the radar reflectivity was programmed to conform to the WRF-3DVar data format.

The absorption of radar reflectivity data by the WRF-3DVar module presupposes that the total water mixing ratio is used as a control variable and that a warm rainfall process is simultaneously introduced into the assimilation module. Assuming a Marshall-Palmer raindrop size distribution and no ice relative reflectivity effect, the following equation can be derived for radar reflectivity Z by introducing a rainwater mixing ratio q_r, in which ρ is the density of air in kg m^{-3}:

$$Z = 43.1 + 17.5\log(\rho q_r) \tag{3}$$

For the four storm events, we selected a uniform 24 h time window that completely described the entire process of rainfall. The duration of the WRF is required to cover the start and end of the rainfall time window; therefore, the assimilation data started to be merged 6 h before the storm events, then assimilated every 6 h until five assimilation cycles were completed, for a total running duration of 36 h (6 × 6 h). Figure 3 presents a schematic diagram of the assimilation using Event 4 for further explanation. As illustrated in Figure 3, there was a seven day warm-up time period (indicated with a dashed line) and a 6 h spin-up time period (indicated with dotted lines) for WRF, and run1 presents the initial WRF run with no data assimilation. Data assimilation started at 00:00 on 21 July 2012 and, subsequently, run2, run3, run4, run5, and run6 were executed at 6 h intervals (i.e., at 00:00, 06:00, 12:00, and 18:00 on 21 July 2012 and 00:00 on 22 July 2012, respectively). The first output file generated in the previous run was used to provide the initial and lateral boundary conditions for subsequent runs and was treated as a benchmark reflecting improvements in rainfall simulation after data assimilation.

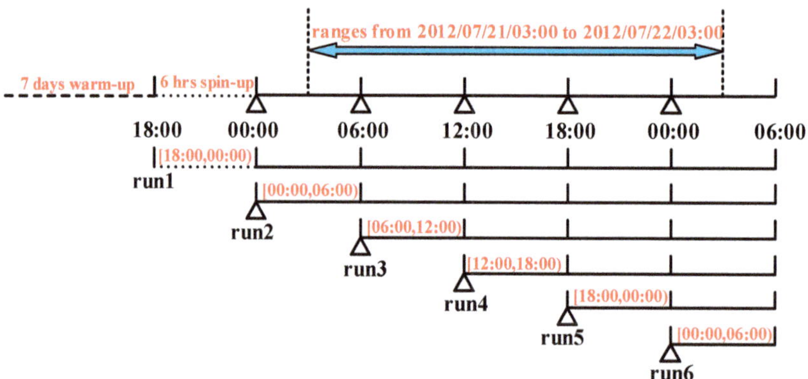

Figure 3. The cycling data assimilation process for Event 4.

3.2. Hydrological Models with Differing Complexity

3.2.1. The Lumped Hebei Model

The lumped Hebei model was developed based on rainfall-runoff mechanisms identified in semi-humid and semi-arid catchments in northern China and has been applied to real-time flood forecasting in Hebei Province. The model comprehensively considers the two inflow mechanisms of infiltration excess and saturation excess (shown in Figure 4). W' and WMM represent the point storage capacity and the storage capacity of the maximum point in catchments. γ and λ are the proportion of the infiltration capability beyond a certain value and the proportion of runoff generation in the catchment. When rainfall intensity was greater than infiltration intensity during the studied period, surface runoff occurs, whereupon continuous infiltration supplements the soil moisture deficiency until the system reaches the point soil water capacity, generating underground runoff. The model divides soil into two layers, in which soil depth should be determined according to the conditions of the catchment. The volume of infiltration (f_t), surface (r_s), and groundwater (r_g) can be calculated using Equations (2)–(4) as follows:

$$f_t = \sum_{i=1}^{t} \overline{f}_i = \left(i - \frac{i^{(1+n)}}{(1+n)f_m^n} \right) e^{-um} + f_c \tag{4}$$

$$r_s = p_t - f_t \tag{5}$$

where p_t (mm/h) is the rainfall rate during the ith hour and $\overline{f}_i$ (mm/h) is the ith hour empirical infiltration rate computed by a variant of Horton equation; m (mm) is the surface soil moisture; and u is an index accounting for the decreasing infiltration rate with increasing soil moisture. f_c (mm/h) and f_m (mm/h) are the stable infiltration rate and the infiltration capacity, respectively. It follows that:

$$r_g = \begin{cases} f_t + p_a - e_t - w_m & p_a + f_t \geq w_m \\ f_t + p_a - e_t - w_m + w_m \left(1 - \frac{f_t + p_a}{w_m}\right)^{1+b} & p_a + f_t < w_m \end{cases} \tag{6}$$

where e_t is the evaporation volume; p_a (mm) represents the rainfall influenced; w_m (mm) is the average storage capacity; and b is a coefficient of the water storage capacity curve.

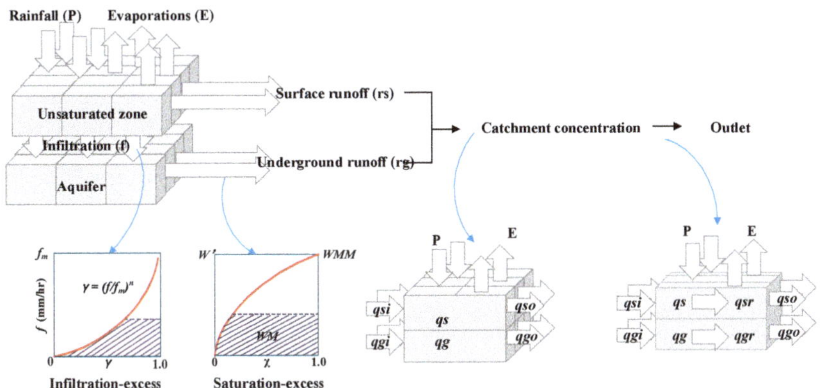

Figure 4. Rainfall-runoff mechanisms for the grid-based Hebei model.

3.2.2. The Grid-Based Hebei Model

The grid-based Hebei model retains the runoff generation concepts behind the lumped Hebei model. Furthermore, it develops terrain index T [48,49] to obtain a grid-type representation of soil moisture storage and infiltration capacity. $T = \ln(\alpha / \tan\beta)$, where α and β are the grid scale and shape parameters, respectively. The value of T in the Fuping and Zijingguan catchments were statistically analyzed. Experimentation demonstrated that the cumulative distribution curves of T values between different regions are always similar [50]. The T values of different grids in the same region can be expressed by a parabolic empirical formula. Hence, the soil water storage capacity and infiltration capacity of different grids can be determined as followed:

$$\frac{W_i}{WMM} = \exp\left\{-\left[\frac{\ln(T_i - T_{\min} + 1)}{\alpha}\right]^{\beta}\right\} \tag{7}$$

$$\frac{f_i}{f_m} = \left\{1 - \left[1 - \exp[-\frac{1}{\alpha}[\ln(T_i - T_{\min} + 1)]^{\beta}]^{b}\right\}^{1/n} \tag{8}$$

where i is a certain grid cell, and T_i and W_i (mm) are the corresponding moisture storage capacity and terrain index of the ith grid. $T_{\min}$ is the minimum value of the grid terrain index. In a non-channel grid, surface runoff outflow $q_{so} = q_{si} + q_s$, where q_s is the grid surface runoff. q_s occurs when the rainfall intensity exceeds f_m. q_{si} denotes the runoff values generated via the surrounding upstream grid. Similarly, the generated groundwater runoff from upstream grid (i.e., q_g, is transmitted to the groundwater runoff outflow according to q_{go}, $q_{go} = q_{gi} \cdot (1 - \mu) + q_g$, where μ is the ratio coefficient). The grid-based Hebei model adopts a simplified form of the Saint Venant equations for confluence calculation. Due to perennial channel water shortage and substantial channel seepage in the study area, an additional Horton infiltration equation was considered in the grid-based Hebei model [50].

3.2.3. The WRF-Hydro Modeling System

The open-code WRF-Hydro model is a multi-core parallel, multi-physics, multiscale, fully distributed hydraulic model. It can be operated only with meteorological data or coupled with the WRF model to constitute the atmospheric-hydrological model system. Its multiscale 3D land surface hydrological simulation mode can improve the one-dimensional vertical generalization of water transport using the original LSM Noah and Noah-MP. WRF-Hydro can use additional modules to achieve lateral flow exchange between the surface and subsurface, thereby improving upon the "column-only" one-dimensional vertical structure. It uses a disaggregation-aggregation solution module between the land model and terrain routing grid, which enables convergence processes occurring at a smaller resolution. WRF-Hydro can set the variables of the scale factor including soil moisture and excess infiltration into the fine grid values.

Noah or Noah-MP LSM are connected to WRF-Hydro to calculate water and heat flux exchange processes. In Noah-MP, a column cylinder structure is used to substitute soil layers into thicknesses, from top to bottom, of 0–10 cm, 10–40 cm, 40–100 cm, and 100–200 cm. The inputs required by WRF-Hydro include meteorological forcing data such as rainfall, air temperature, relative air humidity, surface pressure, wind speed, downward longwave radiation, and downward solar radiation. Gochis and Chen [51] described a sub-grid, spatially weighted disaggregation-aggregation solution to coordinate the land model and terrain routing grid in WRF-Hydro. Its runoff generation method uses a simple water balance (SWB) [52]. Similar to the Hebei model, when the rainfall capacity exceeds the infiltration capacity, a surface infiltration excess occurs in the top soil layer and the corresponding change in surface water depth h (m) is:

$$\frac{\partial h}{\partial t} = \frac{\partial p_e}{\partial t} \left\{ 1 - \frac{\left[\sum\limits_{k=1}^{4} Z_i(\delta_s - \delta_k)\right]\left[1 - \exp\left(-S\frac{R_{dt}}{R_{fd}}\frac{\Delta t}{86400}\right)\right]}{p_e + \left[\sum\limits_{k=1}^{4} Z_i(\delta_s - \delta_k)\right]\left[1 - \exp\left(-S\frac{R_{dt}}{R_{fd}}\frac{\Delta t}{86400}\right)\right]} \right\} \tag{9}$$

where h (m) refers to the change in surface water depth; Δt (s) is the model time step; k is the integer number of the soil layer (i.e., 1–4), Z_k (m) and δ_k (m^{3}m^{-3}) are the depth and grid soil moisture of the kth soil layer; δ_s (m^{3}m^{-3}) is the maximum soil moisture content; S is a coefficient given by Richards' equation to regulate runoff infiltration; and R_{dt} and R_{fd} represent the tunable surface infiltration coefficient and saturated hydraulic conductivity, respectively.

Subsurface routing followed the approach of Wigmosta and Lettenmaier [53] by using a quasi-3D flow taking into account topography, saturated soil depth, and saturated hydraulic conductivity with soil depth. Overland routing is a fully unsteady finite-difference and diffuse wave approach, implemented as described in Downer et al. [54]. River routing is similar to the overland case, using an explicit, one-dimensional, variable time-step diffusion wave equation. Details of specific rainfall-runoff processes can be obtained from the official user guide [55].

3.3. Establishment of Three Coupled Atmospheric-Hydrological Systems

Figure 5 shows a flood prediction diagram for coupled systems comprising three types of hydrological models, WRF models, and WRF-3DVar data assimilation modules. Rainfall in the WRF model was enhanced through optimal parameter scenarios and assimilation. GTS data and radar reflectivity were assimilated every 6 h to generate 3 km grid data in the inner layer Dom2. To eliminate the deviation of rainfall, for each storm, the best performance in the assimilated rainfall ensemble was selected as the forecast forcing data. Subsequently, predicted rainfall was converted into discharges using the aforementioned three models, with parameters calibrated using 17 historical rainfall-runoff events from the two studied catchments since the 1980s. For the calibration of WRF-Hydro, previous parameter sensitivity analysis in the study area noted several key parameters that can cause large fluctuations in flooding prediction including the runoff infiltration parameter (*REFKDT*), the channel Manning roughness parameter (*MannN*), the surface retention depth scaling parameter (*RETDEPRTFAC*), and the overland flow roughness scaling parameter (*OVROUGHRTFAC*). To evaluate the impact of data assimilation on the obtained runoff, WRF outputs under the selected optimal parameterization scheme were also used to drive the model for runoff forecasting; the results of the three systems before and after data assimilation were compared as described below.

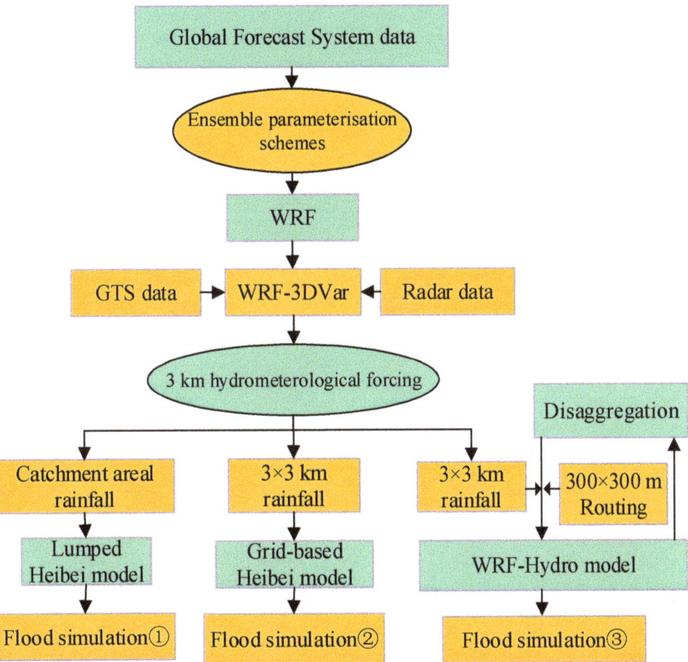

Figure 5. Framework for the coupled atmospheric-hydrologic modeling systems used for rainfall-runoff prediction.

The three coupling systems tested showed three types of rainfall input at different resolutions. The lumped model gives selected grid-averaged rainfall data across the sub-basins, whereas the grid-based Hebei model and WRF-Hydro directly pass the grid coordinates to locate forcing data. The former has a rainfall resolution of 3 km, and the latter has a grid division factor of 10 to allow further downscaling from the WRF assimilation data to 300 m of routing data.

The Nash efficiency coefficient (*NSE*), relative flood peak (R_p), and relative flood volume (R_v) were used to analyze the flow discharge forecasts as follows:

$$NSE = 1 - \frac{\sum_{i=1}^{n} (q\prime_i - q_i)^2}{\sum_{i=1}^{n} (q_i - \overline{q})^2} \tag{10}$$

$$R_p = (q\prime_p - q_p) / q_p \tag{11}$$

$$R_v = (r\prime_v - r_v) / r_v \tag{12}$$

where *i* is the time step; *n* is the total *i* of flood processes; $q\prime_i$, q_i, and $\overline{q}$ are the simulated, observed, and averaged flood discharge, respectively; $q\prime_p$ and q_p are the simulated and observed discharge, respectively; and rl_v and r_v are the simulated and observed flood volumes, respectively.

4. Results

The forecasting results for the three coupled systems before and after data assimilation for the four studied rainfall-runoff events are shown in Figure 6. In each single subfigure, the black bars and black solid curve indicate observed rainfall and runoff, respectively. Red and blue bars and curves indicate rainfall and runoff before and after data assimilation. For the studied coupling systems, assimilation had different degrees of enhancement on rainfall and runoff. The following analyses were conducted separately for data assimilation on the rainfall, runoff, and model systems.

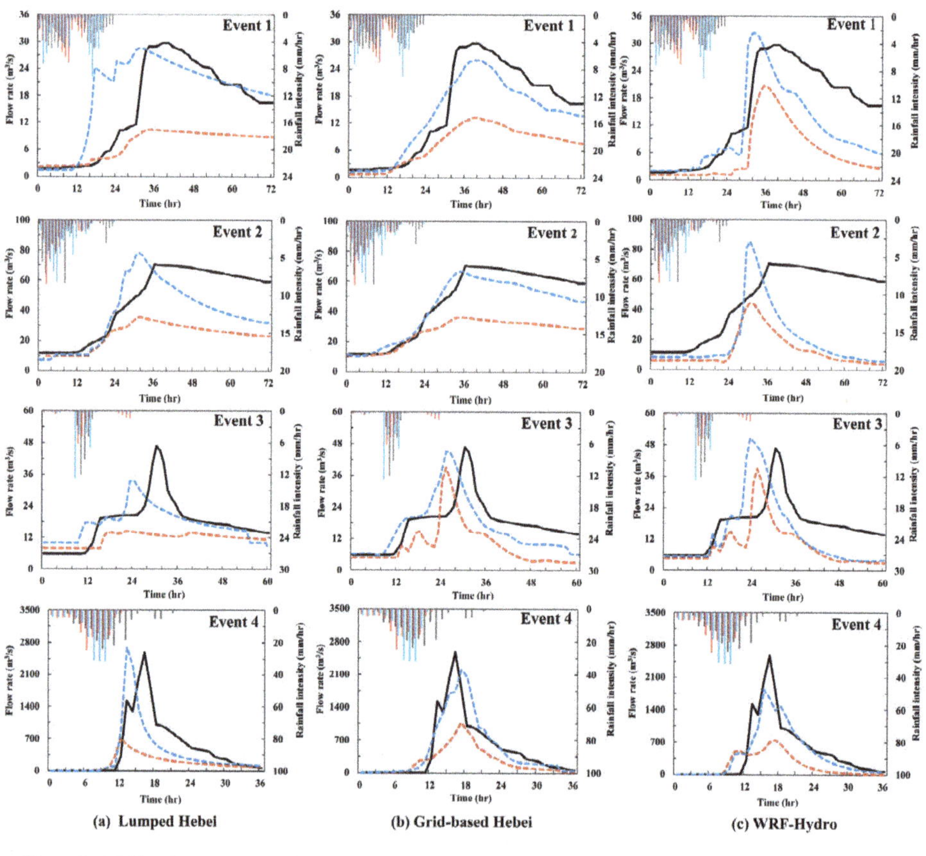

Figure 6. Rainfall-runoff predictions for the three atmospheric-hydrologic coupled systems before and after data assimilation: (**a**) Lumped Hebei; (**b**) Grid-based Hebei; (**c**) WRF-Hydro.

4.1. Effect of Data Assimilation on Rainfall Prediction

Figure 7 shows the cumulative precipitation variation over the simulated period caused by the cycling assimilation processes. The black solid curve represents the observed precipitation, while the other colors represent different assimilation periods. As above-mentioned, run1 was a non-assimilated precipitation process. Under conditions of data assimilation every 6 h, it was found that the cumulative rainfall gradually approached the observed values by the end of run6. For Event 2 and Event 3, the performance of run2 and run3 relative to run1 was impressive. For Event 1, run4 and run5 changed and improved the temporal distribution of precipitation within a few hours after data assimilation, while run2 and run3 seemed to show slightly worse results than run1, if no further assimilation took place. Generally, forecasts of cycling data assimilation after five cycles were largely stable for all events and the final curve integrated by the data assimilation runs (run2 to run6) was enhanced relative to the original run (run1). In addition to showing good performance for selected typical precipitation events, the cycling data assimilation gradually improved the temporal variability.

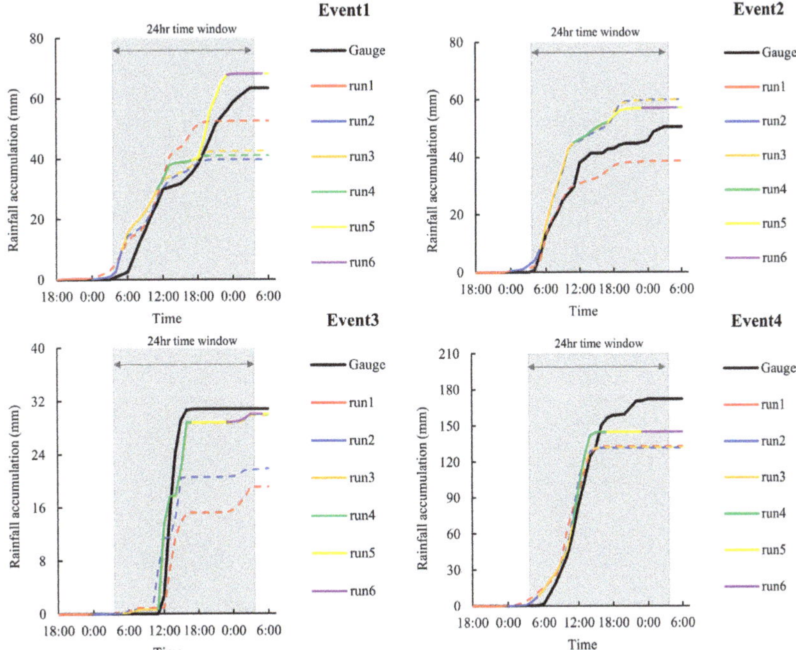

Figure 7. Cumulative rainfall curves before and after data assimilation under different assimilation run conditions.

Table 4 presents further statistics regarding the performance of data assimilation in improving cumulative rainfall. This shows that precipitation significantly increased compared to the period of no data assimilation. Overall, the relative error (*RE*) before and after assimilation was reduced by 0.26; Event 3 errors resulted in the most significant change, with a reduction in deviation of 0.342, while Event 4 rainfall increased most after assimilation by 19.76 mm.

Table 4. Observed and forecasted rainfall accumulation before and after data assimilation.

Storm Event	Observed	No Data Assimilation Forecasted	*RE*	Data Assimilation Forecasted	*RE*	Improvement Forecasted	*RE*
Event 1	63.38	49.35	−0.221	67.46	0.064	18.11	0.157
Event 2	50.48	37.22	−0.263	52.18	0.034	14.96	0.229
Event 3	30.82	19.05	−0.382	29.58	−0.040	10.53	0.342
Event 4	172.17	128.36	−0.254	148.12	−0.140	19.76	0.114
Average	79.21	58.49	−0.280	74.34	−0.020	15.85	0.260

The normalized Taylor diagrams of cumulative rainfall (in which the horizontal and vertical coordinates are normalized by dividing by the standard deviation (*SD*) of the observed series) before and after assimilation are shown in Figure 8. The variations of the assimilated results were closer to the actual observations. The correlation coefficients (*CC*) for both cumulative rainfall before and after assimilation were above 0.9 and the correlation coefficient and root mean square error (*RMSE*) after data assimilation showed a significant improvement compared to the corresponding values before assimilation, especially for Event 1, in which the *CC* increased from 0.93 to 0.99. In the case of Event 3, a decrease in *CC* from 0.98 to 0.96 occurred after assimilation; a similar trend was noted during Event 4. The previous calculations of the temporal *Cv* values for precipitation events showed that the two precipitation events were more heterogeneous in time and space than Event 1 and Event 2. In Event 3, for example, the temporal *Cv* value of 2.3925 was much higher than that of Event 1 (0.6011). This may explain the increased bias in assimilation, since the improved effectiveness of the rainfall forecast after assimilation is determined by the amount of effective information contained in the data. It is clearly easier for radar and GTS to capture data during periods of rainfall that are homogeneously distributed in space and time.

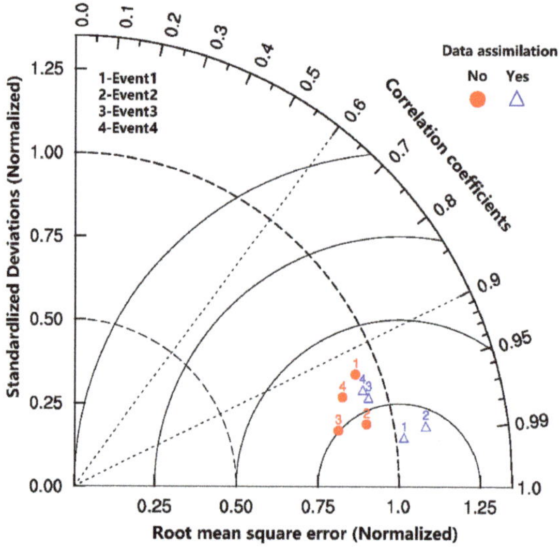

Figure 8. Taylor diagrams of forecasted hourly cumulative rainfall before and after data assimilation.

The above results demonstrate that WRF-3DVar effectively improved the consistency of simulated precipitation. Specifically, cycling assimilations of radar reflectivity and GTS data in the study area were able to improve both initial and lateral boundary conditions, providing a basis for future research into the accurate modeling of atmospheric-hydrological systems.

4.2. Effect of Data Assimilation on Runoff Prediction

In addition to the above analyses of the precipitation process, the performance of the data assimilation on the runoff process was also of interest. We found that runoff forecasts were relatively effective when data assimilation was used. For example, the coupling system from the lumped model simulated Event 2 had significant improvements in flood peak after assimilation (Figure 6).

The evaluation of the *NSE*, R_v, and R_p indices heat map among the flood processes for the four studied events are given in Figure 9. To evaluate the effects of data assimilation on runoff prediction, the indices of events were averaged from the results of the three coupled systems with different complexities. Figure 9 also shows the degree of improvement of the three types of indices after assimilation, demonstrating an overall improvement in *NSE*, R_v, and R_p of 0.386, 0.474, and 0.252. Event 1 presented a relatively homogeneous distribution in space and time and showed the most significant improvements in R_p and *NSE* after assimilation compared to the case of no assimilation, by 0.502 and 0.597, respectively. In contrast, although Event 3 showed the largest improvement in assimilating the *RE* in cumulative rainfall (Table 4), it resulted in the smallest improvement in both R_v and *NSE* of 0.293 and 0.322, respectively. This improved mitigation performance may stem from the poor spatial and temporal homogeneity of Event 3 of the studied storms, which poses difficulties for coupled systems prediction, even if the overall rainfall input does not differ significantly from the actual observations. The indices measured are thus a reflection of the complexity of rainfall-runoff processes.

	Data assimilation		Improvement (%)	
	No	Yes		
R_p	−0.501	−0.022	48.0	
R_v	−0.553	−0.051	50.2	Event1
NSE	−0.211	0.386	59.7	
R_p	−0.452	0.088	54.0	
R_v	−0.547	−0.275	27.2	Event2
NSE	0.608	0.753	14.5	
R_p	−0.405	−0.072	33.2	
R_v	−0.386	−0.093	29.3	Event3
NSE	0.667	0.699	3.1	
R_p	−0.672	−0.128	54.4	
R_v	−0.570	−0.092	47.8	Event4
NSE	0.516	0.749	23.4	
R_p	−0.507	−0.033	47.4	
R_v	−0.514	−0.128	38.6	All systems
NSE	0.395	0.647	25.2	

R_p and R_v Range: −0.672　−0.331　0.088

NSE and Improvement Range: −0.211　0.478　0.753

Figure 9. Averaged runoff prediction indices for *NSE*, R_v, and R_p of the four storm events.

Figure 10 shows the normalized Taylor diagrams of runoff events for the atmospheric-hydrological coupling systems before and after assimilation. The assimilated runoff processes corresponded to smaller *RMSE* and *SD* results closer to the mean observation series, with the exception of Event 3. Assimilated flood discharge may have a higher *CC* than the case of no data assimilation, but this was not always the case; indeed, parts of the spatio-temporally heterogeneous rainfall-runoff events in all three coupling systems were slightly smaller. The *CC* values after rainfall data assimilation were mostly above 0.6, such that only WRF-Hydro simulations of Event 2 and Event 3 had smaller *CC*, as shown in Figure 8. This corresponds to the rainfall processes of Event 2 (where there were two rainfall peaks) and Event 3 (featuring a short and concentrated rainfall process).

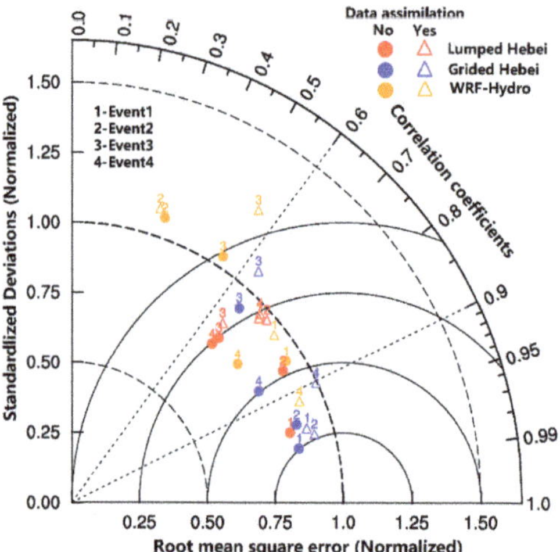

Figure 10. Taylor diagrams of forecasted hourly runoff before and after data assimilation.

4.3. Effect of Data Assimilation on Coupled Systems with Variable Complexity

Smaller catchments are particularly vulnerable to uncertainties and spatial shifts in rainfall patterns that may result in poor streamflow performance [27]. Figure 10 illustrates the coupling systems' stability from the grid-based model in blue, WRF-Hydro in yellow, and the lumped model in red. It can be observed that the *CC* for the grid-based model both before and after assimilation reached above 0.9 and that the values of *RMSE* were smaller than those of the other coupling systems, with the exception of Event 3. The lumped model had the next highest *CC* values, between 0.6 and 0.9, whereas WRF-Hydro exhibited a more scattered *CC* distribution. Nonetheless, after assimilation, the latter captured a better flood peak for Event 4. Detailed indices are given below for storm events, followed by further distinctions in between the effects of varying coupled systems under different types of rainfall before and after assimilation, as shown in Figure 11.

Data assimilation

	No	Yes	No	Yes	No	Yes	
R_p	−0.650	−0.035	−0.552	−0.121	−0.302	0.091	
R_v	−0.536	0.212	−0.519	−0.087	−0.605	−0.279	Event1
NSE	−0.158	0.396	0.023	0.869	−0.500	−0.107	
R_p	−0.497	0.112	−0.485	−0.058	−0.373	0.211	
R_v	−0.498	−0.185	−0.426	−0.068	−0.717	−0.571	Event2
NSE	0.691	0.874	0.764	0.979	0.370	0.406	
R_p	−0.689	−0.271	−0.361	−0.031	−0.164	0.085	
R_v	−0.341	−0.013	−0.370	−0.060	−0.448	−0.206	Event3
NSE	0.734	0.867	0.753	0.801	0.516	0.427	
R_p	−0.724	0.048	−0.584	−0.145	−0.708	−0.288	
R_v	−0.639	−0.222	−0.484	0.041	−0.588	−0.095	Event4
NSE	0.391	0.639	0.643	0.874	0.512	0.734	
	Lumped Hebei		Grid-based Hebei		WRF-Hydro		

R_p and R_v Range: −0.724   −0.295   0.212

NSE Range: −0.500   0.641   0.979

Figure 11. Runoff prediction indices of *NSE*, R_v, and R_p for different atmospheric-hydrologic modeling systems.

4.3.1. Results with the Lumped Hebei Model

Most lumped models lack the spatial information required to describe hydrological processes [56]. The lumped Hebei model consistently responds to spatial variability with probability functions (i.e., infiltration excess and saturation excess curves), thus ignoring the true spatial distribution. For the studied storms, the lumped model generally obtains an early flood peak for rainfall-runoff events, being 8 h and 4 h earlier for Event 3 and Event 4, respectively (Figure 6). Both Event 3 and Event 4 exhibited inhomogeneous spatial and temporal rainfall distributions. For the studied events, forecasts from the lumped model increased the inaccuracy of the peak present time. This may be due to the fact that when the catchment-averaged rainfall is used as an input, the effect of spatial variability in the underlying surface layer on runoff generation can only be considered when the spatial distribution of the rainfall is homogeneous (i.e., neither the combined effect of spatially heterogeneous rainfall distribution and underlying surface layer variability, nor the effect of net rainfall processes as multiple input sources when the spatial distribution of rainfall is heterogeneous).

4.3.2. Results with the Grid-Based Hebei Model

Based on the grid-based Hebei model, an atmospheric-hydrological coupling system was constructed for different rainfall resolutions and applied to the study area. This approach further demonstrated that descriptions of rainfall-runoff generation are compatible with local rainfall and flood forecasting, and that the grid-based Hebei model is more stable for forecasts both before and after assimilation compared with the other coupling systems tested. Flood forecasts before assimilation were slightly less accurate than WRF-Hydro for Event 1 and Event 2, but were generally better than the lumped model. Particularly, for the flood processes of Event 4, the grid-based Hebei model obtained the best *NSE* results (0.643 before assimilation and 0.874 after assimilation), demonstrating that this model is well-adapted to modeling flash floods. Although there was no clearly defined division of soil in the grid-based hydrological model, the influence of spatial heterogeneity due to soil type was somewhat reduced due to the relatively homogeneous nature of the study area.

4.3.3. Results with the WRF-Hydro Modeling System

WRF-Hydro exhibited the opposite prediction accuracy compared with the lumped model. A better forecast for the spatio-temporally heterogeneous Event 3 and Event 4 was noted than for Event 1. With the exception of Event 4, flooding processes were found to exhibit faster surface runoff recession, which may be related to rainfall-runoff generation and interactions between land surface that occurs at every short integration time step in the WRF-Hydro. This increased the volume of infiltrated precipitation, producing higher soil moisture and reducing runoff [57]. For Event 1, which had a small flood magnitude and long flood duration, the flood process was subject to a rapid recession, resulting in a poor NSE (-0.5 before assimilation and -0.107 after assimilation). Nevertheless, WRF-Hydro provided a more favorable forecast for Event 4, demonstrating its potential ability to predict flash floods. Overall, however, the accuracy of the WRF-Hydro forecasts was poor, supporting the findings of previous studies by Wang et al. [58] and Sharma et al. [31].

4.4. Improvement with Different Coupled Systems after Data Assimilation

Figure 12 provides further statistics on the extent to which the three coupled systems contribute to an overall improvement in response to flood events. The indices of systems were averaged from the results of the four studied storm events with different spatial and temporal characteristics.

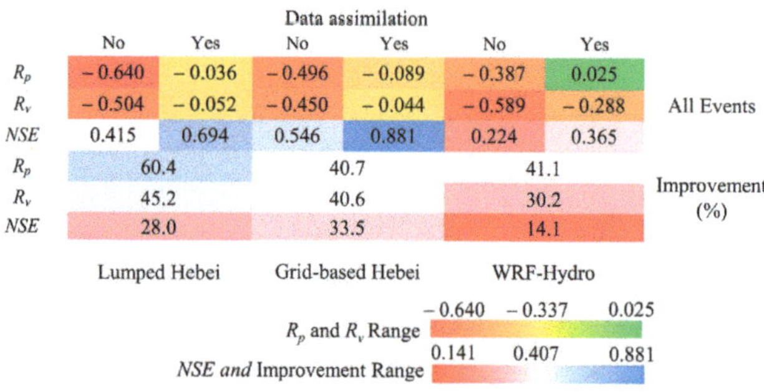

Figure 12. Improvement of the forecasted runoff indices of *NSE*, R_v, and R_p for the three coupled systems.

Data assimilation promoted flooding in all three coupled systems. The coupling system using the lumped model had a weak response with no data assimilation, denoted by low R_p (−0.640) and R_v (−0.504) values, and therefore the most significant enhancement in R_p and R_v after assimilation, especially R_p, with an enhancement of 0.604. Before assimilation, the grid-based system was found to be more stable than the other two systems, so that the improvement was moderate among the three systems. For the WRF-Hydro system, responses to the R_p of Event 3 and Event 4, which were spatially and temporally heterogeneous rainfall events, was better before assimilation, but problems such as rapid surface runoff recession remain a concern. The improvements in R_p and R_v were more obvious after assimilation, whereas *NSE* was less improved.

5. Discussion

The forecasting accuracy of the flood peaks of the lumped model seems to be more demanding in terms of the accuracy of the input rainfall than the two other coupling systems, since the flood peaks of the lumped model were more moderate when unassimilated rainfall was poorly simulated such as in Event 4, where less than a quarter of the observed flood peaks were simulated before assimilation. Although the results of the corresponding indices were better following assimilation, this characteristic greatly increased the uncertainty of the atmospheric-hydrological coupling process, which can easily lead to errors when studying flash floods since it is difficult to guarantee the accuracy of the rainfall forecasting.

Although data assimilation improves the precipitation input required by WRF-Hydro, it is still insufficient for complex model systems due to the need to input more meteorological variables. To improve the performance of the coupled system in the prediction of hydrological processes, research on WRF-Hydro also includes the assimilation of soil moisture variables [59] and real-time flood assimilation. Indeed, the coupling system of the WRF-Hydro model has a stronger basis in physical processes than the former two coupled systems; however, the complexity of parameter estimation that emerges from the model also poses a greater challenge, which is a typical problem in many complex physics-based models such as the variable infiltration capacity model and the community land model (CLM) [60]. In addition to data assimilation, more precise expressions of regional rainfall-runoff mechanisms also need to be further explored. The hydrodynamic parameters developed for local areas may not necessarily be applicable to mesoscale areas; therefore, although the model structure is feasible, the parameters of WRF-Hydro in terms of soil properties are not fully calibrated for northern China. There is an urgent need to find easily available parameters and expression equations that reflect the spatial heterogeneity of local infiltration processes across the region as an alternative to model application.

6. Conclusions

In this study, WRF-3DVar data assimilation experiments were conducted, in which radar reflectivity and GTS data were assimilated with the involvement of coupled hydrological structures of different complexity for rainfall-runoff prediction. The performance of three atmospheric-hydrological systems, established by coupling WRF with the lumped Hebei model, the grid-based Hebei model, and the fully distributed WRF-Hydro, were compared and analyzed for storms with different temporal and spatial distribution characteristics before and after data assimilation. We further explored model potentials and limitations in the localization of flood events. Focusing on the impact of data assimilation on flood forecasting after improving different types of rainfall and coupling systems of varying complexity, we found that WRF-3DVar produces more accurate rainfall forecasts, and that the assimilated model system provides higher confidence in the flood forecasts.

When the lumped model was coupled, its input rainfall was averaged over all grid points at the catchment scale, which may conceal the potential advantages of high-resolution rainfall datasets. The grid-based Hebei model obtained better flood forecasting results, but it did not provide a more comprehensive description of the spatial and temporal processes of the land-surface hydrology. The WRF-Hydro system, on the other hand, is built on the basis of water balance and heat balance in terms of the physical processes, thus clearly necessary for future flood research. The main reasons for the lack of accuracy of the WRF-Hydro predictions might be the preciseness of the input meteorological elements and the structure of the modeling system. Demonstrating the former requires further exploration of the transition from multi-source data assimilation to multi-process data assimilation. The coupling system of WRF-Hydro may differ from actual regional characteristics in its representation of the rainfall-runoff mechanisms, and thus its spatial scale and applicability need to be further explored in the future, especially in relation to the high-resolution land surface and hydrological processes that are essential for flash flood forecasting. There is also a need for future work to build on the strengths of this model and tailor atmospheric-hydrological coupling systems to the study area.

Author Contributions: Conceptualization, W.W. and J.L.; Methodology, W.W., J.L., and F.Y.; Software, W.W. and Y.L.; Validation, W.W. and Y.L.; Formal analysis, W.W. and J.L.; Investigation, W.W., C.L., and F.Y.; Writing-original draft preparation, W.W. and J.L.; Writing-review and editing, W.W., J.L., and C.L.; Visualization, W.W.; Funding acquisition, J.L. and C.L. All authors have read and agreed to the published version of the manuscript.

Funding: This research was funded by the National Natural Science Foundation of China (51822906), the Major Science and Technology Program for Water Pollution Control and Treatment (2018ZX07110001), the National Key Research and Development Project (2017YFC1502405), and the IWHR Research & Development Support Program (WR0145B732017).

Conflicts of Interest: The authors declare no conflict of interest.

References

1. Wang, W.; Seaman, N.L. A comparison study of convective parameterization schemes in a mesoscale model. *Mon. Weather Rev.* **1997**, *125*, 252–278. [CrossRef]
2. Yousef, L.A.; Taha, B. Adaptation of water resources management to changing climate: The role of Intensity-Duration-Frequency curves. *Int. J. Environ. Dev.* **2015**, *6*, 478. [CrossRef]
3. Yang, J.; Zhenming, J.I.; Chen, D.; Kang, S.; Congshen, F.U.; Duan, K.; Shen, M. Improved Land Use and Leaf Area Index Enhances WRF-3DVAR Satellite Radiance Assimilation: A Case Study Focusing on Rainfall Simulation in the Shule River Basin during July 2013. *Adv. Atmos. Sci.* **2018**, *35*, 628–644. [CrossRef]
4. Yucel, I.; Onen, A.; Yilmaz, K.K.; Gochis, D.J. Calibration and evaluation of a flood forecasting system: Utility of numerical weather prediction model, data assimilation and satellite-based rainfall. *J. Hydrol.* **2015**, *523*, 49–66. [CrossRef]
5. Bao, H.; Li, Z.; Yu, Z. Development of coupled atmospheric-hydrologic-hydraulic flood forecasting system driven by ensemble weather predictions. *AGUFM* **2009**, *2009*, H51G-0833. [CrossRef]
6. Ferretti, R.; Lombardi, A.; Tomassetti, B.; Sangelantoni, L.; Colaiuda, V.; Mazzarella, V.; Maiello, I.; Verdecchia, M.; Redaelli, G. Regional ensemble forecast for early warning system over small Apennine catchments on Central Italy. *Hydrol. Earth Syst. Sci. Discuss.* **2019**, 1–25. [CrossRef]
7. Sokol, Z. Assimilation of extrapolated radar reflectivity into a NWP model and its impact on a precipitation forecast at high resolution. *Atmos. Res.* **2011**, *100*, 201–212. [CrossRef]
8. Mohanty, U.C.; Routray, A.; Osuri, K.K.; Prasad, S.K. A Study on Simulation of Heavy Rainfall Events over Indian Region with ARW-3DVAR Modeling System. *Pure Appl. Geophys.* **2012**, *169*, 381–399. [CrossRef]

9.   Routray, A.; Mohanty, U.; Niyogi, D.; Rizvi, S.; Osuri, K.K. Simulation of heavy rainfall events over Indian monsoon region using WRF-3DVAR data assimilation system. *Meteorol. Atmos. Phys.* **2010**, *106*, 107–125. [CrossRef]
10.   Kumar, P.; Varma, A.K. Assimilation of INSAT-3D hydro-estimator method retrieved rainfall for short-range weather prediction. *Q. J. R. Meteorolog. Soc.* **2016**, *143*, 384–394. [CrossRef]
11.   Fierro, A.O.; Clark, A.J.; Mansell, E.R.; MacGorman, D.R.; Dembek, S.R.; Ziegler, C.L. Impact of storm-scale lightning data assimilation on WRF-ARW precipitation forecasts during the 2013 warm season over the contiguous United States. *Mon. Weather Rev.* **2015**, *143*, 757–777. [CrossRef]
12.   Yesubabu, V.; Srinivas, C.V.; Langodan, S.; Hoteit, I. Predicting extreme rainfall events over Jeddah, Saudi Arabia: Impact of data assimilation with conventional and satellite observations. *Q. J. R. Meteorolog. Soc.* **2016**, *142*, 327–348. [CrossRef]
13.   Lin, C.A.; Wen, L.; Lu, G.; Wu, Z.; Zhang, J.; Yang, Y.; Zhu, Y.; Tong, L. Atmospheric-hydrological modeling of severe precipitation and floods in the Huaihe River Basin, China. *J. Hydrol.* **2006**, *330*, 249–259. [CrossRef]
14.   Lu, G.; Wu, Z.; Wen, L.; Lin, C.A.; Zhang, J.; Yang, Y. Real-time flood forecast and flood alert map over the Huaihe River Basin in China using a coupled hydro-meteorological modeling system. *Sci. China Ser. E Technol. Sci.* **2008**, *51*, 1049–1063. [CrossRef]
15.   Lin, C.A.; Wen, L.; Lu, G.; Wu, Z.; Zhang, J.; Yang, Y.; Zhu, Y.; Tong, L. Real-time forecast of the 2005 and 2007 summer severe floods in the Huaihe River Basin of China. *J. Hydrol.* **2010**, *381*, 33–41. [CrossRef]
16.   Wu, J.; Lu, G.; Wu, Z. Flood forecasts based on multi-model ensemble precipitation forecasting using a coupled atmospheric-hydrological modeling system. *Nat. Hazards* **2014**. [CrossRef]
17.   Moser, B.A.; Gallus, W.A., Jr.; Mantilla, R. An initial assessment of radar data assimilation on warm season rainfall forecasts for use in hydrologic models. *Weather Forecast* **2015**, *30*, 1491–1520. [CrossRef]
18.   Yang, D.; Herath, S.; Musiake, K. Comparison of different distributed hydrological models for characterization of catchment spatial variability. *Hydrol. Process.* **2000**, *14*, 403–416. [CrossRef]
19.   Essou, G.R.C.; Arsenault, R.; Brissette, F. Comparison of climate datasets for lumped hydrological modeling over the continental United States. *J. Hydrol.* **2016**, *537*, 334–345. [CrossRef]
20.   Solomatine, D.P.; Dulal, K.N. Model trees as an alternative to neural networks in rainfall—runoff modelling. *Hydrol. Sci. J.* **2003**, *48*, 399–411. [CrossRef]
21.   Bray, M.; Han, D. Identification of support vector machines for runoff modelling. *J. Hydroinf.* **2004**, *6*, 265–280. [CrossRef]
22.   Karpatne, A.; Atluri, G.; Faghmous, J.H.; Steinbach, M.; Banerjee, A.; Ganguly, A.; Shekhar, S.; Samatova, N.; Kumar, V. Theory-guided data science: A new paradigm for scientific discovery from data. *IEEE Trans. Knowl. Data Eng.* **2017**, *29*, 2318–2331. [CrossRef]
23.   Herath, H.M.V.V.; Chadalawada, J.; Babovic, V. Hydrologically Informed Machine Learning for Rainfall-Runoff Modelling: Towards Distributed Modelling. *Hydrol. Earth Syst. Sci. Discuss.* **2020**, *487*, 1–42. [CrossRef]
24.   Cho, Y.; Engel, B.A. NEXRAD Quantitative Precipitation Estimations for Hydrologic Simulation Using a Hybrid Hydrologic Model. *J. Hydrometeorol.* **2017**, *18*, 25–47. [CrossRef]
25.   Song, X.; Zhang, J.; Zhan, C.; Xuan, Y.; Ye, M.; Xu, C. Global sensitivity analysis in hydrological modeling: Review of concepts, methods, theoretical framework, and applications. *J. Hydrol.* **2015**, *523*, 739–757. [CrossRef]
26.   Fatichi, S.; Vivoni, E.R.; Ogden, F.L.; Ivanov, V.Y.; Mirus, B.B.; Gochis, D.J.; Downer, C.W.; Camporese, M.; Davison, J.H.; Ebel, B.A. An overview of current applications, challenges, and future trends in distributed process-based models in hydrology. *J. Hydrol.* **2016**, *537*, 45–60. [CrossRef]
27.   Wagner, S.; Fersch, B.; Yuan, F.; Yu, Z.; Kunstmann, H. Fully coupled atmospheric-hydrological modeling at regional and long-term scales: Development, application, and analysis of WRF-HMS. *Water Resour. Res.* **2016**. [CrossRef]
28.   Liu, J.; Wang, J.; Pan, S.; Tang, K.; Li, C.; Han, D. A real-time flood forecasting system with dual updating of the NWP rainfall and the river flow. *Nat. Hazards* **2015**, *77*, 1161–1182. [CrossRef]
29.   Li, J.; Chen, Y.; Wang, H.; Qin, J.; Li, J.; Chiao, S. Extending flood forecasting lead time in a large watershed by coupling WRF QPF with a distributed hydrological model. *Hydrol. Earth Syst. Sci.* **2017**, *21*, 1279. [CrossRef]
30.   Rogelis, M.C. Streamflow forecasts from WRF precipitation for flood early warning in mountain tropical areas. *Hydrol. Earth Syst. Sci.* **2018**, *22*, 853–870. [CrossRef]
31.   Sharma, S.; Siddique, R.; Reed, S.; Ahnert, P.; Mejia, A. Hydrological Model Diversity Enhances Streamflow Forecast Skill at Short- to Medium-Range Timescales. *Water Resour. Res.* **2019**, *55*, 1510–1530. [CrossRef]
32.   Haddeland, I.; Clark, D.B.; Franssen, W.; Ludwig, F.; Voß, F.; Arnell, N.W.; Bertrand, N.; Best, M.; Folwell, S.; Gerten, D. Multimodel estimate of the global terrestrial water balance: Setup and first results. *J. Hydrometeorol.* **2011**, *12*, 869–884. [CrossRef]
33.   Wood, E.F.; Roundy, J.K.; Troy, T.J.; Van Beek, L.; Bierkens, M.F.; Blyth, E.; de Roo, A.; Döll, P.; Ek, M.; Famiglietti, J. Hyperresolution global land surface modeling: Meeting a grand challenge for monitoring Earth's terrestrial water. *Water Resour. Res.* **2011**, *47*. [CrossRef]
34.   Wang, W.; Liu, J.; Li, C.; Liu, Y.; Yu, F.; Yu, E. An Evaluation Study of the Fully Coupled WRF/WRF-Hydro Modeling System for Simulation of Storm Events with Different Rainfall Evenness in Space and Time. *Water* **2020**, *12*, 1209. [CrossRef]
35.   Fiori, E.; Comellas, A.; Molini, L.; Rebora, N.; Siccardi, F.; Gochis, D.; Tanelli, S.; Parodi, A. Analysis and hindcast simulations of an extreme rainfall event in the Mediterranean area: The Genoa 2011 case. *Atmos. Res.* **2014**, *138*, 13–29. [CrossRef]

36. Verri, G.; Pinardi, N.; Gochis, D.; Tribbia, J.; Navarra, A.; Coppini, G.; Vukicevic, T. A meteo-hydrological modelling system for the reconstruction of river runoff: The case of the Ofanto river catchment. *Nat. Hazards Earth Syst. Sci.* **2017**, *17*, 1741–1761. [CrossRef]
37. Moeng, C.H.; Dudhia, J.; Klemp, J.; Sullivan, P. Examining Two-Way Grid Nesting for Large Eddy Simulation of the PBL Using the WRF Model. *Mon. Weather Rev.* **2007**, *135*, 2295–2311. [CrossRef]
38. Shin, H.H.; Hong, S.-Y. Analysis of resolved and parameterized vertical transports in convective boundary layers at gray-zone resolutions. *J. Atmos. Sci.* **2013**, *70*, 3248–3261. [CrossRef]
39. Tian, J.; Liu, J.; Yan, D.; Li, C.; Yu, F. Numerical rainfall simulation with different spatial and temporal evenness by using a WRF multiphysics ensemble. *Nat. Hazards Earth Syst. Sci.* **2017**, *17*, 563–579. [CrossRef]
40. Tian, J.; Liu, J.; Wang, J.; Li, C.; Yu, F.; Chu, Z. A spatio-temporal evaluation of the WRF physical parameterisations for numerical rainfall simulation in semi-humid and semi-arid catchments of Northern China. *Atmos. Res.* **2017**, *191*, 141–155. [CrossRef]
41. Lin, Y.-L.; Farley, R.D.; Orville, H.D. Bulk parameterization of the snow field in a cloud model. *J. Clim. Appl. Meteorol.* **1983**, *22*, 1065–1092. [CrossRef]
42. Hong, S.Y.; Noh, Y.; Dudhia, J. A new vertical diffusion package with an explicit treatment of entrainment processes. *Mon. Weather Rev.* **2006**, *134*, 2318–2341. [CrossRef]
43. Kain, J.S. The Kain–Fritsch convective parameterization: An update. *J. Appl. Meteorol.* **2004**, *43*, 170–181. [CrossRef]
44. Grell, G.A.; Freitas, S.R. A scale and aerosol aware stochastic convective parameterization for weather and air quality modeling. *Atmos. Chem. Phys.* **2014**, *13*, 5233–5250. [CrossRef]
45. Janjić, Z.I. The Step-Mountain Eta Coordinate Model: Further Developments of the Convection, Viscous Sublayer, and Turbulence Closure Schemes. *Mon. Weather Rev.* **1994**, *122*, 927–945. [CrossRef]
46. Tian, J.; Liu, J.; Yan, D.; Li, C.; Chu, Z.; Yu, F. An assimilation test of Doppler radar reflectivity and radial velocity from different height layers in improving the WRF rainfall forecasts. *Atmos. Res.* **2017**, *198*, 132–144. [CrossRef]
47. Liu, J.; Tian, J.; Yan, D.; Li, C.; Yu, F.; Shen, F. Evaluation of Doppler radar and GTS data assimilation for NWP rainfall prediction of an extreme summer storm in northern China: From the hydrological perspective. *Hydrol. Earth Syst. Sci.* **2018**, *22*, 4329–4348. [CrossRef]
48. Beven, K.J. Physically based, variable contibution area model of basin hydrology. *Hydrol. Ences. Bull.* **1979**, *24*. [CrossRef]
49. Beven, K. Surface water hydrology-runoff generation and basin structure. *Rev. Geophys.* **1983**, *21*, 721–730. [CrossRef]
50. Tian, J.; Liu, J.; Wang, Y.; Wang, W.; Li, C.; Hu, C. A coupled atmospheric-hydrologic modeling system with variable grid sizes for rainfall-runoff simulation in semi-humid and semi-arid watersheds: How does the coupling scale affects the results? *Hydrol. Earth Syst. Sci.* **2020**, *24*, 3933–3949. [CrossRef]
51. Gochis, D.J.; Chen, F. *Hydrological Enhancements to the Community Noah Land Surface Model: Technical Description*; NCAR Tech. Note, TN-454+STR, Boulder, Colo.; University Corporation for Atmospheric Research: Boulder, CO, USA, 2003; 77p. [CrossRef]
52. Schaake, J.C.; Koren, V.I.; Duan, Q.-Y.; Mitchell, K.; Chen, F. Simple water balance model for estimating runoff at different spatial and temporal scales. *J. Geophys. Res. Atmos.* **1996**, *101*, 7461–7475. [CrossRef]
53. Wigmosta, M.S.; Lettenmaier, D.P. A comparison of simplified methods for routing topographically driven subsurface flow. *Water Resour. Res.* **1999**, *35*, 255–264. [CrossRef]
54. Downer, C.W.; Ogden, F.L.; Martin, W.D.; Harmon, R.S. Theory, development, and applicability of the surface water hydrologic model CASC2D. *Hydrol. Process.* **2002**, *16*, 255–275. [CrossRef]
55. Gochis, D.; Barlage, M.; Dugger, A.; FitzGerald, K.; Karsten, L.; McAllister, M.; McCreight, J.; Mills, J.; RefieeiNasab, A.; Read, L. The WRF-Hydro modeling system technical description (Version 5.0). *NCAR Tech. Note* **2018**, *107*. [CrossRef]
56. Moore, R. The probability-distributed principle and runoff production at point and basin scales. *Hydrol. Sci. J.* **1985**, *30*, 273–297. [CrossRef]
57. Cuntz, M.; Mai, J.; Samaniego, L.; Clark, M.; Wulfmeyer, V.; Branch, O.; Attinger, S.; Thober, S. The impact of standard and hard-coded parameters on the hydrologic fluxes in the Noah-MP land surface model. *J. Geophys. Res. Atmos.* **2016**, *121*, 10676–10700. [CrossRef]
58. Wang, Y.; Lahmers, T.; Hazenberg, P.; Gupta, H.; Castro, C.; Gochis, D.; Dugger, A.; Yates, D.; Read, L.; Karsten, L. Enhancements to the WRF-Hydro Hydrologic Model Structure for Semi-arid Environments. *AGUFM* **2018**, *2018*, H41P-2347.
59. Zarekarizi, M. Ensemble Data Assimilation for Flood Forecasting in Operational Settings: From Noah-MP to WRF-Hydro and the National Water Model. *J. Hydrol.* **2018**. [CrossRef]
60. Xi, C.; Yang, T.; Wang, X.; Xu, C.Y.; Yu, Z. Uncertainty Intercomparison of Different Hydrological Models in Simulating Extreme Flows. *Water Resour. Manag.* **2012**, *27*, 1393–1409. [CrossRef]

Article

The Effect of Water Transfer during Non-growing Season on the Wetland Ecosystem via Surface and Groundwater Interactions in Arid Northwestern China

Shufeng Qiao, Rui Ma *, Ziyong Sun, Mengyan Ge, Jianwei Bu, Junyou Wang, Zheng Wang and Han Nie

School of Environmental Studies and MOE Key Laboratory of Biogeology and Environmental Geology, China University of Geosciences, Wuhan 430074, China; sqiao@cug.edu.cn (S.Q.); ziyong.sun@cug.edu.cn (Z.S.); myge@cug.edu.cn (M.G.); jwbu@cug.edu.cn (J.B.); jywang@cug.edu.cn (J.W.); wangzheng@cug.edu.cn (Z.W.); niehan1997@cug.edu.cn (H.N.)
* Correspondence: rma@cug.edu.cn

Received: 23 June 2020; Accepted: 3 August 2020; Published: 5 August 2020

Abstract: The use of ecological water transfer to maintain the ecological environment in arid or semiarid regions has become an important means of human intervention to alleviate vegetation ecosystem degradation in arid and semiarid areas. The water transfer to downstream in a catchment is often carried out during the non-growing season, due to the competitive water use between the upper and middle reaches and lower reaches of rivers. However, the impacts and mechanism of artificial water transfer on vegetation and wetland ecosystem restoration have not been thoroughly investigated, especially in northwest China. Taking the Qingtu Lake wetland system in the lower reaches of the Shiyang River Catchment as the study area, this study analyzed the spatial and temporal distribution surface area of Qingtu Lake and the surrounding vegetation coverage before and after water transfer, by interpreting remote sensing data, the variation of water content in the vadose zone, and the groundwater level by obtaining field monitoring data, as well as the correlation between the water body area of Qingtu Lake and the highest vegetation coverage area in the following year. The conclusion is that there is a positive correlation between the water body area of Qingtu Lake in autumn and the vegetation coverage in each fractional vegetation coverage (FVC) interval in the next summer, especially in terms of the FVC of 30–50%. The groundwater level and soil water content increase after water transfer and remain relatively high for the following months, which suggests that transferred water from upstream can be stored as groundwater or soil water in the subsurface through surface water and subsurface water interaction. These water sources can provide water for the vegetation growth the next spring, or support plants in the summer.

Keywords: ecological water transfer; wetland vegetation ecosystem; surface and groundwater interaction; northwestern China; remote sensing

---

1. Introduction

The arid and semiarid regions in the world usually have fragile ecological environments due to low precipitation and a lack of water resources. Excessive use of water resources leads to ecological system degradation, including wetland degradation and plant ecosystem deterioration [1,2]. For example, climate change and water resource exploitation currently threaten groundwater-dependent ecosystems and put vegetation at risk of degradation in the Nalenggele River Basin, located in the southwest Qaidam Basin in the Qinghai province of northwest China [3]. Due to intense water usage in the Junggar Basin of northwestern China over the last few decades, the water flowing into Ebinur Lake has been greatly reduced. Thus, Ebinur Lake, which lies on the southwest margin of the basin, has been continuously shrinking; the water area had shrunk from over 1000 km^2 to less than 500 km^2 in 2011. The water reduction of Ebinur Lake has led to a serious recession of lakeshore vegetation, with 60% of the desert forest around the lake vanishing, and the desertification area is expanding at a speed of 39.8 km^2 per year on average [4]. Similar situations have also occurred in other countries, such as Australia [5], Mexico [6,7], Nepal [8], the North Basin in Kenya [9], Iran, and Afghanistan [10]. In the Narran Lakes in Australia, one of the most important Ramsar-listed wetlands due to its provision of habitat for wetland fauna during key life history stages, reduced ibis breeding due to water resource exploitation has been reported, from 1 in 4.2 years to 1 in 11.4 years [5]. The increase of water withdrawal and introduction of exotic herbage species have aggravated the transformation of the ecosystem in San Miguel and Zanjon River in the northwest of Mexico, such that the range of crops and grasslands have been significantly reduced and desert shrub species significantly increased [7]. The excessive abstraction of Ewaso Ng'iro River water in the Upper Ewaso Ng'iro North Basin in Kenya has greatly affected downstream water users, and led to the deterioration of the vegetative cover and a reduction in water flow in the Ewaso Ng'iro River and its major tributaries [9]. Hence, the excessive use of water resources can lead to a decline of the groundwater level. Problems associated with this include salinization, land subsidence, and deterioration of water quality, which could be harmful to vegetation and cause ecosystem degradation [11].

Given the current situation, the restoration of wetlands has become an important option for the international community. However, ecological reversal to the natural status is a long-term dynamic change process [12]. People are working to accelerate the restoration of the ecosystem through manual intervention, and ecological water transportation to ecologically fragile areas is one of the most important methods to achieve this [13,14]. In northwest China, this measure has been employed to maintain the ecological system balance of downstream waters, especially for wetland vegetation systems. For example, Xinjiang's Tarim River received water from a water transfer project during ecological emergency periods eight times between 2001 and 2006; the transfer improved groundwater quality and restored natural grasslands [15]. Similarly, a total of 4.512 billion m^3 of water was transferred to the downstream area in the Heihe River Basin from 2000 to 2008, which led to extensive vegetation restoration [16]. Maintaining the ecosystem in downstream waters through ecological water transportation has become the focus of ecological research of arid areas throughout the world [13,14]. Due to the demand for water resources in the upper and middle reaches of these arid river basins in spring and summer for agriculture, little water tends to be available to transfer downstream. Normally, waters in autumn, when the water demand is alleviated in the upstream, are transferred downstream. However, vegetation normally stops growing during this period. Will the water transferred in autumn help maintain the vegetation ecosystem in spring and summer, and if so, how? The quantity and timing of water transfer can determine the effect of water transfer on ecological restoration. However, few studies have been reported on this issue.

Shiyang River is China's third-largest inland river. Its lower reaches are located in the Minqin Basin, the wedge of which is at the cross-point of the Badain Jaran Desert and Tengger Desert. This oasis is an important barrier to prevent inland movement of big sandstorms from the desert. Qingtu Lake, located at the northern edge of Minqin Basin, is the terminal lake of the Shiyang River. It has important ecological significance in preventing the connection of the two big deserts [17]. Because of the construction of a water conservancy project in the upper reaches, and the Hongyashan Reservoir in the middle reaches of Shiyang River, the main channel of Shiyang River in the lower reaches had been dry for a long time, which led to Qingtu Lake, the terminal lake of the Shiyang River, drying in 1959. As a result, the wetland ecosystem and vegetation was transformed into desert, and most areas near the wetland were covered by quicksand, while the regional ecosystem continued to deteriorate [18]. In order to improve the ecological environment in the lower reaches of Shiyang River, the government promoted a water transfer project that transports ecological water into Qingtu Lake downstream via a channel. This started in September 2010 and has continued in autumn every year since. By November

2016, a 25.16 km^2 area of water body surface had formed in the Qingtu Lake [19]. Due to conflict around the ecological water demand in the lower reaches and the agricultural water needs in the middle and upper reaches of Shiyang River in spring and summer, ecological water transportation has been implemented in autumn, when the vegetation has almost stopped growing. However, the impacts and mechanism of artificial water transfer on vegetation and wetland ecosystem restoration remain poorly understood. Taking the Qingtu Lake wetland ecosystem in the lower reaches of the Shiyang River Basin as an example, this study investigated the effect of the water transfer in autumn on the wetland vegetation ecosystem, and explored the mechanism of how water input during the non-growing season improves the vegetation ecosystem.

2. Study Area

Qingtu Lake, the terminal lake of the Shiyang River, is located on the northeast edge of Minqin County, Gansu Province, between 39°04′ and 39°09′ N latitude and 103°36′ and 103°39′ E longitude (Figure 1). The area has a temperate continental arid desert climate. The mean annual temperature is 8.87 °C; the maximum temperature is 37.8 °C and the minimum temperature is −29.5 °C. The mean annual precipitation is 110 mm, with 73% of this occurring in June, July, and August. The mean annual evaporation is 2644 mm [20]. The main vegetation species include *Phragmites australis*, *Nitraria tangutorum*, *Suaeda glauca*, *Haloxylon ammodendron*, and *Kalidium foliatum*. Accompanying shrubs include *Lycium ruthenicum*, *Artemisia sphaerocephala*, and *Kalidium foliatum*. Herbaceous plants include those such as *Salsola ruthenica*, *Zygophyllum fabago*, *Cynanchum sibiricum*, *Salsola collina*, and *Agriophyllum squarrosum*. They are all typical desert vegetation types [21]. The water of Qingtu Lake is mainly used to maintain the ecological system balance around the wetland, especially for improving the vegetation ecosystem [20,22]. The water in Qingtu Lake is weakly alkaline and Na–SO$_4$ (Cl) type, with high salinity and total dissolved solids (TDS) [23]. The lithology is shown in the cross section (Figure 2). The sediments in this area consist of mainly sand, silt, and loam [24].

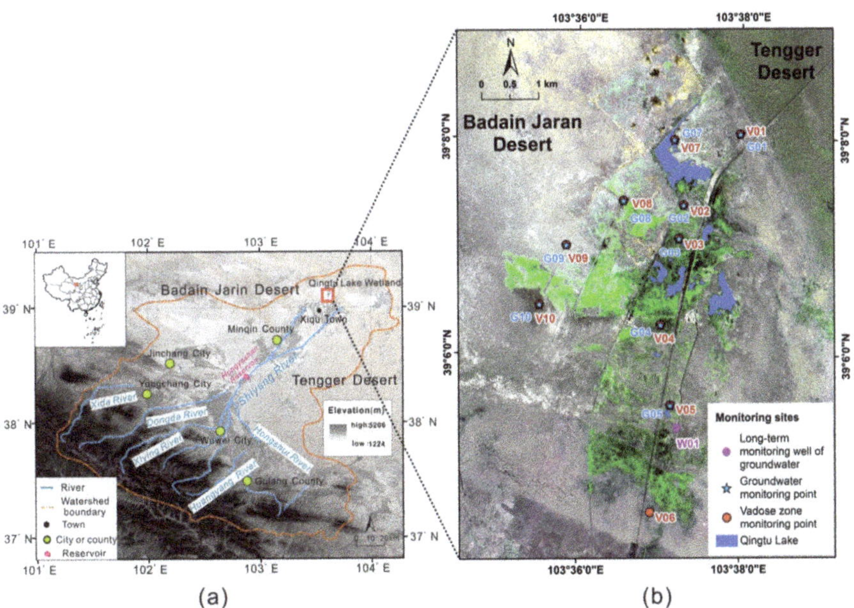

Figure 1. Location of the study area (**a**) and monitoring locations in the Qingtu Lake wetland (**b**).

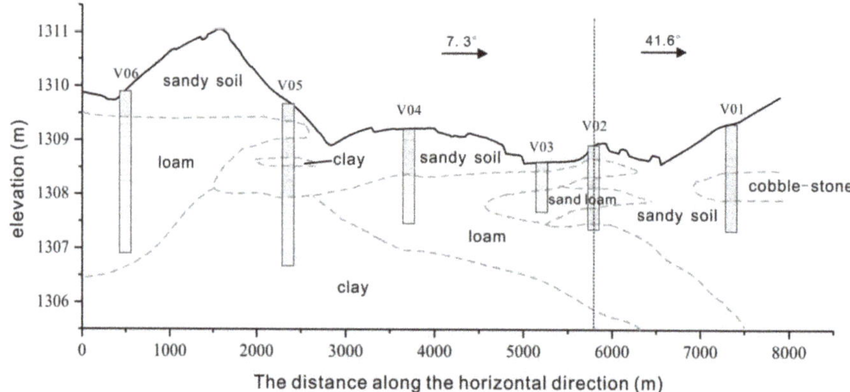

Figure 2. Simplified geological cross section from V01 to V06 (locations are shown in Figure 1b).

During the Western Han Dynasty, Qingtu Lake covered an area of 4000 km^2. Due to the increase of water consumption in the upper reaches of Qingtu Lake, the surface area of the water body of Qingtu Lake began to gradually diminish after 1924. In the 1950s, the construction of the Hongyashan Reservoir caused the Shiyang River to dry quickly [25]. Since the 1960s, there have been a series of ecological problems in the lower reaches of Shiyang River, such as the decline of the groundwater level, vegetation degradation, and soil salinization [23]. In order to restore the regional ecological environment, the local government decided to begin transferring ecological water from Hongyashan Reservoir to Qingtu Lake in September 2010. The water is transferred through irrigation channels from September to October of each year. It is a typical ecological water transportation system, carried out during the non-growing season.

3. Data and Methods

3.1. Data Sources and Preprocessing

Ten monitoring sites were constructed in July and August 2018. Instrumentation at the V01–V10 locations was implemented using 5TE probes developed by the Decagon Company, with a precision of 3% for monitoring the soil water content at different depths along the soil profile (Figure 1b). The G01–G10 wells with screens at depths between 1 and 3.6 m were installed with Canadian Solinst3001 titanium gold edge water-level loggers with a precision of 0.05% for recording the groundwater level, water salinity, and temperature (Figure 1b). The soil water content and parameters in the groundwater were recorded every 30 min. Meteorological data in the study area (temperature, precipitation, air pressure) were downloaded from the China Meteorological Data Sharing Network (http://data.cma.cn).

Considering the timing of water transfer from Hongyashan Reservoir to Qingtu Lake and vegetation evolution, the remote sensing data used in this study were acquired from Landsat series satellite data, including Landsat 8 OLI, Landsat 7 OFF, and Landsat 4-5 TM from 2009 to 2018, with a resolution of 30 m, a synthetic image resolution of 15 m, and a temporal resolution of 16 d. The remote sensing data from after May 2013 were from the Landsat 8 OLI series, while data from before May 2013 were from the Landsat 7 OFF or Landsat 4-5 TM series. The data from the Landsat 7 OFF series were damaged; the images showed data overlap and about 25% of the data was lost. All of the satellite data of the Landsat 7 OFF series used in this study were repaired by the strip removal method [26]. The available Landsat data from before May 2013 were limited, and there were nine images from 2009 to 2018.

All the image data were preprocessed with ENVI 5.2 for radiometric calibration, layer stacking, atmospheric correction, geometric correction, and cropping [27].

3.2. Methods

3.2.1. Water Extraction Methods

The NDWI (normalized difference water index) proposed by McFeeters was initially designed to enhance the difference between water and nonwater bodies, in order to generate initial water body maps. However, it cannot efficiently suppress the signal noise coming from the land cover features of built-up areas [28,29]. Hence, Xu [30] proposed a modified NDWI (MNDWI), which is defined as:

$$MNDWI = \frac{(Green - MIR)}{(Green + MIR)} \tag{1}$$

where Green is the green band, which corresponds to the TM/ETM + image band 2 (0.52–0.60 μm) and the OLI image band 3 (0.525–0.600 μm). MIR represents the mid-infrared band, which corresponds to the TM/ETM + image band 5 (1.55–1.75 μm) and OLI image band 6 (1.560–1.660 μm) [30,31].

Water bodies have positive values in the MNDWI, while soil, vegetation, and built-up classes tend to have negative values. Hence, the normal empirical threshold is zero [29].

3.2.2. Method for FVC Estimation

FVC (fractional vegetation cover) is defined as the fraction of green vegetation seen from nadir, which can characterize the growth conditions and horizontal density of land surface live vegetation. There are three major algorithms for FVC estimation, which include empirical methods, pixel unmixing models, and machine learning methods [32]. The dimidiate pixel model, which was used in this study, is a method of pixel unmixing and widely used for FVC estimation [33]. It is assumes the NDVI (normalized difference vegetation index) in a pixel consists of soil and vegetation, and the NDVI can be derived by [34,35]:

$$NDVI = FVC \times NDVI_{veg} + (1 - FVC) \times NDVI_{soil} \tag{2}$$

where NDVIveg represents the NDVI value of a pure vegetation pixel and NDVIsoil represents the NDVI value of a pure soil pixel. The NDVIsoil selected an NDVI value with a cumulative frequency of 0.5%, and the NDVIveg selected an NDVI value with a cumulative frequency of 99.5% [36,37].

Hence, the FVC can be derived by modifying Equation (2) as:

$$FVC = \frac{(NDVI - NDVI_{soil})}{(NDVI_{veg} - NDVI_{soil})} \tag{3}$$

3.2.3. Classification Method of FVC

For the entire Minqin Basin, there are few vegetation areas and vegetation types. The vegetation coverage types of the Minqin Basin can be divided into five main vegetation coverage categories [38]: extremely low, low, medium, medium and high, and high (Table 1).

Table 1. Vegetation Coverage Classification Standards [38].

Class	Vegetation Coverage	Status of Fractional Vegetation Coverage (FVC)	Description
1	extremely low	FVC ≤ 10%	Intense desertified land, bare rock, bare soil, and waters
2	low	10% < FVC ≤ 30%	Moderately desertified land, low yield grassland, and sparse woodland
3	medium	30% < FVC ≤ 50%	Slightly desertified land, medium grassland, low canopy woodland, and cultivated land
4	medium and high	50% < FVC ≤ 70%	Medium and high yield grassland, forest, and cultivated land
5	high vegetation	FVC > 70%	High yield grassland, forest, and cultivated land

3.2.4. Correlation Analysis

This analysis aims to explore the influence of the former expansion of Qingtu Lake on the secondary vegetation coverage around Qingtu Lake after ecological water transfer. The two variables are one-to-one, corresponding, and relatively continuous, so the Pearson correlation coefficient was selected for calculation. The equation is given below [39].

$$r = \frac{\sum (X - \overline{X})(Y - \overline{Y})}{\sqrt{\sum(X-\overline{X})^2 \sum(Y-\overline{Y})^2}} \tag{4}$$

where X and Y represent the two groups of variables, and X and Y represent the average values of the two groups of variables. If r is greater than 0, then a positive correlation between the two variables is indicated; if the contrary, there is a negative correlation. In addition, the greater the absolute value of r, the stronger the correlation between the two variables.

4. Result

4.1. Spatial and Temporal Variations of the Water Body Area of Qingtu Lake from 2009 to 2018

4.1.1. Temporal Change

Since September 2010, the ecological waters have been transferred from the reservoir to Qingtu Lake from September to October each year [40]. The surface area of the water body of Qingtu Lake exhibited a dynamic change trend (Figure 3). The lake water body reached the largest surface area from November to the following January, with a maximum area of 13.7 km^2. It normally exhibits the lowest area from July to August each year.

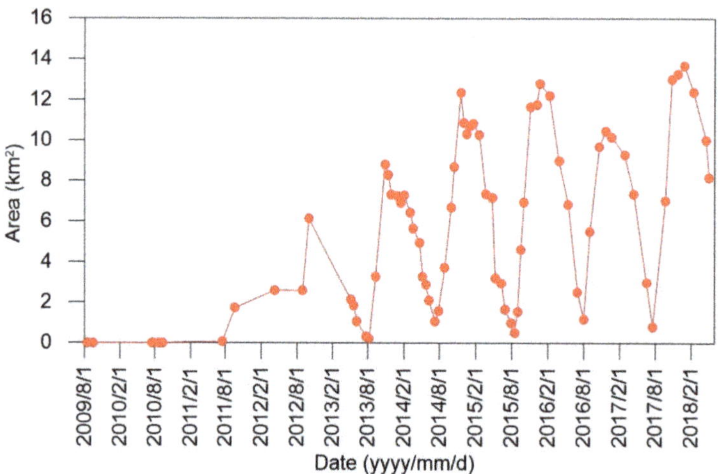

Figure 3. The change in the surface area of the water body of Qingtu Lake over time.

4.1.2. Spatial Change

Due to the similar temporal trend of the Qingtu Lake water surface area each year, only the spatial distributions of Qingtu Lake water area in different months of 2017 are shown in Figure 4. The water surface area of Qingtu Lake changed little between January and February, ranging between 9.2 km^2 and 9.6 km^2 (Figure 4). After April, the water area began to shrink rapidly, and then reached the lowest level (0.8226 km^2) in July. From September, the water area began to expand again, since the water transfer to Qingtu Lake was started in September. The water body expanded to the south and north, and then gradually spread to the west, reaching the highest lake level in November and December. The lake water level then almost stabilized from November to the following February.

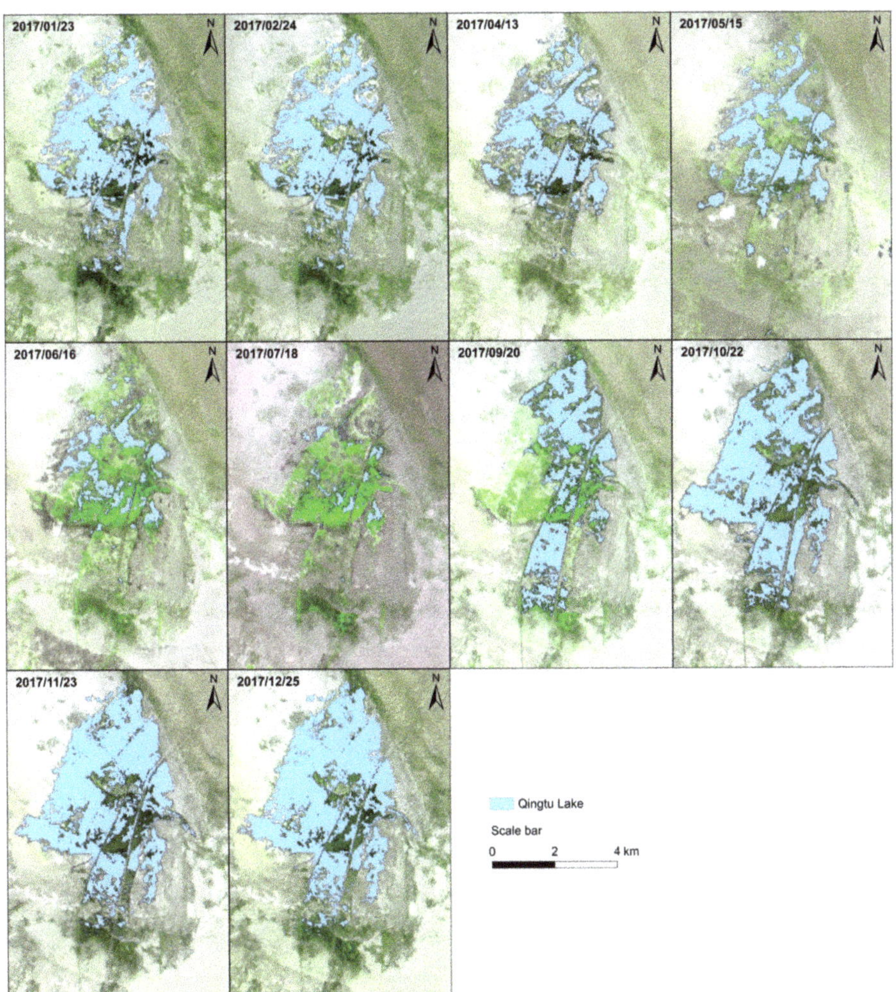

Figure 4. Spatial distribution of the Qingtu Lake water body from January to December 2017. The spatial trend of the lake water body distribution was similar in other years after 2010.

4.2. Variation of Groundwater Level and Soil Water Saturation in Qingtu Lake Wetland Area

4.2.1. Variation of the Groundwater Level

During the frozen period, the groundwater monitoring data loggers at locations G02, G08, and G10 were taken out from the wells in December 2018, and thus the data between December 2018 and May 2019 were missing. As shown in Figure 5, the groundwater level at each monitoring location showed a significant rise after 25 August, and then gradually stabilized. After November, the groundwater level at G05, which is located farther away from the lake center, first declined, and after January, the water level at G04 also showed a downward trend because of the groundwater recharge surface water. After March, the groundwater level at G04 and G05 rose obviously. It is speculated that this phenomenon was caused by the rise of the soil surface temperature, which made the frozen surface soil water melt and recharge the groundwater. In winter, the groundwater at G03, G07, and G09 overflowed the surface, and the groundwater level remained relatively stable for a long time. It can also be seen that the direction of the regional groundwater flow is from G04 to the west and north, and from G9 to G08 and G10.

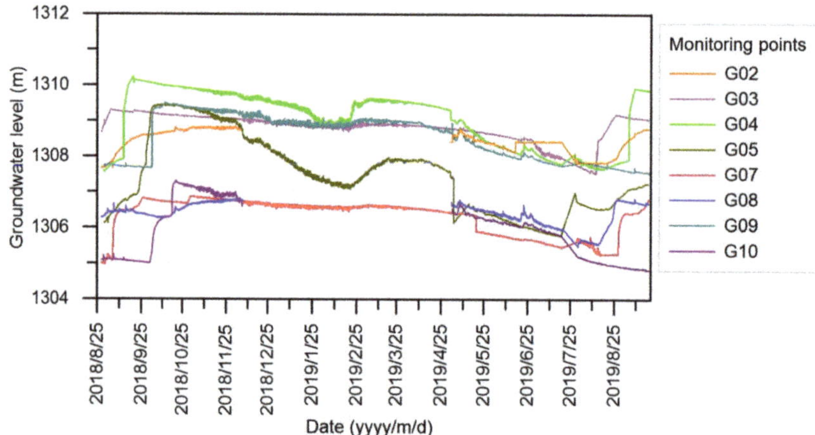

Figure 5. Groundwater-level hydrographs at various monitoring points in the Qingtu Lake wetland area from 2018 to 2019.

Long-term groundwater depth monitoring data at well W01 showed that the groundwater depth gradually decreased from July 2010 to December 2016 under the impact of the water transfer in autumn. The infiltrated surface water greatly increased the groundwater level. Within the same year, the overall groundwater depth fluctuated with greater depth from June to August (Figure 6).

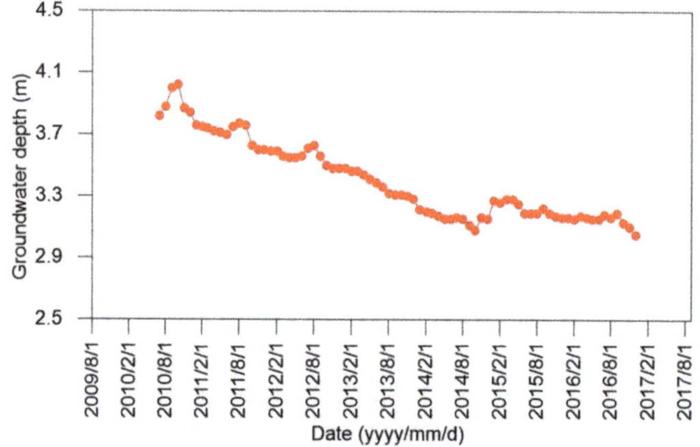

Figure 6. Long-term groundwater depth at location W01 near the Qingtu Lake wetland.

4.2.2. Variation in Soil Water Saturation

The water contents were converted to saturation according to Equation (5) [41]:

$$S_r = \frac{\theta}{n} \tag{5}$$

where Sr represents the soil water saturation, θ is the soil water content, and n is the porosity.

The monitored water contents were normalized by dividing by soil porosity to more clearly show the change in soil water content (Figure 7). The normalized data shown in Figure 7 refer to the soil water saturation. The V02 and V04 locations are close to the center of the lake, and the saturation at V02 and V04 was between 0.10 and 1. The V07, V08, and V09 locations are farther away, and the groundwater overflowed the surface after the water transfer each year. The saturation at V07 was between 0.388 and 1, the saturation at V08 was between 0.155 and 1, and the saturation at V09 was between 0.078 and 1. The V01, V05, V06, and V10 locations are farthest from the lake, and the groundwater could not submerge the surface; the saturation at V01 was between 0.02 and 1, the saturation at V05 was between 0 and 1, the saturation at V06 was between 0 and 0.471, and the saturation at V10 was between 0 and 0.791.

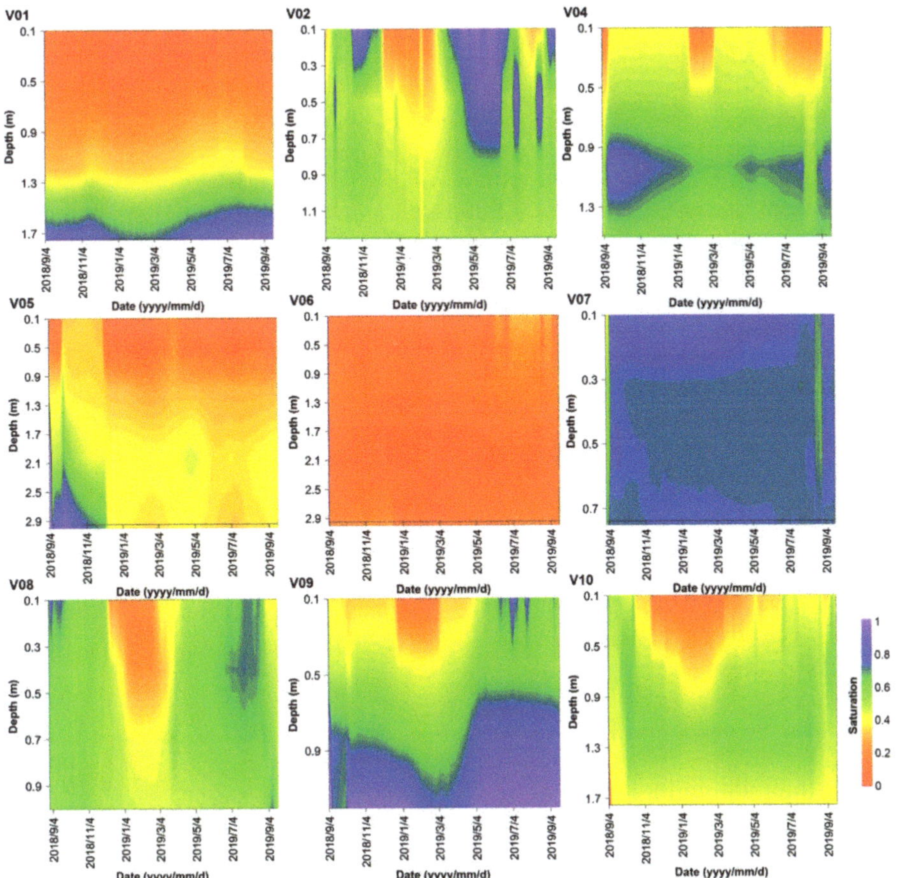

Figure 7. Variation of soil water saturation at various monitoring points.

As is shown in Figure 7, the variation of saturation at each location (except V06) was similar. The saturation of surface soil at each location was low, while the saturation of deep soil at each location was high. The saturation at the V02, V04, V07, V08, and V09 locations, which are closer to the lake, was low from January to March, and increased after March as the surface soil water melted. The conditions described are consistent with the variation characteristics of the groundwater level.

4.3. Variation in Vegetation Coverage around Qingtu Lake Wetland and Mingqin Basin

The vegetation classification of the entire Minqin Basin was calculated by referring to the vegetation coverage classification criteria in Table 1. Due to the influence of agricultural crops in the central basin [42], the interpreted vegetation classification with satellite data may not reflect the restoration of the vegetation cover area caused by water transfer. Therefore, this study interpreted the vegetation cover both in the entire Mingqin Basin and the natural vegetation within 10 km around Qingtu Lake. By comparing and analyzing the vegetation coverage degree within 10 km around Qingtu Lake and the entire Minqin Basin, we found that the characteristics of vegetation coverage were similar, and an obvious seasonal change trend was shown in the vegetation coverage areas for each category. Thus, we only showed the temporal and spatial vegetation coverage change within a 10 km area around Qingtu Lake in Figure 8. Generally, the lowest vegetation cover area occurred from the former November to the latter February, and the highest vegetation cover area occurred from June to September. However, there were some differences in vegetation cover area for each category. For example, the areas of bare rock, bare soil, and water, with the range of 0–10% vegetation coverage, varied little with the seasons, and the area was basically of the same order of magnitude. However, in the range of vegetation coverage above 30%, the vegetation coverage area varied greatly with the seasons, and the order of magnitude was not uniform. Sometimes, a season with an area of 0 may even occur. It is worth noting that, whether considering the entire Minqin Basin or a 10 km area around Qingtu Lake, the vegetation area with vegetation coverage above 70% was abnormal on 15 July 2010 and 18 July 2011. Compared with the data of the same period in other years, the magnitude difference was large. After analysis, it was found that the remote sensing images of these two days were from Landsat 4-5 TM and there were clouds, so it is speculated that the reason for this finding may be cloud interference in the remote sensing images. Meanwhile, the total vegetation coverage within a 10 km area around Qingtu Lake showed an obvious increasing trend since 2009 (Figure 8). The vegetation coverage in each category had increased, and the effect was most obvious in the vegetation coverage between 30 and 50%.

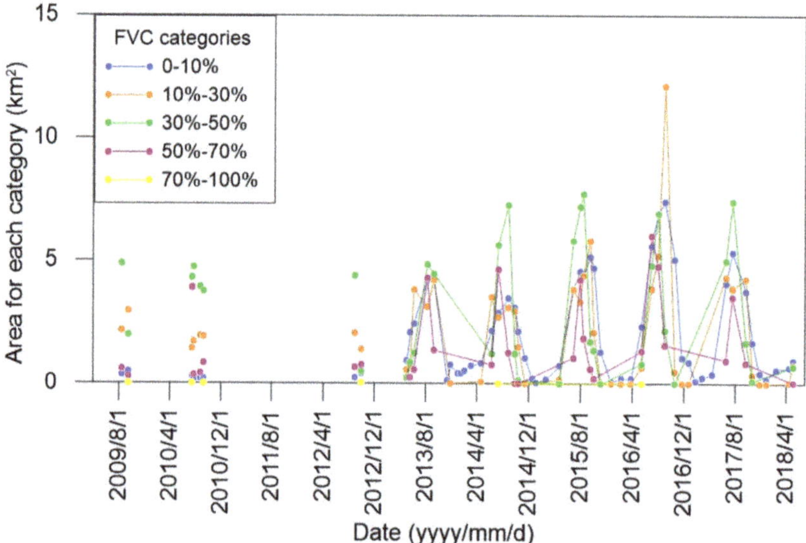

Figure 8. Variations over time of the vegetation coverage for different fractional vegetation coverage (FVC) categories in Qingtu Lake and its surrounding area within 10 kilometers. (Note: There is a lack of Landsat series satellite data from 2009 to 2013, among which there are only two data sets from August and September in 2009, four data sets from July, August and September in 2010, one data set from July, 2011, and two data sets from August and September in 2012).

Figure 9 shows the variation of land use from 1970 to 2017. The obvious changes were in the marshes and low coverage of grass of the natural oasis, which was influenced by the lake area. The area of low coverage grass of the natural oasis increased while the sand area decreased, which was consistent with the variation in vegetation coverage around the Qingtu Lake wetland (Figure 8).

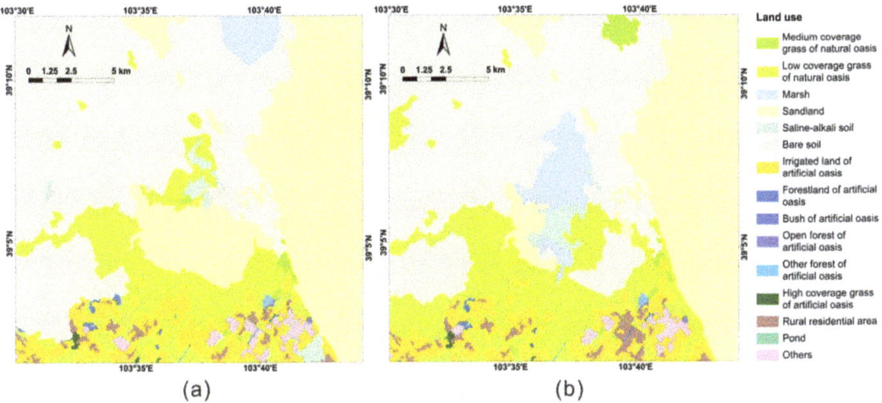

Figure 9. Variation of land use from 1970 (**a**) to 2017 (**b**).

4.4. Spatial Distribution of Vegetation Coverage in Qingtu Lake Wetland Area

The temporal change pattern of vegetation coverage is similar for different years, and thus we only show the spatial distribution of vegetation coverage from January to December 2017 in Figure 10. The areas with high FVC are mainly concentrated in locations close to Qingtu Lake and the south side of the study area, and the FVC of these areas varies greatly with the seasons. In the entire study area, the maximum FVC reached 18.6% from January to April. The maximum FVC reached 67.7% in June. The maximum FVC reached 68.7% in July. The vegetation coverage began to decline after September. In October, the FVC was low, and only sporadic areas had an FVC of more than 30%, while other areas had an FVC of less than 20%. After that, vegetation coverage continued to decline. The FVC, except at Qingtu Lake and the south side of the study area, was lower than 15% for a long time.

5. Discussion

Since September 2010, the water transfer to Qingtu Lake each year began at the end of August [40]. In order to explore the influence of this water transfer on the vegetation growth in the next year around Qingtu Lake, we analyzed the correlation between the water body area of Qingtu Lake in autumn and the maximum vegetation coverage in each FVC range in the following year, combining multiyear data. As shown in Table 2, the correlation coefficients varied among the vegetation areas, with different FVC ranges, and the area of the Qingtu Lake water body. The vegetation area with an FVC of 70 to 100% was small, making it difficult to show any correlation. The lake water body area and vegetation areas in other FVC ranges exhibited a positive correlation, suggesting that the more water transferred into Qingtu Lake in the previous autumn season, the larger the vegetation coverage in the following summer season. Compared with other FVC ranges, the vegetation coverage between 30 and 50% had a stronger correlation with the lake water body area. The correlation between the total vegetation coverage area and lake water body area was also strong, with an overall correlation coefficient of 0.839. As shown by the correlation coefficients, the transferred water had largest impact on the vegetation coverage, from 0 to 50% (Figures 3 and 8). Thus, we concluded that an increase in the vegetation coverage can contribute to the water transfer.

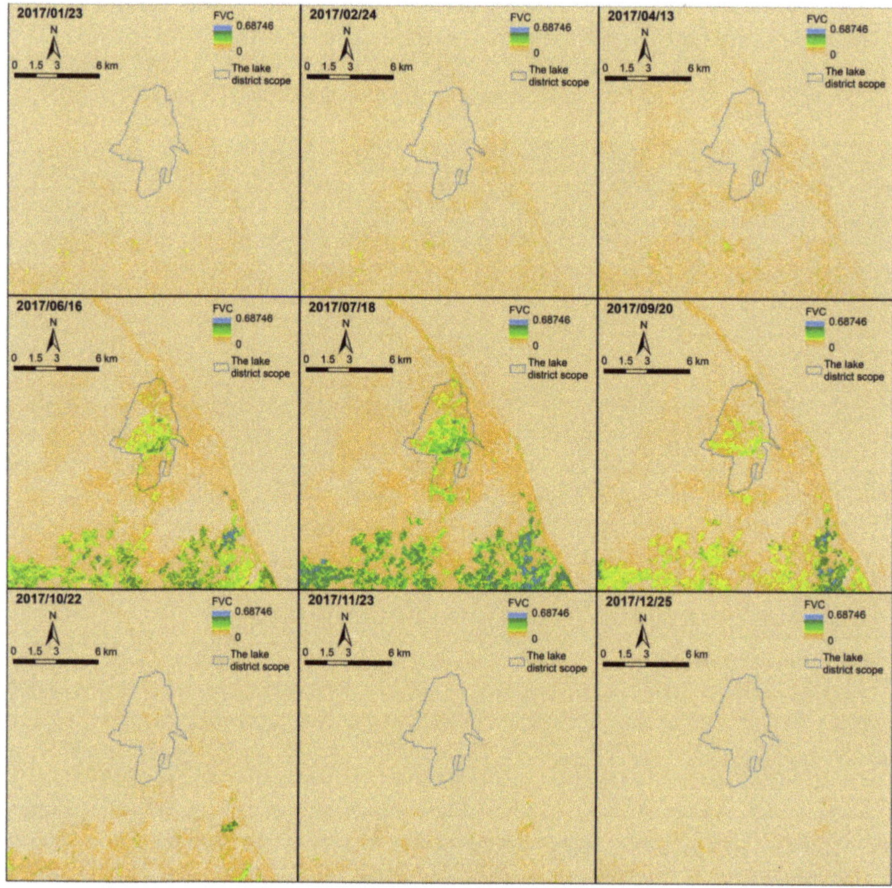

Figure 10. Variation of the spatial distribution of vegetation coverage near the Qingtu Lake wetland.

Table 2. Correlation Coefficients Between Different FVC Ranges and the Surface Area of the Water Body Formed by the Water Transfer.

FVC	0–10%	10–30%	30–50%	50–70%	All
Correlation coefficient	0.853	0.788	0.982	0.437	0.839

The rapid increase of the groundwater level in response to the water transfer suggested that the transfer surface water recharged groundwater (Figure 5). Then, the groundwater level slowly decreased, indicating the dissipation of infiltrated water. Following this, the thawing of frozen water also led to an increase of the groundwater level in spring. Long term, the groundwater-table depth has been slowly recovering since 2010, which could be ascribed to the infiltration from transferred surface water since 2010 (Figure 6). This process is similar to that which occurred in the Tarim River, where the water table near the riverbank was raised by 2–4 m during ecological water transfer during 2001 to 2006 [43]. Many studies reported that the decreased water table led to severe degradation of groundwater-dependent ecosystems in the arid and semiarid area [44], and thus, the increased water table in the study area has benefitted and improved the ecological environment for the vegetation system. A similar conclusion was reported in Heihe River Basin, northwest China [45], where the terminal East Juyan Lake has accumulated 6.19×10^8 m^3 water. The groundwater table rose by an average 0.56 m downstream in the basin. This occurred after receiving ecological water transferred from upper and middle reaches from 2002 until about 2012, which led to an increase in areas of forest and grassland.

Responding to the surface water infiltration and groundwater level increase, increases in soil water content were observed at different monitoring locations, and higher water saturation began to be maintained from April to May (Figure 7). Thus, the soil water can be continuously supplied to the root zone and supports plant growth during spring [46]. In soil that undergoes seasonal freezing and thawing, reduced evaporation and seepage could be beneficial to conserve the soil water and maintain the high water content, and the thawing of frozen soil in spring also increased the water content (Figure 7) [47]. These soil water conditions in arid and semiarid areas are important for supporting the vegetation system. Thus, soil freeze–thaw processes are important for local ecohydrological processes, such as plant germination and growth in spring [48,49]. The transferred waters flowing over the ground surface could also infiltrate, to be either stored as soil water or recharged to groundwater, especially for the shallow distribution of clay soil [50]. The role of fine-textured soil in retaining water recharged by intermittent ecological water conveyances or prior floods as a lasting legacy to sustain riparian plant species over extended drought periods were also reported in the middle Heihe River, China [51] and the lower Rio Grande River, USA [52]. Through the interaction between surface water, groundwater and soil water, ecological water transfer in autumn increased the groundwater level and supported the relatively higher soil water content, providing essential water for vegetation during spring and summer in the following year.

6. Conclusions

To improve the degraded vegetation ecosystem in the arid areas of northwest China, caused by human activities and natural factors, artificial water delivery has become the main method of human intervention. Due to the contradiction between water use for agriculture and vegetation water demand in spring and summer, water transfer is generally carried out in autumn. However, the impact and mechanism of artificial water transfer on vegetation and wetland ecosystem restoration have not been thoroughly investigated, especially in northwest China. Taking the Qingtu Lake wetland system in the lower reach of Shiyang River in northwestern China as the study area, this study illustrated how water transfer affects the surface water, groundwater, and soil water interaction, and improves the vegetation ecosystem.

The surface area of Qingtu Lake expanded to the maximum in autumn and winter after water was transferred, and largely decreased the next summer of each year after 2010. The coverage of vegetation around Qingtu Lake area the next spring and summer also increased significantly each year from 2010. A positive correlation was found between the surface area of the lake body area in autumn and the vegetation coverage in each FVC interval the following summer, suggesting that water transfer improved the vegetation ecosystem in the study area. Further, we also explored the mechanism of how the water transfer in autumn, when the vegetation stops growing, improves the vegetation coverage in the following year. We found that the groundwater level increased after water transfer, and the soil water content increased and remained relatively high for the following months, which suggests that transferred water from upstream can be stored as groundwater or soil water in the subsurface through surface water. These water sources can provide the water for vegetation growth the next spring, or support the plants in the summer. Thus, ecological water transfer plays an important role in improving the vegetation ecosystem, even when the water is transferred in seasons when vegetation does not grow.

Author Contributions: Conceptualization, R.M.; methodology, S.Q.; validation, S.Q., R.M. and Z.S.; formal analysis, S.Q.; investigation, M.G., J.B., J.W., Z.W. and H.N.; resources, R.M. and Z.S.; data curation, M.G. and J.B.; writing—Original draft preparation, S.Q.; writing—Review and editing, R.M. and Z.S.; visualization, S.Q.; supervision, R.M.; project administration, R.M. and Z.S.; funding acquisition, R.M. and Z.S. All authors have read and agreed to the published version of the manuscript.

Remote Sens. **2020**, *12*, 2516

Funding: This research was financially supported by the National Key Research and Development Program of China (2017YFC0406105), National Natural Science Foundations of China (NSFC-41722208 and 41907177), and the Fundamental Research Funds for the Central Universities (CUGL180817).

Acknowledgments: We thank the Institute of Hydrogeology and Environmental Geology, Chinese Academy of Geosciences for their support for field trip and project coordination.

Conflicts of Interest: The authors declare no conflict of interest.

References

1. Hou, X.; Zhang, J.; Huang, T. Rational exploitation of large wellfields based on ecological water demand in arid inland basins. *Fresenius Environ. Bull.* **2016**, *25*, 3003–3011.
2. Xiu, L.; Yan, C.; Li, X.; Qian, D.; Feng, K. Changes in wetlands and surrounding land cover in a desert area under the influences of human and climatic factors: A case study of the Hongjian Nur region. *Ecol. Indic.* **2019**, *101*, 261–273. [CrossRef]
3. Xu, W.; Su, X. Challenges and impacts of climate change and human activities on groundwater-dependent ecosystems in arid areas—A case study of the Nalenggele alluvial fan in NW China. *J. Hydrol.* **2019**, *573*, 376–385. [CrossRef]
4. Yu, X.; Zhuge, Y.; Li, G.; Du, Q.; Zhang, D.; Tan, H.; Zhang, S. A Study of the Ecological Effects of Water Supplement Conditions in Ebinur Lake. *Appl. Ecol. Environ. Res.* **2018**, *16*, 7777–7790. [CrossRef]
5. Brandis, K.J.; Bino, G.; Spencer, J.A.; Ramp, D.; Kingsford, R.T. Decline in colonial waterbird breeding highlights loss of Ramsar wetland function. *Biol. Conserv.* **2018**, *225*, 22–30. [CrossRef]
6. Lopez-Porras, G.; Stringer, L.C.; Quinn, C.H. Corruption and conflicts as barriers to adaptive governance: Water governance in dryland systems in the Rio del Carmen watershed. *Sci. Total Environ.* **2019**, *660*, 519–530. [CrossRef]
7. Mendez-Estrella, R.; Romo-Leon, J.; Castellanos, A.; Gandarilla-Aizpuro, F.; Hartfield, K. Analyzing Landscape Trends on Agriculture, Introduced Exotic Grasslands and Riparian Ecosystems in Arid Regions of Mexico. *Remote Sens.* **2016**, *8*, 664. [CrossRef]
8. Lamsal, P.; Atreya, K.; Ghosh, M.K.; Pant, K.P. Effects of population, land cover change, and climatic variability on wetland resource degradation in a Ramsar listed Ghodaghodi Lake Complex, Nepal. *Environ. Monit. Assess.* **2019**, *191*, 415. [CrossRef]
9. Mutiga, J.K.; Mavengano, S.T.; Zhongbo, S.; Woldai, T.; Becht, R. Water Allocation as a Planning Tool to Minimise Water Use Conflicts in the Upper Ewaso Ng'iro North Basin, Kenya. *Water Resour. Manag.* **2010**, *24*, 3939–3959. [CrossRef]
10. Maleki, S.; Soffianian, A.R.; Koupaei, S.S.; Pourmanafi, S.; Saatchi, S. Wetland restoration prioritizing, a tool to reduce negative effects of drought; An application of multicriteria-spatial decision support system (MC-SDSS). *Ecol. Eng.* **2018**, *112*, 132–139. [CrossRef]
11. Alibrahim, A. Excessive use of groundwater resources in Saudi Arabia: Impacts and policy options. *Ambio* **1991**, *20*, 34–37.
12. Schlesinger, W.H.; Reynolds, J.F.; Cunningham, G.L.; Huenneke, L.F.; Jarrell, W.M.; Virginia, R.A.; Whitford, W.G.J.S. Biological Feedbacks in Global Desertification. *Science.* **1990**, *247*, 1043–1048. [CrossRef] [PubMed]
13. Huang, F.; Chunyu, X.; Zhang, D.; Chen, X.; Ochoa, C.G. A framework to assess the impact of ecological water conveyance on groundwater-dependent terrestrial ecosystems in arid inland river basins. *Sci. Total Environ.* **2020**, *709*, 136155. [CrossRef] [PubMed]
14. Liao, S.; Xue, L.; Dong, Z.; Zhu, B.; Zhang, K.; Wei, Q.; Fu, F.; Wei, G. Cumulative ecohydrological response to hydrological processes in arid basins. *Ecol. Indic.* **2020**, *111*, 106005. [CrossRef]
15. Shi, L.; Tuerxun, H.; Han, G.H. Background, Benefits and Problems of Water Conveyance to the Lower Reaches of the Tarim River. *Bull. Soil Water Conserv.* **2008**, *28*, 176–180.
16. Zhang, Y.; Yu, J.; Qiao, M.; Yang, H. Effects of eco-water transfer on changes of vegetation in the lower Heihe River Basin. *J. Hydraul. Eng.-ASCE.* **2011**, *42*, 757–765.
17. Chen, Z.R.; Liu, S.Z.; Liu, S.J.; Sun, T. Response of Form. Phragmites australis and Form. Nitraria tangutorum after ecological water delivery to Qingtu Lake. *Pratacult. Sci.* **2015**, *32*, 32–40.

18. Zhang, B.; Wang, K.; Li, F. Study of Shiyang River water resources scheduling and Qingtu Lake Ecological Restoration. *Gansu Water Resour. Hydrop. Tech.* **2017**, *53*, 11–14.

19. Shi, W.; Liu, S.; Liu, S.; Yuan, H.; Ma, J.; Liu, H.; An, F. Influence analysis of artificial water transfer on the regional ecological environment of Qingtu Lake in the lower reaches of the Shiyang River. *Acta Ecol. Sin.* **2017**, *37*, 5951–5960.

20. Chunyu, X.; Huang, F.; Xia, Z.; Zhang, D.; Chen, X.; Xie, Y. Assessing the Ecological Effects of Water Transport to a Lake in Arid Regions: A Case Study of Qingtu Lake in Shiyang River Basin, Northwest China. *Int. J. Environ. Res. Public Health.* **2019**, *16*, 145. [CrossRef]

21. Liu, S.; Yuan, H.; Liu, S.; Ma, J. The desert vegetation succession at bank of Qintu Lake with water body formation. *Ecol. Sci.* **2017**, *36*, 64–72.

22. Zhang, X.; Tao, H. Evolution and Conservation Policy of Wetland in Qingtuhu of the Minqin County. *Gansu Sci. Technol.* **2011**, *27*, 7–9.

23. Zhang, Y.; Zhu, G.; Ma, H.; Yang, J.; Pan, H.; Guo, H.; Wan, Q.; Yong, L. Effects of Ecological Water Conveyance on the Hydrochemistry of a Terminal Lake in an Inland River: A Case Study of Qingtu Lake in the Shiyang River Basin. *Water.* **2019**, *11*, 1673. [CrossRef]

24. Zhao, Q. Grain-size characteristics of qingtu lake sediments and its paleaoenvironment explaination. *Arid Land Geogr.* **2003**, *26*, 1–5.

25. Yang, Y.; Li, J.; Chen, F.; Burgess, J.; Li, D.; Chang, G.; Li, Y. The human mechanism research of Minqin Oasis change in the lower reaches of the Shiyang River. *Geogr. Res.* **2002**, *21*, 449–458.

26. Miao, J.; Zhou, X.; Huang, T.-Z.; Zhang, T.; Zhou, Z. A Novel Inpainting Algorithm for Recovering Landsat-7 ETM+ SLC-off Images Based on the Low-Rank Approximate Regularization Method of Dictionary Learning With Nonlocal and Nonconvex Models. *IEEE Trans. Geosci. Remote Sensing.* **2019**, *57*, 6741–6754. [CrossRef]

27. Yang, J.; Zhao, J.; Zhu, G.; Wang, Y.; Ma, X.; Wang, J.; Guo, H.; Zhang, Y. Soil salinization in the oasis areas of downstream inland rivers—Case Study: Minqin oasis. *Quat. Int.* **2020**, *537*, 69–78. [CrossRef]

28. Du, Y.; Zhang, Y.; Ling, F.; Wang, Q.; Li, W.; Li, X. Water Bodies' Mapping from Sentinel-2 Imagery with Modified Normalized Difference Water Index at 10-m Spatial Resolution Produced by Sharpening the SWIR Band. *Remote Sens.* **2016**, *8*, 354. [CrossRef]

29. Yang, X.; Qin, Q.; Grussenmeyer, P.; Koehl, M. Urban surface water body detection with suppressed built-up noise based on water indices from Sentinel-2 MSI imagery. *Remote Sens. Environ.* **2018**, *219*, 259–270. [CrossRef]

30. Xu, H. A study on information extraction of water body with the modified normalized difference water index (MNDWI). *J. Remote Sens.* **2005**, *9*, 589–595.

31. Mamat, A.; Halik, M.; Keram, A.; Song, Z. Remote Sensing Monitoring of Bosten Lake Water Resources and Its Driving Factor Analysis. *Xinjiang Agric. Sci.* **2017**, *54*, 184–192.

32. Liu, D.; Jia, K.; Wei, X.; Xia, M.; Zhang, X.; Yao, Y.; Zhang, X.; Wang, B. Spatiotemporal Comparison and Validation of Three Global-Scale Fractional Vegetation Cover Products. *Remote Sens.* **2019**, *11*, 2524. [CrossRef]

33. Jia, K.; Liang, S.; Gu, X.; Baret, F.; Wei, X.; Wang, X.; Yao, Y.; Yang, L.; Li, Y. Fractional vegetation cover estimation algorithm for Chinese GF-1 wide field view data. *Remote Sens. Environ.* **2016**, *177*, 184–191. [CrossRef]

34. Zhang, X.; Liao, C.; Li, J.; Sun, Q. Fractional vegetation cover estimation in arid and semi-arid environments using HJ-1 satellite hyperspectral data. *Int. J. Appl. Earth Obs. Geoinf.* **2013**, *21*, 506–512. [CrossRef]

35. Zhao, J.; Wang, X.; Li, D. Quantitative analysis on the vegetation coverage changes of Minqin Oasis based on MODIS. *J. Arid Land Resour. Environ.* **2012**, *26*, 91–96.

36. Li, M.; Wu, B.; Yan, C.; Zhou, W. Estimation of Vegetation Fraction in the Upper Basin of Miyun Reservoir by Remote Sensing. *Resour. Sci.* **2004**, *26*, 153–159.

37. Jun, Z.; Jian-xia, Y.; Guo-feng, Z. Effect of Ecological Water Conveyance on Vegetation Coverage in Surrounding Area of the Qingtu Lake. *Arid Zone Res.* **2018**, *35*, 1251–1261.

38. Tao, H.; Wang, D. Change of Vegetation Coverage in Shiyang River Basin 2000–2014. *Chin. Agric. Sci. Bull.* **2017**, *33*, 66–74.

39. Xu, H.; Deng, Y. Dependent Evidence Combination Based on Shearman Coefficient and Pearson Coefficient. *IEEE Access.* **2018**, *6*, 11634–11640. [CrossRef]

40. Zhang, J. Problems and countermeasures of water resources schedule in hongyashan reservoir. *J. Gansu Agric. Univ.* **2015**, 23–24.

41. Ochsner, T.; Horton, R.; Ren, T. Simultaneous water content, air-filled porosity, and bulk density measurements with thermo-time domain reflectometry. *Soil Sci. Soc. Am. J.* **2001**, *65*, 1618–1622. [CrossRef]

42. Xue, X.; Liao, J.; Hsing, Y.; Huang, C.; Liu, F. Policies, Land Use, and Water Resource Management in an Arid Oasis Ecosystem. *Environ. Manage.* **2015**, *55*, 1036–1051. [CrossRef] [PubMed]

43. Chen, Y.; Chen, Y.; Xu, C.; Ye, Z.; Li, Z.; Zhu, C.; Ma, X. Effects of ecological water conveyance on groundwater dynamics and riparian vegetation in the lower reaches of Tarim River, China. *Hydrol. Process.* **2010**, *24*, 170–177. [CrossRef]

44. Wang, Y.; Zheng, C.; Ma, R. Review: Safe and sustainable groundwater supply in China. *Hydrogeol. J.* **2018**, *26*, 1301–1324. [CrossRef]

45. Cheng, G.; Li, X.; Zhao, W.; Xu, Z.; Feng, Q.; Xiao, S.; Xiao, H. Integrated study of the water–ecosystem–economy in the Heihe River Basin. *Natl. Sci. Rev.* **2014**, *1*, 413–428. [CrossRef]

46. Chen, X.; Hu, Q. Groundwater influences on soil moisture and surface evaporation. *J. Hydrol.* **2004**, *297*, 285–300. [CrossRef]

47. Yang, K.; Wang, C. Water storage effect of soil freeze-thaw process and its impacts on soil hydro-thermal regime variations. *Agric. For. Meteorol.* **2019**, *265*, 280–294. [CrossRef]

48. Wang, T.; Li, P.; Li, Z.; Hou, J.; Xiao, L.; Ren, Z.; Xu, G.; Yu, K.; Su, Y. The effects of freeze-thaw process on soil water migration in dam and slope farmland on the Loess Plateau, China. *Sci. Total Environ.* **2019**, *666*, 721–730. [CrossRef]

49. Yi, J.; Zhao, Y.; Shao, M.a.; Zhang, J.; Cui, L.; Si, B. Soil freezing and thawing processes affected by the different landscapes in the middle reaches of Heihe River Basin, Gansu, China. *J. Hydrol.* **2014**, *519*, 1328–1338. [CrossRef]

50. Duan, L.; Huang, M.; Zhang, L. Use of a state-space approach to predict soil water storage at the hillslope scale on the Loess Plateau, China. *Catena.* **2016**, *137*, 563–571. [CrossRef]

51. Sun, Z.; Long, X.; Ma, R. Water uptake by saltcedar (Tamarix ramosissima) in a desert riparian forest: Responses to intra-annual water table fluctuation. *Hydrol. Process.* **2016**, *30*, 1388–1402. [CrossRef]

52. Moore, G.; Li, F.; Kui, L.; West, J. Flood water legacy as a persistent source for riparian vegetation during prolonged drought: An isotopic study of Arundo donax on the Rio Grande. *Ecohydrology* **2016**, *9*, 909–917. [CrossRef]

remote sensing

Article

Application of the GPM-IMERG Products in Flash Flood Warning: A Case Study in Yunnan, China

Meihong Ma [1,2,3]**, Huixiao Wang** [1,3,*]**, Pengfei Jia** [4,†]**, Guoqiang Tang** [5,6]**, Dacheng Wang** [7]**, Ziqiang Ma** [8,9] **and Haiming Yan** [10]

[1] College of Water Sciences, Beijing Normal University, Beijing 100875, China; mamh@tjnu.edu.cn
[2] School of Geographic and Environmental Sciences, Tianjin Normal University, Tianjin 300387, China
[3] Beijing Key Laboratory of urban hydrologic cycle and sponge City Technology, Beijing 100875, China
[4] CITIC Construction Co., Ltd., Beijing 100027, China; jiapf@citic.com
[5] Center for Hydrology, University of Saskatchewan, Saskatoon, SK S7N 1K2, Canada; guoqiang.tang@usask.ca
[6] Coldwater Laboratory, University of Saskatchewan, Canmore, AL T1W 3G1, Canada
[7] Lab of Spatial Information Integration, Institute of Remote Sensing and Digital Earth, Chinese Academy of Sciences, Beijing 100101, China; wangdc@radi.ac.cn
[8] School of Earth and Space Sciences, Peking University, Beijing 100871, China; ziqma@zju.edu.cn
[9] State Key Laboratory of Resources and Environmental Information System, Beijing 100871, China
[10] School of Land Resources and Urban & Rural Planning, Digital Territory Experiment Center, Hebei GEO University, Shijiazhuang 050031, China; Haiming.yan@hgu.edu.cn
* Correspondence: huixiaowang@bnu.edu.cn
† Co-first author: Pengfei Jia.

Received: 26 April 2020; Accepted: 15 June 2020; Published: 17 June 2020

Abstract: NASA's Integrated Multi-Satellite Retrievals for Global Precipitation Measurement (IMERG) is a major source of precipitation data, having a larger coverage, higher precision, and a higher spatiotemporal resolution than previous products, such as the Tropical Rainfall Measuring Mission (TRMM). However, there rarely has been an application of IMERG products in flash flood warnings. Taking Yunnan Province as the typical study area, this study first evaluated the accuracy of the near-real-time IMERG Early run product (IMERG-E) and the post-real-time IMERG Final run product (IMERG-F) with a 6-hourly temporal resolution. Then the performance of the two products was analyzed with the improved Rainfall Triggering Index (RTI) in the flash flood warning. Results show that (1) IMERG-F presents acceptable accuracy over the study area, with a relatively high hourly correlation coefficient of 0.46 and relative bias of 23.33% on the grid, which performs better than IMERG-E; and (2) when the RTI model is calibrated with the gauge data, the IMERG-F results matched well with the gauge data, indicating that it is viable to use MERG-F in flash flood warnings. However, as the flash flood occurrence increases, both gauge and IMERG-F data capture fewer flash flood events, and IMERG-F overestimates actual precipitation. Nevertheless, IMERG-F can capture more flood events than IMERG-E and can contribute to improving the accuracy of the flash flood warnings in Yunnan Province and other flood-prone areas.

Keywords: flash flood; Integrated Multi-Satellite Retrievals for Global Precipitation Measurement; Rainfall Triggering Index; Yunnan

1. Introduction

Flash floods, triggered by heavy rainfall (i.e., short duration, high intensity), are the rapid flooding of water within minutes up to several hours in small basins (hundred square kilometers or less) [1]. It is one of the most devastating floods in the world, which can cause great economic losses and casualties. From 1 October 2007 to 1 October 2015, the National Weather Service (NWS) of the U.S. reported that

278 people died from flash floods, 10% of which resulted in an average property loss of over $100,000 [2]. China is also suffering from severe flash floods, where 984 people have died from flash floods every year on average since 1950 [3]. With the increase in global precipitation and rapid economic development, the frequency and impacts of flash floods will be further exacerbated [4]. The most crucial approach to adapt to flash floods is accurate warnings, which can leave people with more time to respond to these emergencies. However, flash flood early warning remains challenging, especially given the conditions of a short lead time (less than 1–3 h). Accurate and continuous precipitation datasets with high spatial (i.e., 1–4 km) and temporal resolutions (i.e., 5 min to hourly) are critical to the success of flood warnings [5]. Considering climate change and land degradation processes, new tools for flood disaster monitoring and reduction are strongly required. Satellite precipitation products have wide coverage, a high spatiotemporal resolution, and easy data acquisition, and are not restricted by terrain conditions. These products are extremely important to flood warnings in mountainous areas prone to flash floods, where there is scarcely measured rainfall data, and some satellite precipitation products have been used to assess the floods over the basins. However, satellite precipitation products may underestimate extremely strong precipitation, and the accuracy of the satellite precipitation products in catching flash floods remains poorly understood [6,7].

Nowadays, numerous satellite precipitation products, such as the Tropical Rainfall Measuring Mission (TRMM) and Integrated Multi-Satellite Retrievals for Global Precipitation Measurement (GPM) Mission (IMERG), are available for providing post-real-time (PRT) estimation and near-real-time (NRT) estimation. PRT products generally undergo ground-based gauge adjustments while NRT products do not, indicating the former is usually more accurate than the latter [8]. These satellite precipitation products generally have high spatiotemporal resolutions (finer than a 0.25° spatial resolution and shorter than the daily temporal interval) with a wide quasi-global coverage (broader than the 50° N–50° S latitude band); these products are very useful for hydrological studies, especially in data-sparse or ungauged basins [9]. GPM is a new generation of satellite products inheriting TRMM satellite products, with more comprehensive data and higher spatiotemporal accuracy. IMERG is the Level 3 precipitation estimation algorithm of GPM, which can combine gauged rainfall data and satellite data to obtain global rainfall data. IMERG provides three types of products, including the NRT IMERG Early run (hereafter IMERG-E), Late run (hereafter IMERG-L) and the PRT IMERG Final run (hereafter IMERG-F) [10]. Previous studies have proclaimed that IMERG-F has higher accuracy, especially on land, while the IMERG-E and IMERG-L products have better timeliness, which is attractive for flood prediction and monitoring. However, due to the limitations in observation position, density, topography, etc., it is difficult to directly analyze the actual spatial distribution of rainfall with gauged rainfall data [11]. Meanwhile, satellite products also contain uncertainties from instruments, sampling, and retrieval algorithms, and therefore should be comprehensively validated using local gauge data [12].

An appropriate flash flood warning method is another key factor to ameliorate the flash flood warning accuracy. Typical examples include Flash Flood Guidance (FFG), Soil Water Index (SWI), Rainfall Triggering Index (RTI), etc. [13]. FFG is one of the most widely used methods developed by the US River Forecasting Center in the early 1970s, and it calculates the rainfall required to produce bank-full flood conditions associated with flash floods in a given time and area. The calculation steps of the FFG are as follows: (1) gaining the current soil moisture with the hydrological model; (2) reversing the flood peak flow of the basin outlet section; and (3) obtaining the critical rainfall required when the flow reaches the early warning flow [14]. However, FFG has more data requirements since it considers almost all influencing factors. For example, the soil moisture index is an indicator that can accurately describe the trend in soil moisture in the aeration zone, which is mainly calculated from the total water depth of a three-layer tank model with fixed parameters, and FFG has only one parameter related to the infiltration time; however, it is difficult to obtain this parameter regardless of concentration calculation or distribution calculation [15]. The RTI comprehensively considers the effective accumulated rainfall and rainfall intensity in the prediction of flash floods. It focuses on antecedent conditions and has been put into practice for over 10 years in Taiwan [13]. However, this method relies too much on rainfall stations, fails to fully consider intermittent rainfall, and is greatly affected by rainfall field segmentation. Meanwhile, using the deduction coefficient of "t" days in the RTI model for the antecedent rainfall calculation will result in a higher false alert rate under some rainfall patterns. Therefore, Chen et al. (2017) proposed an improved RTI to calculate the antecedent rainfall and effective accumulated rainfall to solve the abovementioned problems, which achieved good practical application effects in Taiwan's 2017 disaster warning [16].

As satellite precipitation data is widely used in large-scale watershed hydrological simulation or land surface process simulation, it is still difficult to meet the needs of flash flood prediction in small and medium-sized watersheds. Most of the existing research focus on the corrected products, but there scarcely has been application verification for real-time products. Most of the satellite products are applied on a large scale, and little attention has been paid to their applicability in small-scale flash flood warnings. Additionally, flash floods generally occur in small and medium-sized river basins with poor economic conditions, low station network coverage density, and there is great difficulty in obtaining data. Given that IMERG products have high spatiotemporal precision, it is therefore of great practical significance to evaluate the applicability of these products to flash flood warnings. Moreover, regional studies are also very common and popular because precipitation exhibits strong spatial variations and different products show varying performance over different regions. Meanwhile, there are still few studies on the inter-comparison of the IMERG products, especially in China, where the superiority of its application in flash flood warnings still needs further exploration [17]. Therefore, taking Yunnan Province in China as the study area, based on two fifth-generation IMERG products (IMERG-E, IMERG-F) and China Meteorological Administration (CMA) data, this study first evaluated these two products with particular attention paid to their systematic and random errors. Then, the empirical RTI method was utilized to evaluate the applicability of the different IMERG products in flash flood warnings. This paper is further organized as follows: Section 2 describes the materials and data; Section 3 presents an assessment of IMERG-E and IMERG-F based on locally measured data and analyzes the effects of flash flood warnings based on satellite precipitation data; and the conclusions are summarized in Section 4.

2. Materials and Data

2.1. Study Area

Yunnan Province is located at the low latitude plateau of southwest China, with an area of 390,000 km^2, accounting for 4.1% of China's total area [18]. The terrain in Yunnan Province is very complex, coupled with the low economic level of remote mountain areas, resulting in the lack of early disaster data. For example, the main terrains are mountains (84%), plateaus and hills (10%), and basins and valleys (6%). The altitude is high in the northwest and low in the southeast, descending stepwise from north to south in this region, and the average altitude is 1980 m, and 87.21% of the land is located in middle-latitude areas (1000–3500 m). The area with a slope below 25° accounts for 56.46% of the national total area. Besides, Yunnan Province is characterized by the most widely distributed karst topography, accounting for 28.9% of the province's area; this results in low water retention and frequent flood disasters. Yunnan is rich in water resources, but unevenly distributed in time and space, where more than 70% of the water resources are concentrated in remote mountainous areas. The total population is 47.1 million in 2014, accounting for 3.5% of the national total population. Yunnan has a typical monsoon climate, mainly dominated by the east Asian monsoon and southwest monsoon, and 90% of the precipitation is concentrated from May to October, especially in the flood season (June to August). For example, long-term studies have shown that the critical rainfall is 35–200 mm in northwest Yunnan, 50–200 mm in southwest Yunnan, and 100–300 mm in eastern Yunnan [19]. The above reasons led to frequent flash flood disasters, which has caused extremely serious disasters. For example, 72 people died from flash floods in Yunnan Province in 2014, accounting for 22.2% of the national total deaths from flash floods. All the preliminary works have promoted the development of flash flood prevention in Yunnan Province [20], but further exploration is still needed, and therefore Yunnan Province was selected as the study area in this study. Figure 1 shows the map of the topography and flash flood disaster distribution in Yunnan Province.

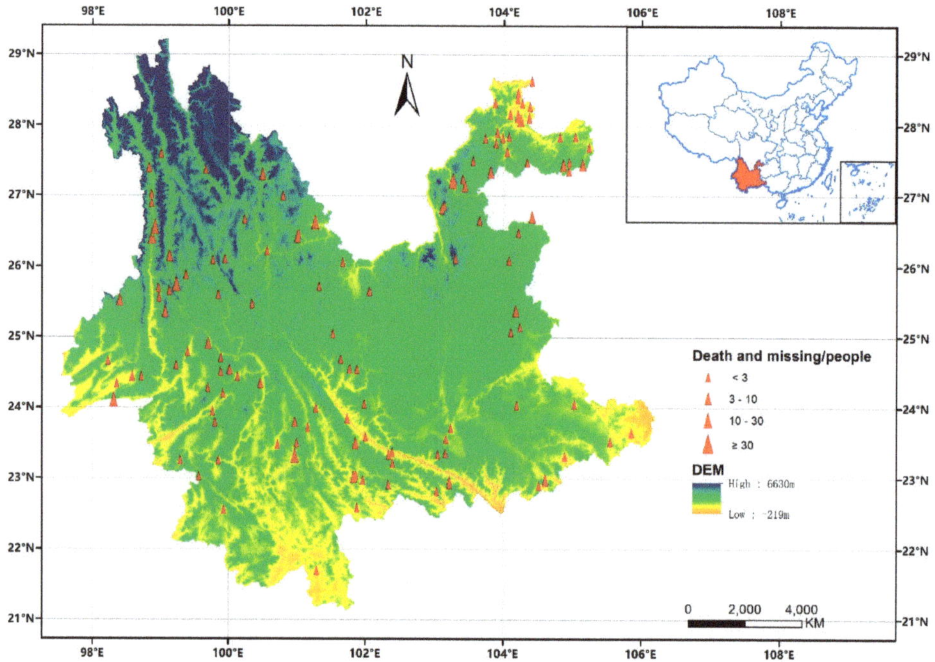

Figure 1. Topographical map and locations of flash flood disasters in Yunnan Province (2011–2018).

2.2. Data

2.2.1. Satellite Data

TRMM Multi-Satellite Precipitation Analysis (TMPA) (50° N–50° S) can retrieve microwave–infrared satellite precipitation estimates with gauge adjustments and can generate rational precipitation estimates at fine spatiotemporal scales (0.25° × 0.25° and 3-hourly temporal intervals) within the global scope (60° N–60° S) [21,22]. As the successor of TRMM, the GPM constellation consists of a core observation platform and 10 cooperative satellites to observe global precipitation through inter-satellite cooperation. The GPM program offers three different levels of data products, all of which are available on the NASA website (http://www.nasa.gov/mission_pages/GPM). This study focuses on the GPM Level 3 products based on the IMERG algorithm. IMERG not only has a high spatial resolution (0.1° × 0.1°, 0.5 h) and fully global coverage (60° S~60° N), but also predicts flood disasters by reducing the uncertainty associated with short-term precipitation accumulation. At present, IMERG data has been developed from the first version to the sixth version, and IMERG-V06B is the latest version of the satellite rainfall data. However, Mohammad et al. (2019) verified that IMERG V05 performs better than V06, and therefore this study has employed IMERG V05 to estimate the precipitation data [23].

IMERG has three different products: Early, Late, and Final. In the real-time phase, the IMERG data generation system generates Early products after running once and then generates Late products after running again. The main difference between them is that the Early product is generated only by the forward propagation algorithm in the cloud mobile vector propagation algorithm, while the Late product is added with the backpropagation algorithm. The time-lag product Final introduces more sensor data sources based on the Late product. Meanwhile, the time of the extracted satellite data is matched with the station's data (08:00 a.m.) [24]. Table 1 shows the detailed information of the three IMERG products, with a study period of 2015–2018. Since IMERG-E and IMERG-L products are NRT products, they are released after 4 h and 12 h after observation, respectively; in turn, the IMERG-F is a PRT product, which has been calibrated for the deviation of the ground rainfall station, so it has a high accuracy and is usually released after two months of observation. Therefore, this study has adopted CMA data as calibration data, pays attention to the accuracy of the IMERG rainfall data to capture flash flood disasters, and combines the precipitation data estimated by IMERG-E and IMERG-F to discuss the accuracy of capturing flash flood, and thereby contributes to obtaining the static early warning thresholds.

Table 1. List of Integrated Multi-Satellite Retrievals for Global Precipitation Measurement (IMERG) products.

	IMERG-E	IMERG-L	IMERG-F
Spatiotemporal resolution	0.1°, 0.5 h	0.1°, 0.5 h	0.1°, 0.5 h
Lag time	6 h	18 h	4 Month
Monitoring range	90° N~90° S	90° N~90° S	90° N~90° S
Data period	Mar. 2015–Dec. 2018	Mar. 2015–Dec. 2018	Mar. 2015–Dec. 2018

2.2.2. Ground Observation Precipitation Data

The regional hourly precipitation integration products of the CMA (http://cdc.cma.gov.cn) is a reference product, which is based on the hourly precipitation observed by more than 30,000 automatic weather stations nationwide and the Climate Prediction Center morphing (CMORPH) satellite. Using the probability density function and optimal interpolation method, the precipitation fusion products with a high spatiotemporal resolution (0.1° × 0.1° and 1-h interval) have been generated in China. Among them, the optimal interpolation method is to combine the semi-hourly infrared data of the Geostationary Earth Orbit (GEO) satellite to interpolate the Passive microwave (PMW) inversion data, and then obtain the relatively fine precipitation. The gauge data involved in the CMA are subject to extreme quality control, including consistency checks, both internal and spatiotemporal. Meanwhile, the complex terrain, the low economic level in remote mountainous areas, and the sparse distribution of the ground stations have led to large errors in the measured data. The CMA data has a wide coverage and high spatiotemporal resolution; the overall error of this product is within 10%, and the error for heavy precipitation and site sparse areas is within 20%, which is more accurate than other similar products in China. Therefore, CMA products are suitable as the calibration data for product evaluation, but their scale should be consistent with the IMERG product in the calibration process [25]. Besides, when using CMA data to estimate the rainfall distribution, we should consider that the dataset may have a delayed phenomenon since it is actually a fusion of measured data and CMORPH satellite rainfall data.

The rainfall level is defined by the CMA, for example, the intraday rainfall is categorized into the rain and light-moderate rain (0–25 mm), heavy rain (25–50 mm), and rainstorm (>50 mm). Meanwhile, the minimum amount of precipitation that the gauges can measure is 0.1 mm. The critical rainfall in Yunnan Province mainly comes from the China rainstorm parameters atlas, which describes the statistical characteristics and laws of China's rainstorms. Based on the study period 2015–2018, the flash flood data were mainly from officially published data. In this study, a total of 120 flash flood events were analyzed, all of which resulted in deaths or missing of people. Additionally, some other data were also employed; for example, the topographic data came from a 1:250,000 digital elevation model (DEM) of the Yunnan Province.

2.3. Methodology

2.3.1. RTI Method

Since 2005, the RTI model has been widely applied in flash flood warnings, which effectively reduced the casualties caused by flash floods. Considering that flash floods are mainly caused by hourly peak rainfall, this method is used to predict flash flood by multiplying the effective cumulative rainfall (R_t) and rainfall intensity (I). Among them, the effective accumulated rainfall mainly uses the accumulated rainfall in the previous 7 days before the flash flood occurs. Based on historical rainfall, this method obtains the flash flood probability under different rainfall conditions by calculating the RTI, and then divides the critical rainfall map into three regions for early warning (low possible, medium, and high) [13]. Meanwhile, it also defines how to segment the rainfall events, i.e., the start time is defined as the rainfall per hour exceeding 4 mm, and the end time is the rainfall per hour falling below 4 mm for six consecutive hours. A single day refers to the period from yesterday at 08:00 a.m. to today at 08:00 a.m. Rainfall data were selected according to the flash flood disaster events. If the flash flood disaster occurs in an area without a monitoring station, calculations are performed using data from the nearest rainfall stations (within a radius of 50 mm). The RTI equations are as follows:

$$RTI_t = I * R_t \tag{1}$$

$$R_t = \sum_{i=0}^{n} \alpha^i R_i \tag{2}$$

where RTI_t is the RTI calculated at time t; I is the rainfall intensity; R_t is the effective cumulative rainfall; i means the antecedent day number from one to n; α is the rainfall attenuation coefficient, mainly taken from the measured value of 0.78 by Cui Peng in Jiangjiagou, Yunnan Province [26]; α^i is the reduction factor of the previous i day; and R_i is the 24-h cumulative rainfall of the previous i-day, where the initial cumulative rainfall R_0 is 50 mm, which is mainly obtained through actual statistical analysis.

Since the previous rainfall was calculated using "t" days of accumulated rainfall, the false alarms rate is higher in certain rainfall patterns (e.g., long-term duration and low rainfall intensity). Besides, this method does not consider the effects of intermittent rainfall and rainfall segmentation, etc.; all of the above results in low accuracy. Therefore, based on the reduction period and the reduction coefficient being unchanged, Chen et al. (2018) proposed an improved RTI method and provides a detailed flowchart describing the method [16]. The specific equation of this method is

$$R_t = I_t + R_{t-1} * (\alpha)^{\frac{1}{24}} \tag{3}$$

where I_t is the current rainfall intensity at time t (mm/h); and R_{t-1} is the effective cumulative rainfall one hour earlier. In Formula (2), each operation needs to calculate R_i separately, multiply it by α^i, and then accumulate; however, in Formula (3), it has only one rainfall intensity and one accumulated rainfall, which greatly reduces the calculation amount and contributes to the future subsequent large-scale grid operations.

2.3.2. Evaluation Metrics

Six indicators were employed to evaluate the accuracy of the satellite precipitation products. Table 2 shows the formulas and optimal values of these indexes. CC (correlation coefficient) represents the correlation between the satellite precipitation data and site precipitation data, the greater the better; BIAS (Relative Bias) and RMSE (Root Mean Square Error) are quantitative indicators, representing the deviation degree between the satellite precipitation data and ground station precipitation data. The method also includes three classification indicators: Probability Of Detection (POD), False Alarm Ratio (FAR), and Critical Success Index (CSI), which can comprehensively reflect the ability of the precipitation products to estimate the probability of precipitation event occurrence. Among them,

POD is the correct forecast rate, while FAR is the wrong forecast rate. The critical success index (CSI) is the function of POD and FAR that provides a more balanced estimate of the satellite products [27].

Table 2. List of the formulas and optimal values of the indexes.

Diagnostic Statistics	Equation	Optimum Value	Value Ranges	Unit
CC	$CC = \dfrac{\sum\limits_{i=1}^{n}(D_i-\overline{D})(M_i-\overline{M})}{\sqrt{\sum\limits_{i=1}^{n}(D_i-\overline{D})^2\sum\limits_{i=1}^{n}(M_i-\overline{M})^2}}$	1	$(-1,1)$	-
RMSE	$RMSE = \sqrt{\dfrac{1}{n}\sum\limits_{i=1}^{n}(D_i-M_i)^2}$	0	$(0,\text{Inf})$	mm
BIAS	$BIAS = \dfrac{\frac{1}{n}\sum\limits_{i=1}^{n}(D_i-M_i)}{\sum\limits_{i=1}^{n}M_i}\times 100\%$	0	$(-\text{Inf},\text{Inf})$	%
POD	$POD = \dfrac{O}{O+U}$	1	$(0,1)$	-
FAR	$FAR = \dfrac{P}{O+P}$	0	$(0,1)$	-
CSI	$CSI = \dfrac{O}{O+U+P} = \dfrac{1}{1/(1-FAR)+1/POD-1}$	1	$(0,1)$	-

Variables: n, total number of flash floods; D_i, the i-th of the evaluated; M_i, the i-th of the reference data; $\overline{D}$, mean of D_i; $\overline{M}$, mean of M_i; O, number of hits; U, number of misses; P, number of false positive.

2.3.3. Systematic or Random Error

Random error is an avoidable error caused by measurement or calculation. Systematic error is the unavoidable error caused by the experimental instrument or accuracy. Systematic and random errors are determined by objective and subjective factors, respectively, both of which can be reduced but not eliminated [28]. The spatiotemporal variability of precipitation, measurement error, and the uncertainty of sampling affects the satellite precipitation data accuracy, where the uncertainties include systematic and random errors (hereafter Syst and Rand), mainly from (1) sensor observation; (2) the algorithm used to estimate rainfall; and (3) sampling error. Reference [29] developed a method for estimating Syst and Rand for satellite precipitation products. The system of mean square error (MSE) and the formula for random error are

$$MSE = MSE_S + MSE_R \tag{4}$$

$$\frac{\sum_n(x-y)^2}{n} = \frac{\sum_n(\hat{x}-y)^2}{n} + \frac{\sum_n(x-\hat{x})^2}{n} \tag{5}$$

where x is the satellite precipitation; y is the CMA precipitation; and n is the time steps number (here days); the formula for calculating $\hat{x}$ is as follows:

$$\hat{x} = ay + b \tag{6}$$

where a and b are the slope and intercept parameters that need to be calibrated, respectively. The calculation formula is

$$Syst = \frac{\sum_n(\hat{x}-y)^2}{n} \Big/ \frac{\sum_n(x-y)^2}{n} \tag{7}$$

$$Rand = \frac{\sum_n(x-\hat{x})^2}{n} \Big/ \frac{\sum_n(x-y)^2}{n} \tag{8}$$

3. Results and Discussion

3.1. Spatial Distribution of Precipitation

Figure 2 shows the spatial distributions of the daily average precipitation in Yunnan Province captured by IMERG-E, IMERG-F, and CMA from March 2015 to December 2018. Using CMA data as the reference calibration data, the rainfall distribution in Yunnan Province is shown in Figure 2a. In Yunnan Province, the precipitation increases from northeast to southwest, with the largest precipitation occurring in the western border region. Referring to Figure 1, Yunnan's terrain is complex and changeable, with a high northwest and low southeast, descending stepwise from north to south. Moreover, flash floods are mainly concentrated in the southwest region and densely distributed in parts of the northeast, reflecting that the flash flood disaster is mainly affected by multiple factors with heavy rainfall as the main trigger. Meanwhile, there is a clear precipitation zone in the western region, with the smallest precipitation in the northwest and more precipitation in the central-western region. Figure 2b,c presents the distribution of the estimated precipitation in Yunnan Province using IMERG-E and IMERG-F, respectively. The overall trend is consistent with the precipitation distribution measured by CMA; that is, the estimated precipitation is relatively large in the southwest area, while low in the northeast relatively. Among them, the estimated rainfall of IMERG-E is less in the northeast region and has a larger coverage area, which is lower than that of CMA. The estimated maximum rainfall area is consistent with the CMA, but its coverage area is much smaller than the CMA. Therefore, IMERG-E underestimates the precipitation, especially in the southwest region. IMERG-F is the opposite of IMERG-E; its estimated rainfall covers a larger area, and the minimum and maximum rainfall are higher than the CMA. IMERG-F overestimates precipitation, especially in the central and western regions.

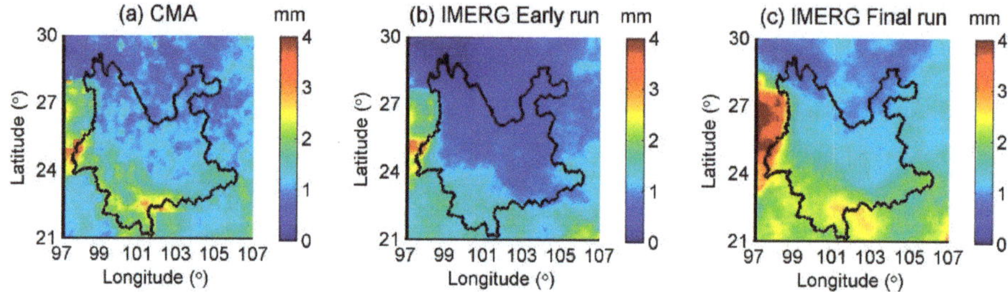

Figure 2. Spatial distributions as per the CMA (**a**), IMERG-E (**b**), and IMERG-F (**c**).

3.2. Evaluation of IMERG-E and IMERG-F

Figure 3 presents the spatial patterns of six indicators obtained from IMERG-E and the CMA using hourly data over the Yunnan Province. Generally, CC was relatively low, changing from 0.2 to 0.5, especially over the northwest regions (CC < 0.2) with the highest altitudes. As for the RMSE and BIAS, their spatial patterns are similar, with an increasing trend for the RMSE and BIAS and a decreasing trend for relative error from northwest to southeast. Many previous studies have confirmed that satellite precipitation estimates are usually low with large errors in mountainous areas. For example, compared with the highest regions in the Northwest, IMERG-E's BIAS is underestimated by about 50% [30]. As for satellite-based precipitation metrics, there is also a significant trend; that is, POD and CSI are increasing, while FAR is decreasing, which is in harmony with the overall trend of the IMERG-E data. Table 3 shows the evaluation metrics for IMERG calculated with the mean value at hourly and daily timescales; the evaluation index calculated by the mean is larger on a daily scale than that on an hourly scale. Given the relatively low accuracy of IMERG-E, the differences between the IMERG-E and CMA were relatively obvious.

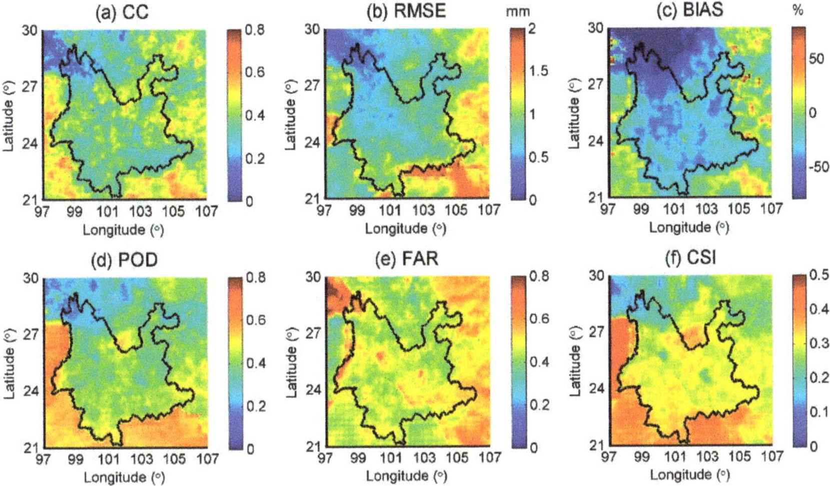

Figure 3. The spatial patterns of six metrics (CC (**a**), RMSE (**b**), BIAS (**c**), POD (**d**), FAR (**e**), and CSI (**f**)) generated from IMERG-E and CMA using hourly data in Yunnan Province.

Table 3. Calculation results for evaluating IMERG.

Timescale	Product	CC	RMSE (mm)	BIAS (%)	POD	FAR	CSI
Hourly	IMERG-E	0.41	1.1	−0.69	0.39	0.53	0.27
	IMERG-F	0.46	0.97	23.33	0.43	0.51	0.29
Daily	IMERG-E	0.63	7.59	−1.18	0.63	0.38	0.45
	IMERG-F	0.73	5.74	28.24	0.66	0.34	0.48

The spatial patterns of the six metrics derived from IMERG-F and CMA (Figure 4) exhibited similar trends as those from IMERG-E and CMA. Table 2 revealed the four indicators CC, BIAS, POD, and CSI of NRL IMERG-E products are lower than these for PRL IMERG-F, regardless of daily and hourly data, while the two indicators RMSE and FAR are higher than that for IMERG-F. Therefore, IMERG-F indicated a better performance in this region. For instance, the POD and CSI of IMERG-F and CMA were also overall higher than those of IMERG-F and CMA, especially for CSI. Besides, in both Figures 3 and 4, the POD and CSI in these two figures show significant differences inside and outside of the Yunnan Province, which is mainly due to the complex and changeable terrain that induces large systematic errors in satellite precipitation. Besides, there is no actual rainfall measurement site outside the Yunnan border, and the rainfall distribution obtained only by interpolation appears discontinuous.

The following is a further analysis of the error source (i.e., systematic error or random error) of EMERG-E. The spatial patterns of systematic and random errors in IMERG-E at the 1 and 24 h temporal scales in Yunnan Province, respectively, demonstrated almost the same spatial patterns (Figure 5). Overall, the system error is mainly distributed in the northwest region of Yunnan Province with a high altitude (about 80%), and the coverage area is relatively small. Meanwhile, except for the northwestern part of Yunnan Province, the systematic error at 1 h is significantly higher than that at 24 h, especially the error at 24 h mostly disappeared. Random error mainly occurs in the southern region of Yunnan Province with relatively high rainfall (about 80%), but a relatively low in the northwest region. Besides, the random error is higher at 24 h than at 1 h. Therefore, for the IMERG-E data, the estimated error in the high-altitude and low-precipitation areas of Yunnan Province is mainly determined by the systematic error; the southern area of Yunnan Province with a low elevation and high precipitation is dominated by random errors.

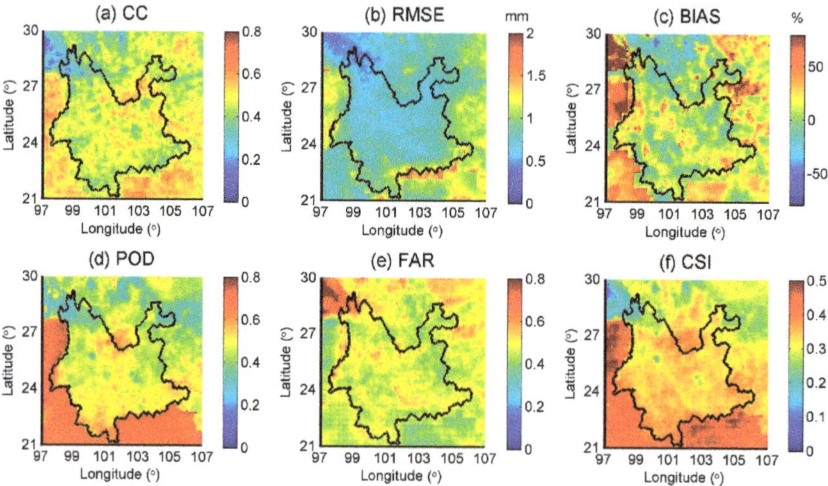

Figure 4. The spatial pattern of six indicators generated from hourly data of IMERG-F and CMA in Yunnan, where the indicators indicated by (**a**), (**b**), (**c**), (**d**), (**e**), and (**f**) are CC, RMSE, BIAS, POD, FAR, CSI.

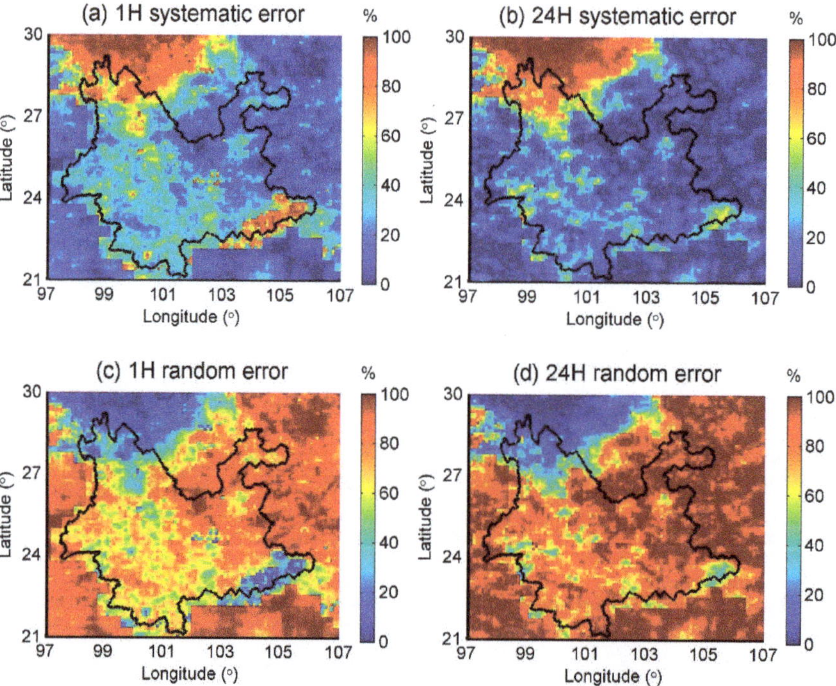

Figure 5. The spatial patterns of the system and random errors for IMERG-E at the 1 h and 24 h temporal scales in Yunnan Province, respectively, where (**a**) and (**b**) are the systematic errors of 1 h and 24 h, respectively; (**c**) and (**d**) are random error of 1 h and 24 h.

As for IMERG-F, systematic errors accounted for less than 20% of the total errors across the whole study area at both the 1 and 24 h temporal scales. Compared with IMERG-E, IMERG-F has been significantly improved because the systematic errors were effectively reduced, especially in mountainous areas with the complex terrain. Figure 6 shows that the IMERG-F data error (≥80%) is primarily a random error, independent of the underlying surface features. Besides, Figures 6 and 7 behave differently inside and outside the Yunnan border. This phenomenon is mainly due to the complex and changeable terrain of Yunnan Province, which further triggers the discontinuous changes in satellite precipitation errors. It is a discontinuous problem in the original satellite's precipitation error, rather than using different spatiotemporal resolution rainfall. Therefore, the overall IMERG-F data is better than the IMERG-E data.

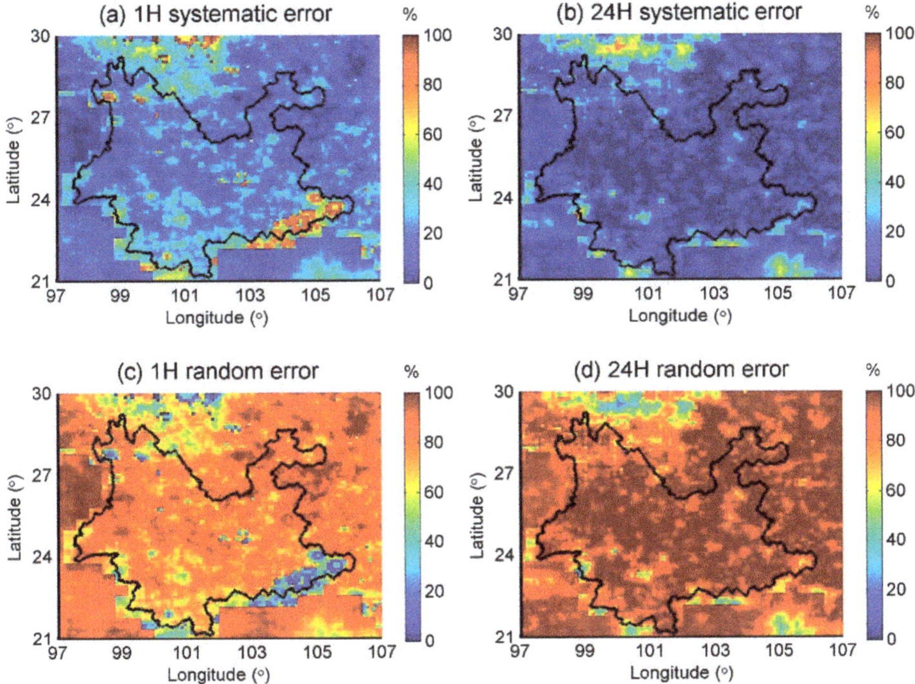

Figure 6. The spatial patterns of the system and random errors in IMERG-F at the 1 and 24 h temporal scales, respectively; where (**a**) and (**b**) are the systematic errors of 1 h and 24 h, respectively; (**c**) and (**d**) are random error of 1 h and 24 h.

3.3. Applicability Analysis of IMERG in Flash Flood Warning

Based on historical flash flood disaster events, combined with three types of rainfall products (IMERG-E and -F, and CMA), the multi-period rainfall (1 h, 3 h, 6 h, and 24 h) is obtained. The resultant effective accumulated multi-period rainfall is calculated by the improved RTI method. Meanwhile, combined with the flash floods' actual frequency and the rainstorm statistical parameter atlas of China, the multi-period critical rainfall (1 h, 3 h, 6 h, and 24 h) (hereafter, CR1, CR3, CR6, and CR24) is obtained; the G (x) early warning model is constructed for effective cumulative rainfall (R_t) and corresponding period critical rainfall (CRt). Since this model does not take into account the potential multiple influencing factors, such as slope, vegetation, and human activity, when issuing the flash flood warning, we should make a comprehensive analysis regarding the flash flood risk distribution to determine whether a flash flood event has been captured. Considering the flash flood risk map obtained by Ma et al. (2019), the obtained results are shown in Figures 7 and 8 [31].

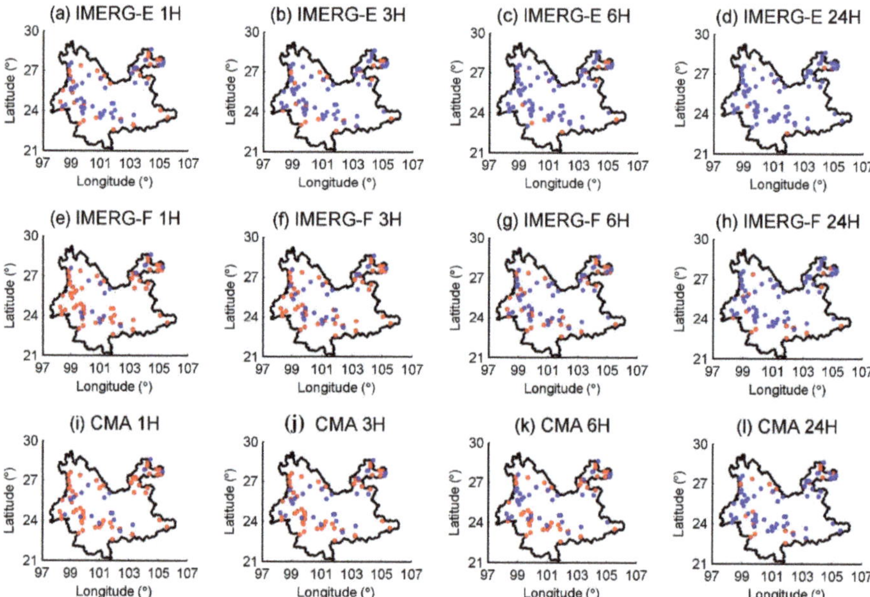

Figure 7. The performance of the flash floods captured by the three products (IMERG-F, EMERGE, and CMA) for different times (1 h, 3 h, 6 h, and 24 h). Note: Red indicates that the warning issued has captured the flash flood event; blue indicates that the issued warning has not captured the flash flood event, where (**a**), (**b**), (**c**), and (**d**) are the distribution of IMERG-E products in 1 h, 3 h, 6 h and 24 h to capture flash flood disasters; (**e**), (**f**), (**g**), and (**h**) are the distribution of IMERG-F products in 1 h, 3 h, 6 h and 24 h to capture flash flood disasters; (**i**), (**j**), (**k**), and (**l**) are the distribution of CMA products in 1 h, 3 h, 6 h and 24 h to capture flash flood disasters.

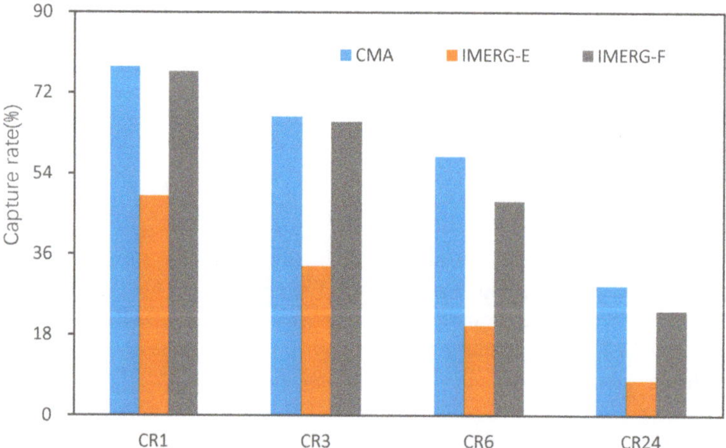

Figure 8. Percentage of flash floods caught by CMA, IMERG-E, and IMERG-F. Note: CRt represents the critical rainfall at time t.

Obviously, for the same product, the captured flash flood events has decreased over time, where the hit rate of both CR6 and CR24 is less than 60%. Among them, the flash floods events captured by these three rainfall products are concentrated in western Yunnan, while there are fewer flash flood events in the relatively flat southwest and central Yunnan. In general, if a flash flood event cannot be captured in a short period, it will not be anyhow captured. Moreover, the effect of IMERG-E products on capturing disasters is significantly lower than that of IMERG-F, and the capture rate in each period is less than 50%. However, the capture accuracy of IMERG-F at CR1 and CR3 is almost comparable to the CMA—only a 1% difference—and the capture rate of CR1 is nearly 80%, with an extremely high accuracy. Nonetheless, with the time marching, the capture effect decreased significantly. Meanwhile, if the flash flood events cannot be captured by IMERG-F and CMA, the same phenomenon occurs in IMERG-F; but, as time goes on, there is an out-of-sync phenomenon in the catching flash flood events by IMERG-F and CMA. Besides, the CMA data is a fusion of measured data and CMORPH, leading to possible delays in data acquisition. Therefore, the hit rate is decreasing with an increasing temporal resolution or increasing the averaging time of the satellite images.

4. Conclusions

This study first quantitatively evaluated the IMERG-E and IMERG-F satellite precipitation products, and then combined the improved RTI method to analyze its application effects in flash flood warnings, which has three major contributions:

(i)   Flash flood warning aspects are integrated, for the first time, with satellite precipitation to account for the applicability of satellite data in flash flood warnings. The result shows that the early warning effect of IMERG-F products is better at the 1 h and 3 h scale than that at the daily scale;

(ii) the study area has not been documented in previous studies. Yunnan Province is characterized by a low latitude but high altitude where satellite precipitation exhibited some new characteristics, including that precipitation in Yunnan Province has increased from northeast to southwest, where the largest precipitation occurred in the flood-prone area in the southwest part;

(iii) this study reveals some interesting phenomena that were not reported in related research [32]. For example, the systematic error of IMERG-E is mainly distributed in areas with high altitudes and low precipitation, and the random error is mainly distributed in areas with low altitudes and high precipitation. The most important thing is that as for the same satellite rainfall product, the flash flood disaster events that can be captured decreases with time. For different satellite rainfall products, the flash flood events captured by the IMERG-E products are significantly lower than IMERG-F. Meanwhile, the capture rate of each period is less than 50%.

To sum up, the above research results will contribute to the application of satellite remote sensing data in flash flood warnings.

Author Contributions: Conceptualization, M.M. and P.J.; methodology, M.M., P.J. and G.T.; software, G.T.; validation, P.J., G.T., and Z.M.; formal analysis, M.M.; investigation, D.W.; resources, M.M. and G.T.; data curation, P.J.; writing—original draft preparation, M.M., P.J. and Z.M.; writing—review and editing, M.M., P.J., H.W. and H.Y.; visualization, G.T. and P.J.; supervision, H.W.; project administration, H.W.; funding acquisition, H.W. All authors have read and agreed to the published version of the manuscript.

Funding: This work was supported by the National Key R&D Program of China (2018YFC1508105) and National Natural Science Foundation of China (NSFC. General Projects: (Grant No. 41471430), the Young Scientists Fund of the National Natural Science Foundation of China ((No. 51909052), Natural Science Foundation of Tianjin (18JCQNJC08800), Research on Key Technologies of Flash Flood Prevention Based on Multi-machine Learning Model (2019KJ086), the National Natural Science Foundation of China (Grant No. 41901343), the Key R&D Program of Ministry of Science and Technology (Grant No. 2018YFC1506500), the State Key Laboratory of Resources and Environmental Information System, the China Postdoctoral Science Foundation (No. 2018M630037, and 2019T120021), the Open Fund of the State Key Laboratory of Remote Sensing Science (Grant No. OFSLRSS201909), and Research on Flash Flood Warning Based on Machine Learning: A Case Study of Yunnan Province (52XB1903).

Acknowledgments: The authors wish to thank the anonymous reviewers for their comments and suggestions.

Conflicts of Interest: The authors declare no potential conflict of interests.

Abbreviations

Acronyms	Full Name
NASA	National Aeronautics and Space Administration
NWS	National Weather Service
DEM	Digital Elevation Model
GEO	Geostationary Earth Orbit
PMW	Passive Microwave
CMORPH	Climate Prediction Center morphing
IMERG	Integrated Multi-Satellite Retrievals for Global Precipitation Measurement
TRMM	Tropical Rainfall Measuring Mission
IMERG-E	IMERG Early run product
IMERG-F	IMERG Final run product
PRT	Post-real-time
NRT	Near-real-time
GPM	Global Precipitation Measurement
TMPA	TRMM Multi-Satellite Precipitation Analysis
FFG	Flash Flood Guidance
RTI	Rainfall Triggering Index
SWI	Soil Water Index
CC	Correlation Coefficient
RMSE	Root Mean Square Error
BIAS	Relative Bias
POD	Probability of Detection
FAR	False Alarm Ratio
CSI	Critical Success Index
CMA	Regional hourly precipitation integration products of the China Meteorological Administration

References

1. Hong, Y.; Adhikari, P.; Gourley, J. Flash Flood. *Encycl. Earth Sci.* **2013**, *18*, 324–325.
2. Gourley, J.J.; Flamig, Z.L.; Vergara, H.; Kirstetter, P.E.; Clark, R.A., III; Argyle, E.; Hong, Y. The flooded locations and simulated hydrographs (flash) project: Improving the tools for flash flood monitoring and prediction across the united states. Bulletin of the American Meteorological Society. *Bull. Am. Meteorol. Soc.* **2017**, *98*, 361–372. [CrossRef]
3. Guo, L.; He, B.; Ma, M.; Chang, Q.; Li, Q.; Zhang, K.; Hong, Y. A comprehensive flash flood defense system in China: Overview, achievements, and outlook. *Nat. Hazards* **2018**, *92*, 727–740. [CrossRef]
4. Gourley, J.J.; Arthur, A. *Rainfall Rate, Use in the Hydrological Sciences*; Encyclopedia of GIS: Norman, OK, USA, 2017.
5. Borga, M.; Boscolo, P.; Zanon, F.; Sangati, M. Hydrometeorological Analysis of the 29 August 2003 Flash Flood in the Eastern Italian Alps. *J. Hydrometeorol.* **2007**, *8*, 1049–1067. [CrossRef]
6. Belabid, N.; Zhao, F.; Brocca, L.; Huang, Y.; Tan, Y. Near-Real-Time Flood Forecasting Based on Satellite Precipitation Products. *Remote Sens.* **2019**, *11*, 252. [CrossRef]
7. Aminyavari, S.; Saghafian, B.; Sharifi, E. Assessment of Precipitation Estimation from the NWP Models and Satellite Products for the Spring 2019 Severe Floods in Iran. *Remote Sens.* **2019**, *11*, 2741. [CrossRef]
8. Tang, G.; Wen, Y.; Gao, J.; Long, D.; Ma, Y.; Wan, W.; Hong, Y. Similarities and differences between three coexisting spaceborne radars in global rainfall and snowfall estimation. *Water Resour. Res.* **2017**, *53*, 3835–3853. [CrossRef]
9. Yuan, F.; Wang, B.; Shi, C.; Cui, W.; Zhao, C.; Liu, Y.; Ren, L.; Zhang, L.; Zhu, Y.; Chen, T.; et al. Evaluation of hydrological utility of IMERG Final run V05 and TMPA 3B42V7 satellite precipitation products in the Yellow River source region, China. *J. Hydrol.* **2018**, *567*, 696–711. [CrossRef]
10. Huffman, G.J.; Bolvin, D.T.; Nelkin, E.J.; Wolff, D.B.; Adler, R.F.; Gu, G.; Hong, Y.; Bowman, K.P.; Stocker, E.F. The TRMM Multi-satellite Precipitation Analysis (TMPA). In *Satellite Rainfall Applications for Surface Hydrology*; Springer: Dordrecht, The Netherlands, 2010.

11. Tang, G.; Zeng, Z.; Ma, M.; Liu, R.; Wen, Y.; Hong, Y. Can Near-Real-Time Satellite Precipitation Products Capture Rainstorms and Guide Flood Warning for the 2016 Summer in South China? *IEEE Geosci. Remote Sens. Lett.* **2017**, *14*, 1208–1212. [CrossRef]

12. Tyralis, H.; Dimitriadis, P.; Koutsoyiannis, D.; O'Connell, P.E.; Tzouka, K.; Iliopoulou, T. On the long-range dependence properties of annual precipitation using a global network of instrumental measurements. *Adv. Water Resour.* **2018**, *111*, 301–318. [CrossRef]

13. Jan, C.D.; Li, M.H. A Debris-Flow Rainfall-Based Warning Model. *J. Chin. Soil Water Conserv.* **2004**, *35*, 275–285. (In Chinese)

14. Clark, R.A.; Gourley, J.J.; Flamig, Z.L.; Hong, Y.; Clark, E. CONUS-Wide Evaluation of National Weather Service Flash Flood Guidance Products. *Weather Forecast.* **2014**, *29*, 377–392. [CrossRef]

15. Zhao, G.; Pang, B.; Xu, Z.; Yue, J.; Tu, T. Mapping flood susceptibility in mountainous areas on a national scale in china. *Sci. Total Environ.* **2018**, *615*, 1133–1142. [CrossRef]

16. Chen, Y.; Wei, Z.; Chia, H. A Rainfall-based Warning Model for Predicting Landslides Using QPESUMS Rainfall Data. *J. Chin. Soil Water Conserv.* **2017**, *48*, 44–55.

17. Zubieta, R.; Getirana, A.; Espinoza, J.C.; Lavado-Casimiro, W.; Aragon, L. Hydrological modeling of the Peruvian–Ecuadorian Amazon Basin using GPM-IMERG satellite-based precipitation dataset. *Hydrol. Earth Syst. Ences Discuss.* **2017**, *21*, 3543–3555. [CrossRef]

18. Zhang, K.; Xue, X.; Hong, Y.; Gourley, J.J.; Lu, N.; Wan, Z.; Wooten, R. iCRESTRIGRS: A coupled modeling system for cascading flood-landslide disaster forecasting. *Hydrol. Earth Syst. Sci.* **2016**, *20*, 1–23. [CrossRef]

19. Tian, Y.; Peters-Lidard, C.D.; Eylander, J.B.; Joyce, R.J.; Huffman, G.J.; Adler, R.F.; Zeng, J. Component analysis of errors in satellite-based precipitation estimates. *J. Geophys. Res. Atmos.* **2009**, *114*. [CrossRef]

20. Zeng, Z.; Tang, G.; Long, D.; Zeng, C.; Ma, M.; Hong, Y.; Xu, J. A cascading flash flood guidance system: Development and application in Yunnan Province, China. *Nat. Hazards* **2016**, *84*, 2071–2093. [CrossRef]

21. Ma, M.; Zhang, J.; Su, H.; Wang, D.; Wang, Z. Update of early warning indicators of flash floods: A case study of hunjiang district, northeastern china. *Water* **2019**, *11*, 314. [CrossRef]

22. Hou, A.Y.; Kakar, R.K.; Neeck, S.; Azarbarzin, A.A.; Kummerow, C.D.; Kojima, M.; Iguchi, T. The Global Precipitation Measurement Mission. *Bull. Am. Meteorol. Soc.* **2014**, *95*, 701–722. [CrossRef]

23. Mohammad, S.; Hosseini, M.; Tang, Q. Validation of GPM IMERG V05 and V06 Precipitation Products over Iran. *J. Hydrometeorol.* **2020**, *5*, 1011–1037. [CrossRef]

24. Sungmin, O.; Foelsche, U.; Kirchengast, G.; Fuchsberger, J.; Tan, J.; Petersen, W.A. Evaluation of GPM IMERG Early, Late, and Final rainfall estimates using WegenerNet gauge data in southeastern Austria. *Hydrol. Earth Syst. Sci.* **2017**, *21*, 6559–6572.

25. Shen, Y.; Zhao, P.; Pan, Y.; Yu, J. A high spatiotemporal gauge-satellite merged precipitation analysis over China. *J. Geophys. Res. Atmos.* **2014**, *119*, 3063–3075. [CrossRef]

26. Cui, P.; Yang, K.; Cui, P. Relationship Between Occurrence of Debris Flow and Antecedent Precipitation: Taking the Jiangjia Gully as an Example. *Sci. Soil Water Conserv.* **2003**, *1*, 11–15.

27. Ma, M.; He, B.; Wan, J.; Jia, P.; Guo, X.; Gao, L.; Maguire, L.; Hong, Y. Characterizing the Flash Flooding Risks from 2011 to 2016 over China. *Water* **2018**, *10*, 704. [CrossRef]

28. Maggioni, V.; Sapiano MAdler, R.F. Estimating Uncertainties in High-Resolution Satellite Precipitation Products: Systematic or Random Error? *J. Hydrometeorol.* **2016**, *17*, 1119–1129. [CrossRef]

29. Ayugi, B.; Tan, G.; Ullah, W.; Boiyo, R.; Ongoma, V. Inter-comparison of remotely sensed precipitation datasets over Kenya during 1998-2016. *Atmos. Res.* **2019**, *225*, 96–109. [CrossRef]

30. Palomino-Angel, S.; Anaya-Acevedo, J.A.; Botero, B.A. Evaluation of 3B42V7 and IMERG daily-precipitation products for a very high-precipitation region in northwestern South America. *Atmos. Res.* **2018**, *217*, 37–48. [CrossRef]

31. Ma, M.; Liu, C.; Zhao, G.; Xie, H.; Jia, P.; Wang, D.; Hong, Y. Flash Flood Risk Analysis Based on Machine Learning Techniques in the Yunnan Province, China. *Remote Sens.* **2019**, *11*, 170. [CrossRef]

32. Rivera, J.A.; Marianetti, G.; Hinrichs, S. Validation of chirp's precipitation dataset along the central Andes of Argentina. *Atmos. Res.* **2018**, *213*, 437–449. [CrossRef]

Article

Performance of an Unmanned Aerial Vehicle (UAV) in Calculating the Flood Peak Discharge of Ephemeral Rivers Combined with the Incipient Motion of Moving Stones in Arid Ungauged Regions

Shengtian Yang [1,2], Chaojun Li [1,2], Hezhen Lou [1,2,*], Pengfei Wang [1,2], Juan Wang [3,4] and Xiaoyu Ren [5]

[1] College of Water Sciences, Beijing Normal University, Beijing 100875, China; yangshengtian@bnu.edu.cn (S.Y.); 201831470002@mail.bnu.edu.cn (C.L.); 201721470024@mail.bnu.edu.cn (P.W.)
[2] Beijing Key Laboratory of Urban Hydrological Cycle and Sponge City Technology, Beijing 100875, China
[3] Beijing Key Laboratory of Environmental Remote Sensing and Digital Cities, Beijing Normal University, Beijing 100875, China; 201731170029@mail.bnu.edu.cn
[4] School of Geography, Faculty of Geographical Science, Beijing Normal University, Beijing 100875, China
[5] Beijing Weather Modification Office, Beijing 550001, China; 201431170007@mail.bnu.edu.cn
* Correspondence: louhezhen@bnu.edu.cn

Received: 31 March 2020; Accepted: 13 May 2020; Published: 18 May 2020

Abstract: Ephemeral rivers are vital to ecosystem balance and human activities as essential surface runoff, while convenient and effective ways of calculating the peak discharge of ephemeral rivers are scarce, especially in ungauged areas. In this study, a new method was proposed using an unmanned aerial vehicle (UAV) combined with the incipient motion of stones to calculate the peak discharge of ephemeral rivers in northwestern China, a typical arid ungauged region. Two field surveys were conducted in dry seasons of 2017 and 2018. Both the logarithmic and the exponential velocity distribution methods were examined when estimating critical initial velocities of moving stones. Results reveal that centimeter-level orthoimages derived from UAV data can demonstrate the movement of stones in the ephemeral river channel throughout one year. Validations with peak discharge through downstream culverts confirmed the effectiveness of the method. The exponential velocity distribution method performs better than the logarithmic method regardless of the amount of water through the two channels. The proposed method performs best in the combination of the exponential method and the river channel with evident flooding ($>20\ \mathrm{m^3/s}$), with the relative accuracy within 10%. In contrast, in the river channel with a little flow (around $1\ \mathrm{m^3/s}$), the accuracies are weak because of the limited number of small moving stones found due to the current resolution of UAV data. The poor performance in the river channel with a little flow could further be improved by identifying smaller moving stones, especially using UAV data with better spatial resolution. The presented method is easy and flexible to apply with appropriate accuracy. It also has great potential for extensive applications in obtaining runoff information of ephemeral rivers in ungauged regions, especially with the quick advance of UAV technology.

Keywords: UAV remote sensing; Ephemeral rivers; flood peak discharge; incipient motion; arid ungauged regions

1. Introduction

Water tensions are particularly acute in arid regions globally due to the infrequency and short periods of rainfall. The urbanization process of expanding populations has exacerbated water scarcity in these areas, which are generally underdeveloped and lack hydrological gauging stations [1–3]. The quantity and quality of water resources in arid ungauged regions remain crucial issues of common concern to international communities [4].

In arid ungauged regions, most water resources are in the form of surface runoff, largely accounted for by ephemeral rivers, especially in upstream mountainous water systems [5]. As one of the least-known aquatic ecosystems, ephemeral rivers are not only crucial as an essential water supply, but also support vegetation and other biota living nearby [6]. The structure, productivity, and spatial distribution of biotic communities in semi-arid and arid regions are strongly affected by ephemeral water flows [7,8]. There is thus an urgent need to monitor the discharge of ephemeral rivers, especially in upstream watersheds. Moreover, the shortage of available data for traditional hydrological research methods severely impedes the study of ephemeral rivers and related ecological-hydrological processes in arid regions, which further limits the rational exploitation and management of water resources.

Hydrological models have long been used to demonstrate hydrological processes and to estimate the discharge of ephemeral rivers [9], such as the lump model [10] and the Soil and Water Assessment Tool (SWAT) model [11]. However, the time-consuming processing of hydrological models requires a great deal of observation data and regional parameterization. Furthermore, results of hydrological models vary significantly because of highly unpredictable flow patterns, complex interactions with and among anthropogenic pressures, and lack of information on the physiographic and environmental conditions of many catchments [11,12]. Satellite remote sensing has been widely applied to extract river hydraulic variables in the calculation of peak discharge because of its low cost and high sequence characteristics [13–15], and most studies have estimated discharge by constructing the relationship between channel geometry and discharge [16–18]. Nevertheless, the extremely high variability of ephemeral river runoff at different spatial and temporal scales has made it hard to estimate the discharge of ephemeral rivers with the abovementioned methods [19]. Specifically, when the river is too narrow or the river only flows for a few days, it is difficult to determine the water level and water surface area, and furthermore, to extract hydraulic variables with coarse remotely sensed data. Additionally, the peak discharge of ephemeral rivers has commonly been calculated by indirect means, such as the slope-area method [20]. The continuous slope-area method for gaging volumetric discharge has been tested in various ephemeral channels [21,22], while the method is still hard to implement in arid ungauged regions where access is limited.

In recent years, the rapid development of unmanned aerial vehicle (UAV) techniques has made it possible to obtain terrain information at centimeter-level precision [23–25]. These advances help to determine the hydraulic geometry accurately [26], and thus to identify riverbed change and sediment movement in a river channel, especially of ephemeral rivers in arid ungauged regions. With the help of UAV data, the study of riverbed sediment could provide a new perspective to estimate flood peak discharge [27]. The most commonly used approach defining the critical hydrodynamic conditions of sediment transport is still regarded as the Shields approach [28]. In subsequent studies, the threshold of the initiation of sediment transport was expressed by either the bed shear stress or the average flow rate, and numerous equations have been published based on simplified assumptions, laboratory data, and field measurements [29–31]. Among the equations, the power law and Prandt–von Karman velocity distribution are widely employed to estimate the critical velocities of sediment transport in open-channel flow, separately representing the logarithmic and exponential velocity distribution. Therefore, it is worth exploring the performance of UAVs in estimating the flood peak discharge of ephemeral rivers, with the logarithmic or the exponential velocity distribution equations applied.

The present paper uses high-resolution UAV data identifying stone movements to explore a fast and convenient method of calculating the flood peak discharge of ephemeral rivers in ungauged arid regions. The research objectives of the study are (1) to identify the movement of stones in the channel of the ephemeral river and to calculate the critical initial velocities of moving stones using high-resolution UAV data, (2) to obtain the flood peak discharge of the ephemeral river, and (3) to compare the performance of the logarithmic and the exponential velocity distribution methods. The proposed method of estimating the flood peak discharge of ephemeral rivers is presented to address the conflict between the shortage of hydrological gauge data and the data needs for research and management of ephemeral rivers.

2. Study Area

The study area comprised two small watersheds, namely Harulin (45°0′22″–45°8′52″N, 80°53′15″–80°59′17″E) and Sailimu (44°47′10″–44°49′2″N, 80°9′15″–81°11′26″E), located in the Bortala River Basin of northwestern China (Figure 1). The two watersheds are in the upstream mountainous area of the Bortala River, separately covering an area of 47.9 km² (Harulin watershed) and 4.0 km² (Sailimu watershed). The climate in Bortala River Basin is a typical temperate continental climate with annual precipitation less than 100 mm and annual evaporation more than 1600 mm. With an arid climate, the primary source of the water system in the basin is seasonal ice and snow meltwater from the top of the mountain and the surface runoff formed by heavy precipitation in summer [32]. Therefore, many streams in the basin are usually ephemeral, and the runoff from April to August often accounts for more than 50% of the annual runoff. Culverts with flood watermarks were only found in the main river channels of Harulin and Sailimu watersheds. Thus, the maximum flood peak flow of ephemeral rivers based on the water level can be calculated in the two watersheds. The main river channels in the two watersheds are separately referred to as river channel H and river channel S.

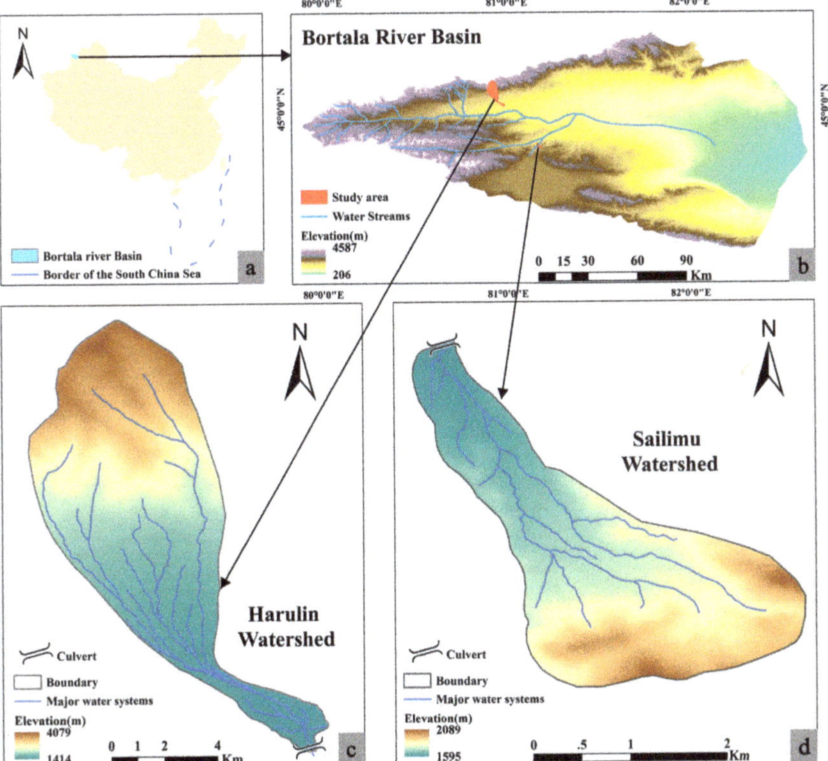

Figure 1. (**a**) Location of Bortala River Basin in China, (**b**) topography and water systems of Bortala River Basin, (**c**) Harulin watershed, and (**d**) Sailimu watershed.

3. Methodology

In this study, a new method is presented to estimate the peak discharge of ephemeral rivers, in which high-resolution UAV data was taken as a primary data source, and the method of initiation of motion was carried out. The detailed workflow of the study is as follows (Figure 2): Step 1, high-resolution UAV data acquired in the field were processed to obtain Digital Orthophoto Maps (DOMs) and Digital Surface Models (DSMs) of the studied channels. Step 2, orthoimages were compared to identify moved stones around selected cross sections, then the critical velocities of these stones were calculated separately using logarithmic and exponential velocity distributions. Step 3, the average velocity of selected cross sections was calculated from the critical velocities of moving stones, and the discharge through cross sections was then obtained with the extracted cross-sectional area taken from the DSM. Step 4, essential hydraulic variables of culverts were extracted from orthoimages and the DSM to calculate the discharge through a culvert for validation.

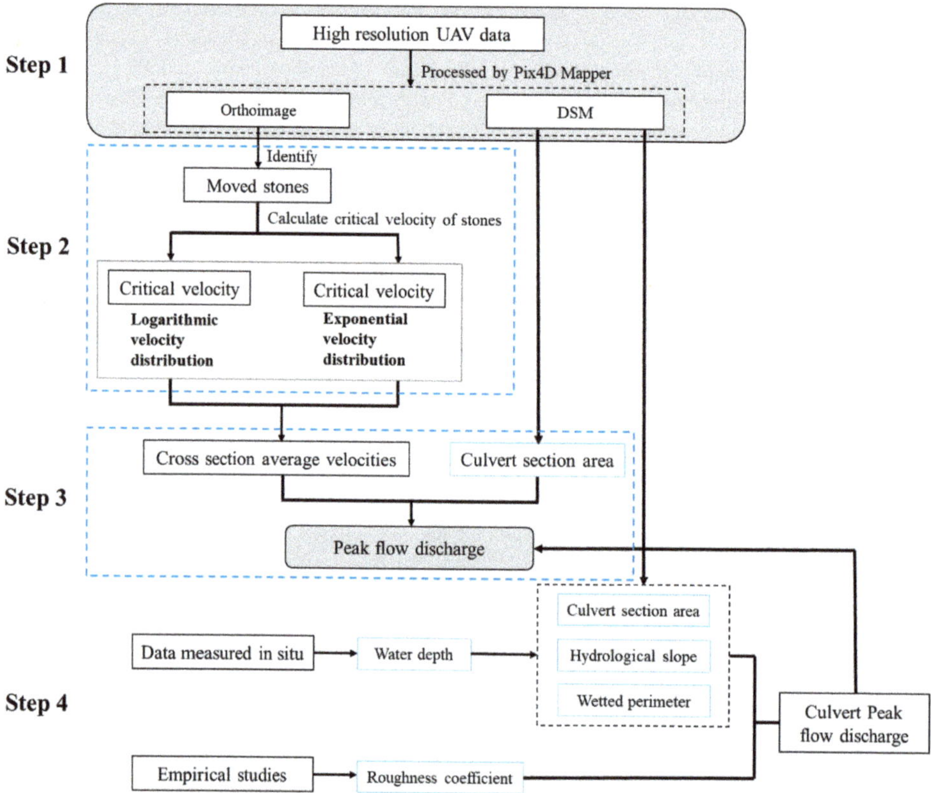

Figure 2. Workflow of the method calculating flood peak discharge of ephemeral rivers.

3.1. Data

3.1.1. UAV Data Collection

Aerial images were acquired via a DJI Phantom 3 Professional UAV (DJI, Shenzhen, China), manually flown by sight and equipped with an independent camera set to take photographs at a fixed interval (Table 1). The DJI Phantom 3 Professional UAV has a vertical takeoff weight of 1280 g and can reach a maximum speed of 16 m/s. Fully charged batteries provide a maximum flight time of 23 min. The camera was a standard lens mounted on a Cardan suspension and based on complementary metal-oxide semiconductor (CMOS) image sensor technology. The FC300X sensor provides images with 12.4 million effective pixels and has separate channels for red, green, and blue light. The spatial and elevation resolution of the acquired UAV data was confirmed to be centimeter-level in our previous study [33].

Table 1. Key parameters of the unmanned aerial vehicle (UAV).

Characteristics	FC300X
Sensor (Type)	1/2.3″ CMOS sensor
Image size (Columns and Rows)	1.2 million (4000 × 3000)
Million effective pixels	12.4
Maximum aperture	f/2.8
Camera focal length	20 mm
Field of view	94°
ISO range	100–1600 (photo)

UAV images were acquired twice, in August of 2017 and 2018, the dry season of the study area with no flow through the channel. In each survey, the same study area was surveyed in two flight missions pre-programmed with Pix4D Capture software for several waypoints to achieve a 90% image overlap along the track. The flight altitude was on average 70 m and was a trade-off between the size of the studied channel and the desired spatial detail of approximately 3 cm per pixel. The two in situ surveys produced 380 and 384 UAV images separately. A total of 8 Ground Control Points (GCPs), 4 for each river channel, were measured by Real-Time Kinematic in situ surveys. The root-mean-square error (RMSE) between the ground measurements and UAV data was ±2.79 cm in the horizontal direction, which is feasible to identify the stone movement in the exposed channel.

3.1.2. UAV Data Processing

The stereoscopic images were processed using the rapid and automatic professional processing software Pix4Dmapper. The software supports not only UAV data but also aerial photography, oblique photography, and close-range photogrammetry. Through data importing, initial processing, and point cloud encryption, Pix4D mapper provides a high-resolution Digital Ortho Map (DOM) and a Digital Surface Model (DSM) [34,35] (Figure 3). The processing steps include (1) determining the internal orientation element and calibrating the camera, (2) searching and fast matching the same name point employing the scale-invariant feature transform (SIFT) algorithm [36], (3) using the position information of the same name point and the image to adjust the regional network and to restore the position and posture of the image, (4) performing aerial triangulation encryption and generating an image point cloud using the control point or the matching point, and (5) generating the DSM from an image point cloud and then generating the DOM.

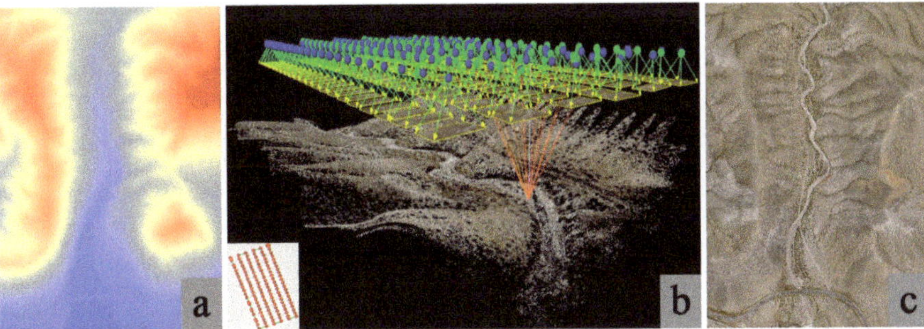

Figure 3. (**a**) The derived Digital Surface Model (DSM), (**b**) the three-dimensional triangulation process by UAV and the flight path, and (**c**) the derived Digital Orthophoto Map (DOM).

3.1.3. In Situ Survey Data

In situ surveys were conducted in August of 2017 and 2018, the same time as UAV data collection. During the surveyed dry season, river channel H and S were completely exposed. Evident watermarks were found both on the bank of some reaches and the downstream culvert of the studied river channels. There is also a clear vegetation growth boundary to help determine the water level. The watermark conditions of the two years on the riverbank were compared and measured by tape to determine the maximum water level of the peak flood throughout the year. Cross sections with the determined maximum water levels were selected from the measured river reaches for subsequent calculation. The watermarks on the culvert were also measured to determine the maximum water level when the peak flood passed through the downstream culvert in each study site for validation.

3.2. Critical Velocity of Particle Transport

In arid ungauged regions, upstream ephemeral rivers rarely flow, and river channels are scarcely affected by human activities. Limited stones scattered in the river channel, especially identifiable moving stones, are regarded as only moved by water flow. When the flood flow promotes a stone to move, the movement of the stone is less likely affected by the surrounding stones and bed materials. Thus, the description of a moving stone targeted at the incipience by a hydraulically rough wall-shear flow can be illustrated in Figure 4. The catalyst forces for the moving stone are the hydrodynamic drag force F_D and the lift force F_L. The submerged weight W brings the stabilizing condition to the stone. The three forces reach equilibrium and have the relationship [37]:

$$F_D = \tan\varnothing(W - F_L), \tag{1}$$

where $\varnothing$ is the angle of repose.

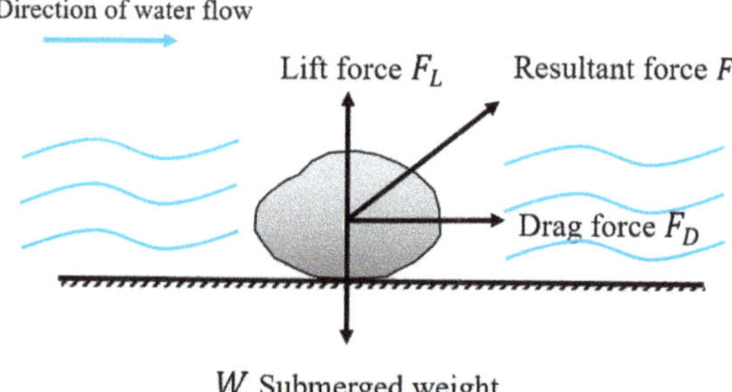

Figure 4. Major forces exerted on stones on the incipient motion.

The submerged weight W, drag F_D, and lift F_L can be expressed as

$$F_D = C_D \frac{\pi d^2}{4} \frac{\rho U_0^2}{2}, \tag{2}$$

$$F_L = C_L \frac{\pi d^2}{4} \frac{\rho U_0^2}{2} \tag{3}$$

$$W = \frac{1}{6}(\rho_s - \rho)\pi d^3, \tag{4}$$

where U_0 is the fluid velocity at the bottom of the channel, C_D and C_L are respectively the drag and lift coefficients, d is particle nominal diameter, ρ_s is stone density, and ρ is liquid density.

In the studied river channels, size compositions of the surface sediment mainly include sand fraction (<2 mm), and gravel fraction (2–20 mm). Cobbles (20–200 mm) and boulders (>200 mm) are scattered on the riverbed, and only some could be identified from UAV images due to the resolution. Most of the stones in the channel are granite with an average empirical density ρ_s = 2.80 g/cm^3 [38,39].

In the studied river channels H and S, four cross sections (sections A, B, C, and D) were analyzed and measured in situ individually (Figure 5). The research reach refers to the range of 5 meters upstream and downstream of each section. In each research reach, moved and unmoved stones were identified by comparing the high-resolution orthoimages taken over two years. For each moved stone, the average flow rate when it starts to move can express the water flow intensity. The power law and Prandt–von Karman logarithmic velocity were used to simulate the velocity profile and to calculate the critical velocity of the incipient motion of each moving stone separately.

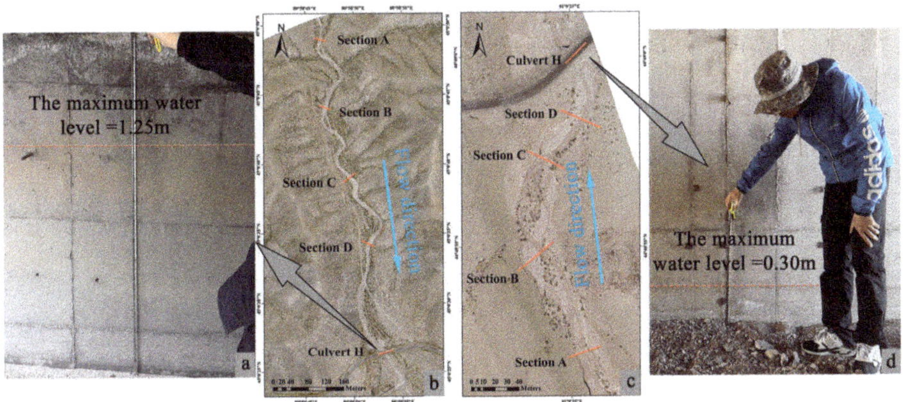

Figure 5. (**b**) and (**c**) present the location of cross sections and the downstream culvert in river channels H and S. (**a**) and (**d**) are fieldwork photos of measuring flood max water level.

3.2.1. Logarithmic Distribution of the Flow Velocity

Assuming that the flow velocity profile in the channel follows the power law representing a logarithmic velocity distribution, the average velocity through a cross section of the open channel with turbulence can be expressed using the Einstein unified formula [40]:

$$\frac{U}{U'_*} = 5.75 \lg\left(\frac{12.27R'\chi}{k_s}\right),\tag{5}$$

where k_s is the height of the protrusion of the boundary roughness, also known as the side-wall roughness or the riverbed roughness. U'_* is the shear velocity corresponding to the sand resistance and $U'_* = \sqrt{gR'J}$. χ is a correction factor related to k_s/δ. δ is the calculated thickness of the viscous bottom layer and $\delta = 11.6\nu/U_*$. R' is the hydraulic radius corresponding to the sand resistance.

The critical initial shear stress can be expressed as:

$$\Theta = \frac{\tau_c}{(\rho_s - \rho)D} = f\left(\frac{U_* D}{\nu}\right).\tag{6}$$

Sand waves have not yet formed, so let $R' = R$ and U_c represent the average vertical velocity of the critical condition of incipient motion. The critical initial condition expressed by the average critical velocity through the cross section U_c can be derived as:

$$\frac{U_c}{\sqrt{\frac{\rho_s - \rho}{\rho} g D}} = 5.75\sqrt{\Theta}\lg\left(\frac{12.75R\chi}{k_s}\right),\tag{7}$$

where $\Theta = f(R_*) \approx 0.06$ under a turbulent condition ($R_* \gg 500$). The density of water $\rho = 1$ g/cm^3. g is the acceleration due to gravity and has a value of 9.8 m/s^2. D is the nominal diameter of stones. In the case of a wide and shallow channel, the hydraulic radius R equals the water depth, the Einstein correction factor $\chi = 1.0$, and the roughness height k$_s$ = D.

3.2.2. Exponential Distribution of the Flow Velocity

Assuming the average velocity on the cross section is exponentially distributed on the vertical line, the critical initial velocity can be written as

$$U_c = K \sqrt{\frac{\rho_s - \rho}{\rho} gD} \left(\frac{h}{D}\right)^{1/6}, \tag{8}$$

where K is a comprehensive coefficient that considers the submerged weight of the stones, lifting force, drag force, and effect of the irregular shape of stones on the three force coefficients. Shamov combined measurement and laboratory data of a river, determining K as 1.14, which has been widely used with good effect [41].

3.3. Discharge Calculation of River Sections

It is assumed that the hydraulic conditions in a relatively short reach of a river do not vary obviously, and the average velocity through each section can then be expressed as the mean of the critical velocity of all scattered moving stones. The DSM acquired in the dry season and the identified riverbank watermarks are used to obtain the profile in each section when the flood peak flows through the river channel. The peak discharge through each section is then calculated as

$$Q = v \times A, \tag{9}$$

where Q is the peak flood discharge through each section, v is the average velocity of the flow through each section, and A is the cross-sectional area when the flood peak flows through culverts.

3.4. Performance Evaluation

Studies have long been conducted to estimate the discharge through culverts on various hydraulic conditions, and flood watermarks have also been used to identify the maximum water level during a specific period [42]. In each river channel, the peak discharge through the downstream culvert (Figure 5) was calculated by Equations (9) and Equation (10) (the Manning formula) to validate the effectiveness of the proposed method and to evaluate the performance of two velocity distributions during the calculation process. The underwater cross-sectional area when the flood peak flows through the culverts is derived from the culvert profile obtained from the DSM and the measured maximum water depth. The Manning formula is normally used to calculate the flow velocity of open channels [43], but its reliability to estimate a peak discharge from the channel cross section has also been confirmed [44]. The general form of the Manning formula is written as

$$v_c = \frac{k}{n_c} R_c^{2/3} J^{1/2}, \tag{10}$$

where v_c is the average velocity of the flow, R_c is the hydraulic radius, and J is the hydraulic gradient. k is the factor of conversion between the International System of Units (SI) and English units, normally regarded as 1 for SI units. n_c is the roughness coefficient of the culvert, determined by the riverbed materials, water bank, and growing plants [45].

The relative accuracy (RA) was calculated to indicate the accuracy of individual calculation results, and the threshold for the relative accuracy of discharge was set to 20% according to empirical studies [46]. Root-mean-square error and mean absolute percentage error (MAPE) were used to evaluate the performance of logarithmic and exponential velocity distribution in the method. These metrics are calculated as:

$$\text{Relative accuracy} = \frac{|Q_i - Q_m|}{Q_m}, \tag{11}$$

$$RMSE = \sqrt{\left(\frac{\sum (Q_e - Q_m)^2}{n}\right)}, \tag{12}$$

$$MAPE = \frac{1}{n}\sum \frac{|Q_e - Q_m|}{Q_m}, \tag{13}$$

where Q_i is the individual result of the estimated peak discharge through each cross section ($i = 1, 2, 3, 4$), Q_e is the estimated peak discharge through each culvert, Q_m is the average peak discharge through all calculated cross sections in each river channel, and n is the number of calculated cross sections.

4. Results

4.1. Stone Movement and Velocity Distribution in the River Channel

Comparing the derived orthoimages in two years, stone movement by water flow was identified in 10 m river reaches over each selected cross section in two river channels (Figure 6). In the magnified view in the white rectangular area in each river reach over the two years, it is possible to distinguish the movement or non-movement of stones in each river channel. Stones found in the 2017 image but that had disappeared in the 2018 image were identified as moving with the flow throughout the year, while stable stones were regarded as unmoved stones. The nominal particle size of the identified stones was at least 8 cm due to the resolution of the UAV images, and most of these stones can be described as cobbles. The particle sizes of some large boulders in the river section are even more than the average water depth of the selected cross section during the large peak flow event. These boulders were excluded from the circle selection in Figure 6 and the following calculation.

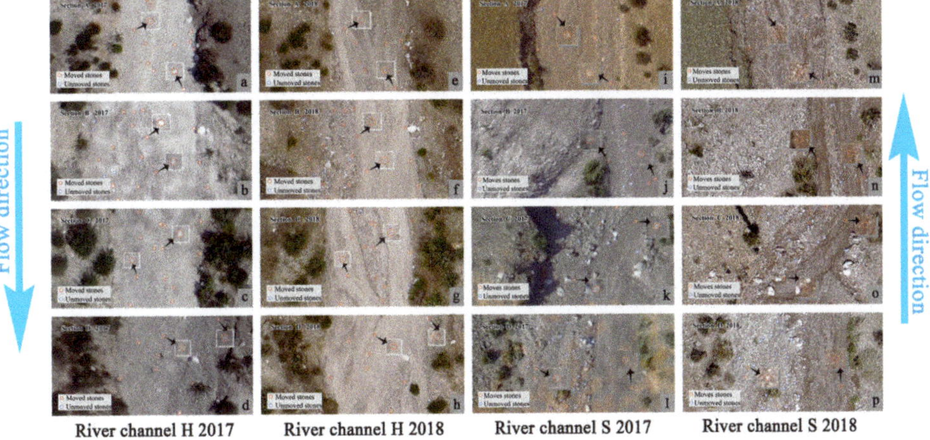

Figure 6. Moved and unmoved cobbles in four river reaches of river channel H and S throughout the year. Magnified views of the moved cobbles are presented in each white rectangular area. (**a–d**) identified stones in river channel H in 2017, (**e–h**) identified stones in river channel H in 2018, (**i–l**) identified stones in river channel S in 2017, and (**m–p**) identified stones in river channel S in 2018.

In both river channel H and S, the movement of cobbles was visible, and cobbles apparently moved by water flow were mainly distributed in the main channel. The main channel was more specific in river channel H with more moved cobbles (8–15) found in red circles (Figure 6a–h). In river channel S (Figure 6i–p), cobbles of various sizes were found, and the main channel was not obviously washed by flow over one year. Consequently, more unmoved cobbles (3-25) in blue circles were identified in river channel S.

Considering that all moving cobbles were mainly moved by water flow, the initial velocities of all identified moving cobbles calculated by the logarithmic or the exponential velocity distribution are presented in Figure 7. In four river reaches of river channel H (Figure 7a–h), the critical initial velocity of the same moving cobble calculated using the logarithmic method is appreciably larger than the velocity calculated using the exponential method. In river reach A, the critical initial velocities of moving cobbles greatly varied, 1.96–2.65 m/s for the exponential method and 3.40–4.39 m/s for the logarithmic method. In addition, there were more cobbles in the higher velocity grading (shown with the longer arrow) for the logarithmic method (Figure 7a) than for the exponential method (Figure 7e). In river reach B, the critical initial velocities of moving cobbles were low, and there were more cobbles with the lowest velocity grading for the logarithmic method (Figure 7b) than for the exponential method (Figure 7f). The channel in river reach C was slowly washed by the water flow and the critical initial velocities of moving cobbles were high (above 4 m/s by the logarithmic method and above 2.4 m/s by the exponential method). There were more cobbles with the highest velocity for the logarithmic method (Figure 7c) than for the exponential method (Figure 7g), with there being more red arrows in Figure 7c.

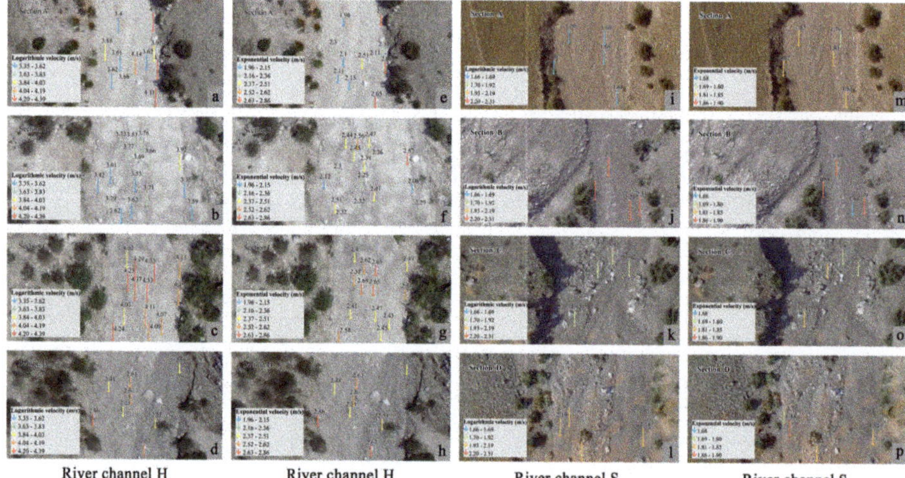

River channel H   River channel H   River channel S   River channel S
Logarithmic velocity distribution   Exponential velocity distribution   Logarithmic velocity distribution   Exponential velocity distribution

Figure 7. The critical initial velocities of all identified moving cobbles in four river reaches of river channel H and S are calculated separately by the logarithmic or the exponential velocity distribution. Each arrow indicates the direction of the water flow when each moving cobble starts to move, and a longer length of the arrow indicates a higher initial velocity. (**a–d**) initial velocities calculated by the logarithmic velocity distribution in river channel H, (**e–h**) initial velocities calculated by the exponential velocity distribution in river channel H, (**i–l**) initial velocities calculated by the logarithmic velocity distribution in river channel S, and (**m–p**) initial velocities calculated by the exponential velocity distribution in river channel S.

Regarding river channel S (Figure 7i–p), velocities obtained by the logarithmic method and exponential method were similar, but the former method produced a slightly broader range of velocities (1.66–2.31 m/s) than the latter method (1.68–1.90 m/s). The logarithmic velocities of moving cobbles in river reach A (Figure 7i) were the lowest among four reaches, all graded at the minimum velocity level, whereas the calculated exponential velocities (Figure 7m) were graded at the average level for the four sections. The critical initial velocities of the moving cobbles in river reach C were similar for the two methods. The condition of river reaches B and D was alike. The maximum velocities calculated using the logarithmic method were concentrated in river reaches B and D (Figure 7j,l) while the maximum and minimum velocities obtained using the exponential method were both in sections B and D (Figure 7n,p).

4.2. Peak Discharge of the River Cross Section

The profiles of each selected cross section at the highest water level are presented in Figure 8. In two river channels, the elevation of the riverbed and riverbank in each cross section continuously decreased downstream. The underwater area of each cross section of river channel H was relatively large, and the section shape was likely rectangular. In contrast, the underwater area of river channel S was relatively small, and the section shape was irregular.

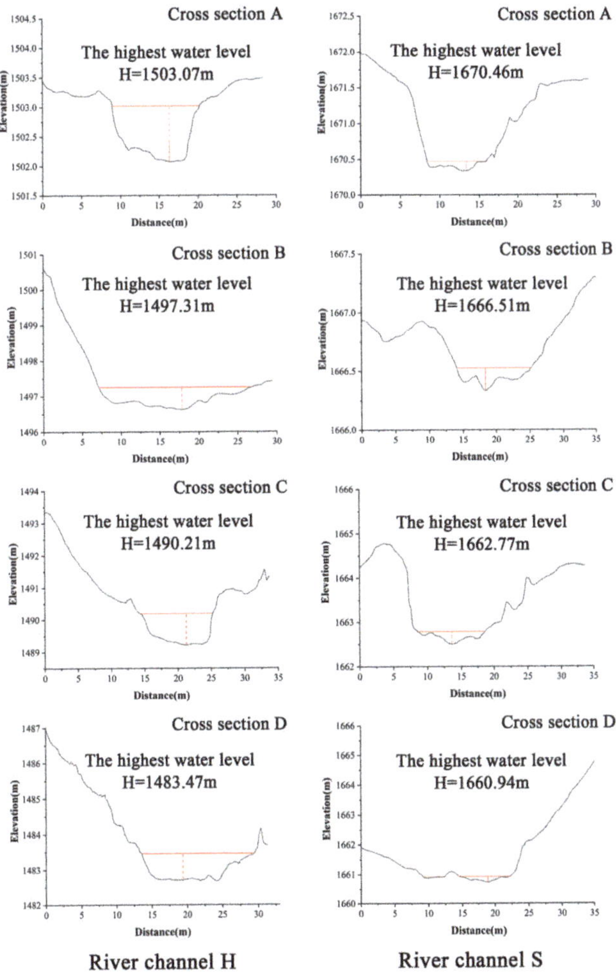

Figure 8. Profile of four selected cross sections in river channel H and S.

When the maximum peak flood flowed through cross sections of the two rivers, the maximum water depth and average water depth of the section greatly varied, while the submerged underwater area was similar (Table 2). At the highest peak water level, the average critical initial velocity of all identified moving cobbles in each river reach was regarded as the cross-sectional flow velocity, and the cross-sectional flow rate and peak discharge calculated using the logarithmic and exponential methods were obtained, as shown in Table 2. The cross-section velocities of the river channel H are different using the two methods, with all logarithmic velocities exceeding 3 m/s and all exponential velocities being around 2 m/s. The difference in the cross-sectional velocity in river channel S is not apparent between the two calculation methods, with all velocities being about 2 m/s. Additionally, the peak discharge calculated by the logarithmic method is larger than that by the exponential method in the case of river channel H. For river channel S, the calculated peak discharges are also similar in four cross sections using the logarithmic and exponential method, with the biggest peak discharges both in section C and the smallest both in section D.

Table 2. Results of the average velocity and discharge through four cross sections in river channels H and S.

River Channel	Cross Section	Cross Section Area A (m^2)	Exponential		Logarithmic	
			Velocity U_c (m/s)	Discharge Q (m^3/s)	Velocity U_c (m/s)	Discharge Q (m^3/s)
	A	8.47	2.33	19.74	3.90	33.03
	B	8.69	2.37	20.60	3.66	31.81
H	C	8.46	2.53	21.40	4.19	35.45
	D	8.84	2.59	22.89	4.08	36.05
	Average	-	2.46	21.15	3.96	34.08
	A	0.89	1.95	1.74	2.17	1.93
	B	0.93	2.00	1.86	1.82	1.69
S	C	1.19	1.95	2.32	2.29	2.73
	D	0.91	1.85	1.68	1.66	1.51
	Average	-	1.94	1.90	1.99	1.96

4.3. Validation of the Estimated River Discharges

The specific profile and key hydraulic variables of each culvert at the time of the maximum peak flood within one year were determined from the topographic information from the DSM and the field measurement (Figure 9). Culvert H is a wide and shallow box-type culvert with a maximum water depth of 1.25 m when the flood peak passed through. The abundant water in river channel H was almost close to the maximum design discharge of the culvert, while culvert S was square with a smaller water flow. The maximum water depth of culvert S was only 0.3 m. The bases of the two culverts were relatively flat laying a large amount of sediment mixed with sporadically distributed gravel stones. The culvert roughness n_c was confirmed as 0.033, a practical value from local hydrological work experience according to conditions of the culvert.

$$v_c = \frac{k}{n_c} R_c^{2/3} J^{1/2} \tag{14}$$

The validation of the peak flood in each river channel is given in Table 3. The flooding in river channel H is evident with a larger peak discharge through culvert H than culvert S, while in river channel S there is little water with larger relative accuracies. The relative accuracies of the logarithmic method and in river channel S all exceed the threshold of 20%. Only the exponential method applied in river channel H indicates an accurate estimation, with the relative accuracies within 10%. The RMSE and MAPE help to further evaluate the performance of different velocity distributions used in the proposed method. RMSE and MAPE are both lower for the exponential method than for the logarithmic method in two river channels, indicating that the former method is more accurate in terms of the incipient motion of moving cobbles regardless of the amount of water in the river channel.

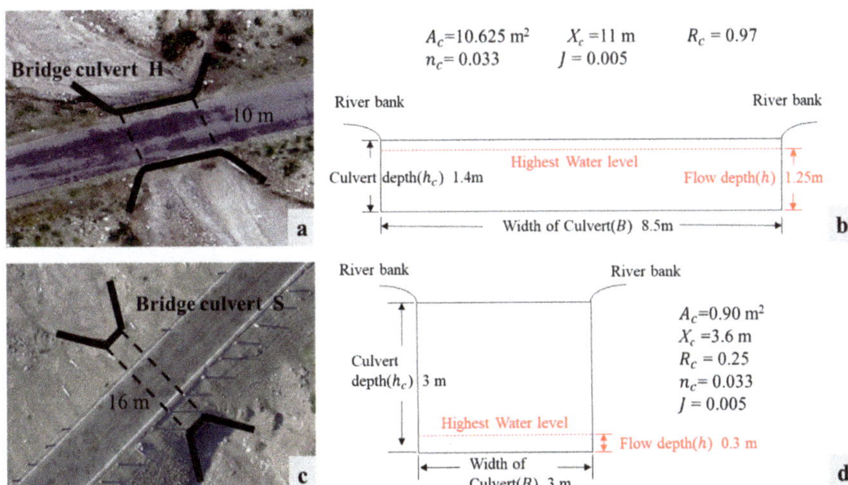

Figure 9. (**a**) and (**c**) are the magnified view of bridge culverts H and S. (**b**) and (**d**) present cross-section profiles and hydraulic parameters used to calculate average velocity and peak discharge through culverts H and S. A_c is the underwater cross-sectional area of the culvert, X_c is the wetted perimeter, J is the hydraulic gradient, R_c is the hydraulic radius, and n_c is the roughness coefficient.

Table 3. Validation of the peak discharge estimation.

River Channel	Cross Section	The Average Peak Discharge (m³/s)	Exponential Velocity Distribution			Logarithmic Velocity Distribution		
			RA	RMSE (m³/s)	MAPE (m³/s)	RA	RMSE (m³/s)	MAPE (m³/s)
H	A	22.02	10%	1.45	0.06	50%	12.18	0.55
	B		6%			44%		
	C		3%			61%		
	D		4%			64%		
S	A	0.76	174%	1.17	1.51	193%	1.29	1.59
	B		186%			169%		
	C		232%			273%		
	D		168%			151%		

5. Discussion

5.1. Value and Extension of the Proposed Method

The present paper provides a convenient, flexible, and reliable method of obtaining the annual peak discharge of an ephemeral river in arid ungauged regions. In empirical studies, statistical methods, hydrological models, and multi-source remote sensing methods are most commonly used to estimate peak discharge (Table 4). In the application of these methods, a variety of ways have been conducted for verification. Most studies used records from nearby gauging stations, such as the water level, the flow rate, and the discharge, to validate discharge estimations [47–53]. A few studies were verified directly by field observations [16,46,54]. Furthermore, flood marks on the deposits, places, cliffs and other facilities were also commonly used to calculate the flood peak discharge with reliable accuracy in regions lacking flood gauging measurements or to reconstruct historical floods [48,55–58]. In arid ungauged regions, flood processes of ephemeral rivers are few and continue to be difficult to directly monitor. Therefore, the identification of clear flood watermarks on construction facilities can be used to verify the existence of flood processes effectively and to validate the maximum peak discharge.

Table 4. Overview of studies to calculate peak discharge.

	Study	Data Used	Approach	Results Verification
Statistical methods	Gallart et al. (2016) [47]	Flow data, interviews and high-resolution aerial photographs	Developing the relationship between different aquatic states and discharge	Flow records measured at the gauging stations
	Yang (2000) [55]	Sediment layers, eyewitnesses accounts	Using slackwater deposits to reveal the magnitude and frequency of palaeofloods	Instrument flood records
	Kimura (2010) [56]	Botanical evidence	Field vegetation investigation, dendrochronological method, Manning formula	Scars and inclinations of the vegetation by flood
	Zha (2009) [48]	Gauging flood level marks, local interviews	Flood level-discharge, slope-area method	Data from gauging stations
Hydrological models	Gallart et al. (1997 & 2002) [49,50]	Gauging reports and soil water content	TOPMODEL	Field Observations & observed streamflow
	Sharma et al. (1994) [10]	Gauging reports and field data	A lumped model	Observed data representing 79 Hydrographs in 15 channel reaches
	Bullard et al. (2007) [51]	Gauging reports and field data	The Urban Runoff and Basin Systems rainfall-runoff model, Hydrologic Engineering Centre-River Analysis System computer model, and empirically-based velocity-area method	Rainfall from gauges in the catchment and streamflow data
	Kim & Shin. (2018) [54]	Gauging reports and field data	The grid-based rainfall-runoff model (GRM), using the relationship between the runoff coefficient, intensity of rainfall, and curve number and the rational method	The observed flow data
Multi-remote sensing methods	Gleason et al. (2014) [16]	Landsat TM	At-Many-Stations Hydraulic Geometry (AMHG)	In situ river gauge observations data of mean daily discharge
	Bjerkie et al. (2003 & 2005) [52,53]	Digital orthophoto quadrangles (DOQs) and ERS-1	Modeled equation based on the resistance equation formulated by Chezy and Manning	Flow measurements database at river sites
	Birkinshaw et al. (2014) [46]	ERS-2, ENVISAT, and Landsat	Substituting Time series of river channel stage levels, channel slope and channel widths into the Bjerklie et al. (2003) [52] equation	Daily in situ discharge measurements data
	Sichangi et al. (2016) [19]	Multiple satellite altimetry data, MODIS, and field data	Using satellite derived parameters: river stages and effective river width to optimize unknown parameters in modified Manning's equation	In situ discharge measurements are used to derive rating curves
	Huang et al. (2018) [59]	Multiple satellite altimetry, Landsat series, Sentinel-1/2, and Google Earth Engine (GEE)	Using river width and water depth derived from the water surface and water level	Obtained high-spatial-resolution images with a UAV

Daily precipitation data of the study area were also collected from the nearest meteorological station during the study period to reveal the characteristics of the precipitation in the study area. According to empirical studies, the average rainfall intensity in the study area is around 22.14 mm/h [60]. Generally, it will take more than one hour to generate surface runoff for continuous precipitation with this rainfall intensity in arid regions [61]. In the study area, considerable precipitation was rare, and the top 10 precipitations (>30 mm) during the study period are presented (Figure 10). Only during concentrated precipitations would the surface flow probably yield and then lead to flooding in the channel. Stones in the ephemeral river channel could have been moved by only one or two flood peaks in September of 2017 or July of 2018. Therefore, the stone movement by flooding in the channel could be used to estimate the discharge of the flood peak.

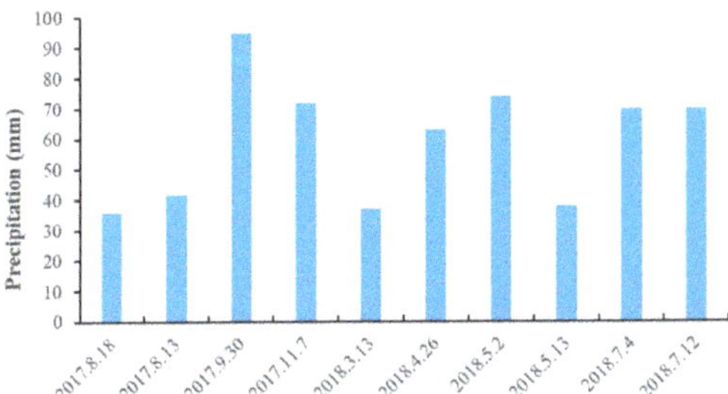

Figure 10. Top 10 precipitations in the study area in 2017.8–2018.8.

The proposed method takes full advantage of the UAV data and the long-term water-free characteristic of ephemeral rivers to estimate peak discharge of ephemeral rivers in ungauged regions. Considering both geographical variables and hydraulic variables, the proposed method is similar to the classic slope-area method, but with the high-resolution UAV data playing a dominant role in the process. On the one hand, the UAV data help to identify terrain information and stone movement. On the other hand, the UAV data are used to derive significant hydraulic variables, such as hydraulic slope, hydraulic radius, and underwater cross-sectional area. The proposed method is vital to the effective long-term monitoring of ephemeral rivers in vast ungauged regions and has great potential to promote at spatial and temporal scales. The primary data required for this method is high-resolution imagery in the dry season. In addition to the UAV data used in this study, other centimeter-level high-resolution images can also be used as data sources. Selecting a typical cross section during the dry season is only needed for in situ measurements, and the water level at the time of the maximum peak flood flow can be determined by comparing the distribution of the soil layer, vegetation, and cobbles. The method established in this paper can also be extended to calculate the maximum peak floods of ephemeral rivers in other typical arid and semi-arid regions with determined dry seasons in these regions.

5.2. Performance Evaluation of the Estimated Velocities

Due to the short and variable flowing time of ephemeral rivers during the year, it remains hard to capture the flooding process in time and to obtain field measurements for validation. According to empirical studies, the culvert peak discharge estimated by flood watermarks was regarded as reliable and could be used to evaluate the estimated peak discharge. In addition to the validation of the peak discharge, we also examined the performance of cross-sectional velocity estimations by the incipient motion of stones. The culvert peak discharge was regarded as the real flood peak discharge and was used to calculate cross-sectional velocities (Equation (15)) for evaluating the estimated velocities by logarithmic and exponential velocity distribution methods:

$$v_i = \frac{Q_c}{A_i}, \tag{15}$$

where v_i represents the cross-sectional velocities calculated by the culvert peak discharge (i = 1,2,3,4,5), Q_c is the culvert peak discharge through each river channel, and A_i is the cross-sectional area (i = 1,2,3,4,5).

Results indicate that the performance of velocity estimation was best using the exponential method in channel H (Figure 11). The overestimation by the logarithmic method was mostly due to the poor performance of original parameters of large grain sediment [27,40]. In contrast, the exponential velocity equation used in this study was the classic formula proposed by Shamov. The formula is concise and generalized, only using K as a comprehensive coefficient to represent three major forces exerted on the stones and the effects of the irregular shape of stones on the incipient motion. The K value used in this study, also proposed by Shamov, has been verified by many experiments and popularization, and is most suitable for inland arid areas [35]. Furthermore, the river channel S only has a small peak discharge of 0.74 m^3/s, which could move only small stones during the flood. Due to the current spatial resolution of UAV data, enough small stones moved by water could not be identified. Therefore, the estimated velocities by both methods in channel S significantly deviated from the velocities calculated by culvert peak discharge.

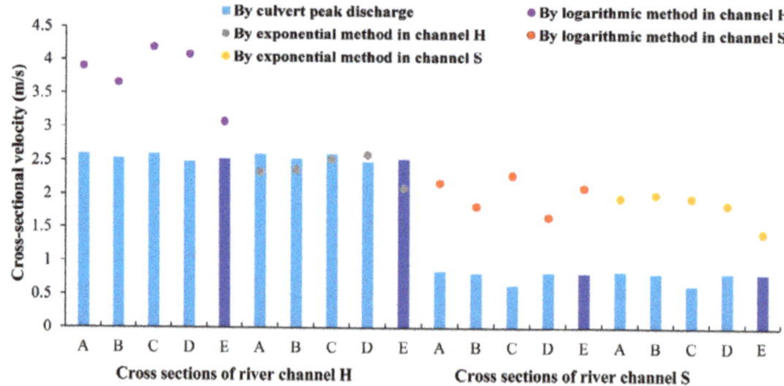

Figure 11. Error analysis of cross-sectional velocity estimation by the incipient motion of moving stones.

We also added one more cross section (cross section E) to each river channel to evaluate the performance of velocity estimation by the proposed method when smaller stones were used. The identified stone movements of cross section E are presented in Figure 12. The main difference of the new cross section E is that moving stones were much smaller than the other four cross sections, with an average nominal diameter of around 8 cm (Table 5). When smaller stones were identified and used, estimated velocities by the exponential method decreased. In channel S, using smaller stones significantly lead to a decrease of the error of cross-sectional velocity, and the error of cross section E is the minimum in channel S. In channel H, with evident flooding, only using small stones in the calculation could not reveal the real peak flood and led to underestimations. Regarding the logarithmic method, using smaller stones also reduced the error of velocity estimations in cross section E of channel H. Regardless of using smaller stones, the logarithmic method still performed poorly in channel S with only a little flow.

Table 5. Comparison of the size of moving stones.

Cross Section	Average Nominal Diameter of Moving Stones (cm)	
	River Channel H	**River Channel S**
A	16.35	14.00
B	18.14	18.56
C	16.25	16.98
D	18.99	19.34
E	7.97	8.30

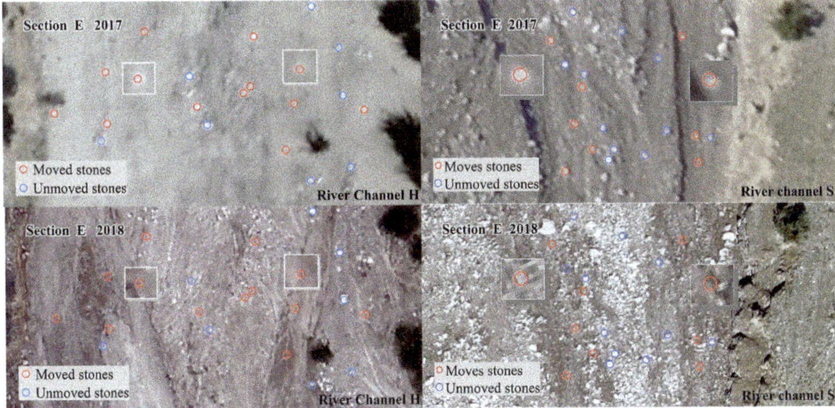

Figure 12. Stone movement in section E in two river channels.

5.3. The Effects of the Selection of Large Boulders on the Estimation of Peak Discharge

Some of the identified moving stones in the river channel are irregular and appreciably larger than other moving stones in the riverbed. They cannot be entirely submerged during the flood peak events, and major forces exerted on them are different when they start to move [62]. Therefore, whether large boulders are selected in the calculation may affect the estimation of peak discharge and is worth exploring [63,64]. Moved boulders were only identified in cross sections A and D of river channel H, and were found in all four cross sections of river channel S. The results of the critical initial velocity and peak discharge in two river channels using two velocity distribution methods are compared in Figure 13.

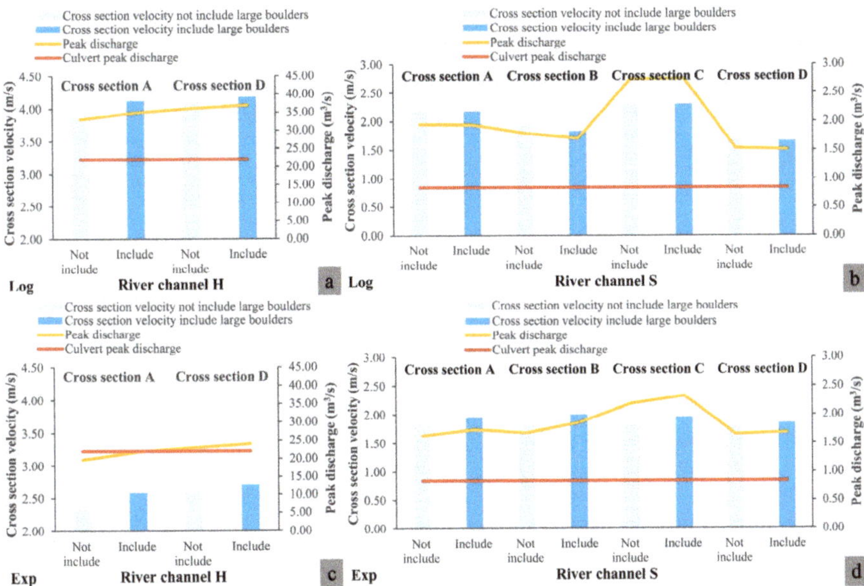

Figure 13. Cross section velocity and peak discharge results on condition of boulder selection: (**a**) logarithmic velocity distribution of river channel H, (**b**) logarithmic velocity distribution of river channel S, (**c**) exponential velocity distribution of river channel H, (**d**) exponential velocity distribution of river channel S.

In the case of river channel H with a large amount of water flow, the selection of large boulders apparently increases the cross-sectional velocity and discharge for both logarithmic and exponential methods (Figure 13a,c). The peak discharge in river channel H obtained using the logarithmic method is initially greater than the culvert flow, and the peak discharge thus deviates further from the culvert peak discharge considering the large boulders. In the case of the exponential method, the calculated peak discharge is similar to the culvert peak discharge, irrespective of whether large boulders are included in the calculation. In the case of river channel S with only a small amount of water, there are many stones with an equal nominal diameter more substantial than the average water depth (around 0.1 m). When large boulders are included, the velocity and peak discharge calculated by the logarithmic method are relatively stable in three cross sections A, C, and D, and only decrease in cross section B with a corresponding decline in the deviation of peak discharge (Figure 13b). Instead, the velocity and peak discharge obtained by the exponential method in four sections of channel S increase, and the deviation from the peak discharge of the culvert further increases (Figure 13d). In all, whether large boulders are considered does affect the discharge estimation, and the effect greatly varies in two river channels with different scales of water flow. With a large amount of water flow, the inclusion of large moved boulders in the calculation significantly increases the estimation of peak discharge in the river channel. Nevertheless, the inclusion of large moved boulders does not significantly affect the results of peak discharge if there is only a small amount of water in the river channel.

5.4. Limitations and Uncertainties of the Present Research

The proposed method of calculating the peak discharge of ephemeral rivers using the critical initial velocity still has some weaknesses and uncertainties. First of all, the resolution of the UAV dramatically affects the number of stones identified moving, and the measurement accuracy of the length and width of the cobbles. With the current image resolution, we failed to distinguish enough small moving stones, which leads to less accurate estimations in river channel S with a small discharge. Topographic data and orthoimages with higher resolution could effectively enhance the performance of the method in the future, especially in ephemeral rivers with a small discharge [65]. Secondly, stone properties, such as density and shape, would also generate uncertainties [66–68]. In the study area, most of the river stones are irregular granite stones. The constant value of density and the generalized size of stones used in critical initial velocity estimations influenced the accuracies of the peak flood discharge. In addition, only classic incipient motion theories were adopted in this study, while the equation to calculate the critical initial velocity of each moved cobble could be modified with local empirical studies for future extension.

6. Conclusions

Ephemeral rivers, as an essential part of surface runoff, maintain the stability of the watershed ecosystem and fulfill the water requirements of humans living in arid ungauged regions. With the advantage of high-resolution data and flexible application, the UAV technique has provided more perspectives for runoff monitoring in addition to the traditional hydrological model and satellite remote sensing methods. In this study, a method was proposed to estimate the flood peak discharge of ephemeral rivers using the sediment mobility and UAV data derived in dry seasons. The main conclusions are as follows:

1. The proposed method performs best in the combination of the exponential method and the river channel with evident flooding (>20 m^3/s), with relative accuracy within 10%. In the river channel with a little flow (around 1 m^3/s), the accuracies are weak because of the limited number of small moving stones found due to the current resolution of UAV data.
2. The exponential velocity distribution method performs better regardless of the amount of water through the two channels, because of the reliable comprehensive coefficient used in the generalized formula.

3. The effects of using small moving stones or large boulders in the proposed method depend on the discharge in the ephemeral river. In the river with a little flow, identifying smaller moving stones would increase the estimation accuracy. In large ephemeral rivers, estimation results are greatly influenced by using smaller stones or large boulders.

The proposed methodology provides a fast and flexible way of estimating the peak discharge of ephemeral rivers in arid ungauged regions with appropriate accuracy. The method is worthy of further development and local calibration for extensive applications in obtaining runoff information of ephemeral rivers in ungauged regions in the future, especially with the quick advance of UAV technology.

Author Contributions: Conceptualization, S.Y., C.L. and H.L.; Methodology, C.L. and P.W.; Software, J.W. and X.R.; Visualization, P.W. and J.W.; Writing – original draft, C.L.; Writing – review & editing, S.Y. and H.L. All authors have read and agreed to the published version of the manuscript.

Funding: This research was funded by National Natural Science Foundation of China (grant number U1603241, 41801334, and U1812401), China Postdoctoral Science Foundation (grant number 2017M620663), and National Young Scientist Natural Science Foundation of China (grant number 41601569).

Acknowledgments: We thank Glenn Pennycook for editing the English text of the original draft of this paper.

Conflicts of Interest: The authors declare no conflict of interest.

References

1. Vorosmarty, C.J.; McIntyre, P.B.; Gessner, M.O.; Dudgeon, D.; Prusevich, A.; Green, P.; Glidden, S.; Bunn, S.E.; Sullivan, C.A.; Liermann, C.R.; et al. Global Threats to Human Water Security and River Biodiversity. *Nature* **2010**, *467*, 555–561. [CrossRef] [PubMed]

2. WWAP (United Nations World Water Assessment Programme). *The United Nations World Water Development Report 3: Water in a Changing World*; UNESCO and Earthscan: Paris, France; London, UK, 2009.

3. WWAP (United Nations World Water Assessment Programme). *The United Nations World Water Development Report 2015: Water for a Sustainable World*; UNESCO: Paris, France, 2015.

4. Chen, Y.N.; Li, B.F.; Li, Z.; Li, W.H. Water resource formation and conversion and water security in arid region of Northwest China. *J. Geogr. Sci.* **2016**, *26*, 939–952. [CrossRef]

5. Leigh, C.; Boulton, A.J.; Courtwright, J.L.; Fritz, K.; May, C.L.; Walker, R.H.; Datry, T. Ecological research and management of intermittent rivers: An historical review and future directions. *Freshw. Biol.* **2016**, *61*, 1181–1199. [CrossRef]

6. Larned, S.T.; Datry, T.; Arscott, D.B.; Tockner, K. Emerging concepts in temporary river ecology. *Freshw. Biol.* **2010**, *55*, 717–738. [CrossRef]

7. Gallart, F.; Prat, N.; García-Roger, E.M.; Latron, J.; Rieradevall, M.; Llorens, P.; Barber´a, G.G.; Brito, D.; De Girolamo, A.M.; Lo Porto, A.; et al. A novel approach to analysing the regimes of temporary streams in relation to their controls on the composition and structure of aquatic biota. *Hydrol. Earth Syst. Sci.* **2012**, *16*, 3165–3182. [CrossRef]

8. Shumilova, O.; Zak, D.; Datry, T.; Von Schiller, D.; Corti, R.; Foulquier, A.; Obrador, B.; Tockner, K.; Allan, D.C.; Altermatt, F.; et al. Simulating rewetting events in intermittent rivers and ephemeral streams: A global analysis of leached nutrients and organic matter. *Glob. Chang. Biol.* **2019**, *25*, 1591–1611. [CrossRef]

9. Skoulikidis, N.T.; Sabater, S.; Datry, T.; Morais, M.M.; Buffagni, A.; Dörflinger, G.; Zogaris, S.; Sánchez-Montoya, M.M.; Bonada, N.; Kalogianni, E.; et al. Non-perennial Mediterranean rivers in Europe: Status, pressures, and challenges for research and management. *Sci. Total Environ.* **2017**, *577*, 1–18. [CrossRef]

10. Sharma, K.D.; Murthy, J.S.R.; Dhir, R.P. Streamflow routing in the Indian arid zone. *Hydrol. Process* **1994**, *8*, 27–43. [CrossRef]

11. De Girolamo, A.M.; Lo Porto, A.; Pappagallo, G.; Tzoraki, O.; Gallart, F. The hydrological status concept. Application at a temporary river (Candelaro, Italy). *River Res. Appl.* **2015**, *31*, 892–903. [CrossRef]

12. Singh, V.P.; Woolhiser, D.A. Mathematical modeling of watershed hydrology. *J. Hydr. Eng.* **2002**, *7*, 270–292. [CrossRef]

13. Birkinshaw, S.J.; O'Donnell, G.M.; Moore, P.; Kilsby, C.G.; Fowler, H.J.; Berry, P.A.M. Using satellite altimetry data to augment flow estimation techniques on the Mekong River. *Hydrol. Process* **2010**, *24*, 3811–3825. [CrossRef]

14. Tarpanelli, A.; Barbetta, S.; Brocca, L.; Moramarco, T. River discharge estimation by using altimetry data and simplified flood routing modeling. *Remote Sens.* **2013**, *5*, 4145–4162. [CrossRef]

15. UNESCO. Detecting water bodies and water related features in the Niger Basin (Niamey area) using SAR data: The ESA TIGER WADE Project. In *Application of Satellite Remote Sensing to Support Water Resources Management in Africa: Results from the TIGER Initiative*; UNESCO: Paris, France, 2010; pp. 108–118.

16. Gleason, C.J.; Smith, L.C.; Lee, J. Retrieval of river discharge solely from satellite imagery and at-many-stations hydraulic geometry: Sensitivity to river form and optimization parameters. *Water Resour. Res.* **2014**, *50*, 9604–9619. [CrossRef]

17. Durand, M.; Gleason, C.J.; Garambois, P.A.; Bjerklie, D.; Smith, L.C.; Roux, H.; Rodriguez, E.; Bates, P.D.; Pavelsky, T.M.; Monnier, J.; et al. An intercomparison of remote sensing river discharge estimation algorithms from measurements of river height, width, and slope. *Water Resour. Res.* **2016**, *52*, 4527–4549. [CrossRef]

18. Wang, S.S.; Zhou, F.Q.; Russell, H.A.J. Estimating Snow Mass and Peak River Flows for the Mackenzie River Basin Using GRACE Satellite Observations. *Remote Sens.* **2017**, *9*, 256. [CrossRef]

19. Sichangi, W.; Wang, L.; Yang, K.; Chen, D.L.; Wang, Z.J.; Li, X.P.; Zhou, J.; Liu, W.B.; Kuria, D. Estimating continental river basin discharges using multiple remote sensing data sets. *Remote Sens. Environ.* **2016**, *179*, 36–53. [CrossRef]

20. Quick, M.C. Reliability of flood discharge estimates. *Can. J. Civ. Eng.* **1991**, *18*, 624–630. [CrossRef]

21. Stewart, A.M.; Callegary, J.B.; Smith, C.F.; Gupta, H.V.; Leenhouts, J.M.; Fritzinger, R.A. Use of the continuous slope-area method to estimate runoff in a network of ephemeral channels, southeast Arizona, USA. *J. Hydrol.* **2012**, *472–473*, 148–158. [CrossRef]

22. Lee, K.; Firoozfar, A.R.; Muste, M. Technical Note: Monitoring of unsteady open channel flows using the continuous slope-area method. *Hydrol. Earth Syst. Sci.* **2017**, *21*, 1863–1874. [CrossRef]

23. Tamminga, A.D.; Eaton, B.C.; Hugenholtz, C.H. UAS-based remote sensing of fluvial change following an extreme flood event. *Earth Surf. Process. Landf.* **2015**, *40*, 1464–1476. [CrossRef]

24. Cook, K.L. An evaluation of the effectiveness of low-cost UAVs and structure from motion for geomorphic change detection. *Geomorphology* **2017**, *278*, 195–208. [CrossRef]

25. Hemmelder, S.; Marra, W.; Markies, H.; De Jong, S.M. Monitoring river morphology & bank erosion using UAV imagery—A case study of the river Buëch, Hautes-Alpes, France. *Int. J. Appl. Earth Obs. Geoinf.* **2018**, *73*, 428–437.

26. Zhao, C.S.; Zhang, C.B.; Yang, S.T.; Liu, C.M.; Xiang, H.; Sun, Y.; Yang, Z.Y.; Zhang, Y.; Yu, X.Y.; Shao, N.F.; et al. Calculating e-flow using UAV and ground monitoring. *J. Hydrol.* **2017**, *552*, 351–365. [CrossRef]

27. Yousef, H. Hydraulics of Sediment Transport. In *Hydrodynamics—Theory and Model*; Zheng, J.H., Ed.; InTech: Shanghai, China, 2012; pp. 23–58.

28. Shields, A. *Application of Similarity Principles and Turbulence Research to Bed-Load Movement*; California Institute of Technology: Pasadena, CA, USA, 1936.

29. Yalin, M.S.; Karahan, E. Inception of sediment transport. *J. Hydraul. Div.* **1979**, *105*, 1433–1443.

30. Yang, C.T. *Sediment Transport: Theory and Practice*; MC Graw-Hill: New York, NY, USA, 1996.

31. Dey, S.; Ali, S.Z. Advances in modeling of bed particle entrainment sheared by turbulent flow. *Phys. Fluids* **2018**, *30*. [CrossRef]

32. Song, S.K.; Bai, J. Increasing Winter Precipitation over Arid Central Asia under Global Warming. *Atmosphere* **2016**, *7*, 139. [CrossRef]

33. Zhang, C.B.; Yang, S.T.; Zhao, C.S.; Lou, H.Z.; Zhang, Y.C.; Bai, J.; Wang, Z.W.; Guan, Y.B.; Zhang, Y. Topographic data accuracy verification of small consumer UAV. *J. Remote Sens.* **2018**, *22*, 185–195.

34. Turner, D.; Lucieer, A.; Wallace, L. Direct georeferencing of ultrahigh-resolution UAV imagery. *IEEE Trans. Geosci. Remote Sens.* **2014**, *52*, 2738–2745. [CrossRef]

35. Ruzgienė, B.; Berteška, T.; Gečyte, S.; Jakubauskiene, E.; Aksamitauskas, V.C. The surface modeling based on UAV Photogrammetry and qualitative estimation. *Measurement* **2015**, *73*, 619–627. [CrossRef]

36. David, G.L. Distinctive image features from scale-invariant keypoints. *Int. J. Comput. Vis.* **2004**, *60*, 91–110.

37. Shao, X.J.; Wang, X.Q. *Introduction to River Mechanics*, 2nd ed.; Tsinghua University: Beijing, China, 2012.

38. Malpas, J.; Fletcher, C.J.N.; Ali, J.R.; Aitchison, J.C. *Aspects of the Tectonic Evolution of China*; The Geological Society: London, UK, 2004.

39. Ma, L.; Wu, J.; Abuduwaili, J.; Liu, W. Geochemical Responses to Anthropogenic and Natural Influences in Ebinur Lake Sediments of Arid Northwest China. *PLoS ONE* **2016**, *11*, e0155819. [CrossRef] [PubMed]

40. Einstein, H.A. *The Bed-Load Function for Sediment Transportation in Open Channel Flows*; Technical Bulletin (No.1026); U.S. Department of Agriculture: Washington, DC, USA, 1950.

41. Shamov, G.I. *Riverine Sediments: River Channel and Their Catchment Erosion (In Russian)*; Gidrometeopizdat: Leningrad, Russia, 1954.

42. Meselhe, E.A.; Hebert, K. Laboratory measurements of flow through culverts. *J. Hydraul. Eng.* **2007**, *133*, 973–976. [CrossRef]

43. Powell, R.W. The Origin of Manning's Formula. *J. Hydraul. Div.* **1968**, *94*, 1179–1181.

44. Wu, Q.; Zhao, Z.; Liu, L.; Granger, D.E.; Wang, H.; Cohen, D.J.; Wu, X.H.; Ye, M.L.; Yosef, O.; Lu, B.; et al. Outburst flood at 1920 BCE supports historicity of China's Great Flood and the Xia dynasty. *Science* **2016**, *353*, 579–582. [CrossRef] [PubMed]

45. William, F.C. *Estimation of Roughness Coefficients for Natural Stream Channels with Vegetated Banks*; U.S. Geological Survey Water-Supply Paper 2441; U.S. Geological Survey: Denver, CO, USA, 1998.

46. Birkinshaw, S.J.; Moore, P.; Kilsby, C.G.; O'Donnell, G.M.; Hardy, A.J.; Berry, P.A.M. Daily discharge estimation at ungauged river sites using remote sensing. *Hydrol. Process* **2014**, *28*, 1043–1054. [CrossRef]

47. Gallart, F.; Llorens, P.; Latron, J.; Cid, N.; Rieradevall, M.; Prat, N. Validating alternative methodologies to estimate the regime of temporary rivers when flow data are unavailable. *Sci. Total Environ.* **2016**, *565*, 1001–1010. [CrossRef]

48. Zha, X.; Huang, C.; Pang, J.; Gu, M. Flood Discharges Reconstructed by Flood Level Marks on Cliff Faces in Middle Reaches of Jinghe River. *Bull. Soil Water Conserv.* **2009**, *29*, 149–153.

49. Gallart, F.; Latron, J.; Llorens, P.; Rabada, D. Hydrological functioning of mediterranean mountain basins in vallcebre, catalonia: Some challenges for hydrological modelling. *Hydrol. Process* **1997**, *11*, 1263–1272. [CrossRef]

50. Gallart, F.; Llorens, P.; Latron, J.; Regues, D. Hydrological processes and their seasonal controls in a small Mediterranean mountain catchment in the Pyrenees. *Hydrol. Earth Syst. Sci.* **2002**, *6*, 527–537. [CrossRef]

51. Bullard, J.E.; Mctainsh, G.H.; Martin, P. Establishing Stage-Discharge Relationships in Multiple-Channelled, Ephemeral Rivers: A Case Study of the Diamantina River, Australia. *Geog. Res.* **2007**, *45*, 233–245. [CrossRef]

52. Bjerklie, D.M.; Dingman, S.L.; Vorosmarty, C.J.; Bolster, C.H.; Congalton, R.G. Evaluating the potential for measuring river discharge from space. *J. Hydrol.* **2003**, *278*, 17–38. [CrossRef]

53. Bjerklie, D.M.; Dingman, S.L.; Bolster, C.H. Comparison of constitutive flow resistance equations based on the Manning and Chezy equations applied to natural rivers. *Water Resour. Res.* **2005**, *41*, W11502. [CrossRef]

54. Kim, N.W.; Shin, M. Estimation of peak flow in ungauged catchments using the relationship between runoff coefficient and curve number. *Water* **2018**, *10*, 1669. [CrossRef]

55. Yang, D.; Yu, G.; Xie, Y.; Zhan, D.; Li, Z. Sedimentary records of large Holocene floods from the middle reaches of the Yellow River, China. *Geomorphology* **2000**, *33*, 73–88. [CrossRef]

56. Taksshi, K.; Sunao, W.; Suk, W.K. Characteristics of past flood behavior reconstructed from botanical evidence in a mountain gorge river system. *Int. J. Eros. Control Eng.* **2010**, *3*, 27–33.

57. Shi, F.; Yi, Y.; Mu, P. *Investigation, Research and Study on the Yellow River Historical Flood*; Yellow River Water Conservancy Press: Zhengzhou, China, 2002.

58. Institute of Hydrology. *Flood Estimation Handbook*; Institute of Hydrology: Wallingford, UK, 1999.

59. Huang, Q.; Long, D.; Du, M.D.; Zeng, C.; Qiao, G.; Li, X.D.; Hou, A.Z.; Hong, Y. Discharge estimation in high-mountain regions with improved methods using multi-source remote sensing: A case study of the Upper Brahmaputra River. *Remote Sens. Environ.* **2018**, *219*, 115–134. [CrossRef]

60. An, W.Z.; Zhu, L.M.; Zhou, W.Q.; Gao, L.L. Analysis on characteristics and change patterns of precipitation in northwest China. *Agric. Res. Arid Areas* **2006**, *24*, 194–199.

61. Wang, Y.; Wang, Z.; Zhou, P. Experiment on rainfall-runoff process on slope of loess plateau. *Acta Conserv. Soli Aquae Sin.* **1991**, *5*, 25–31.

62. Buffington, J.M.; Montgomery, D.R. A systematic analysis of eight decades of incipient motion studies, with special reference to gravel-bedded rivers. *Water Resour. Res.* **1998**, *33*, 1993–2029. [CrossRef]

63. Recking, A. Theoretical development on the effects of changing flow hydraulics on incipient bed load motion. *Water Resour. Res.* **2009**, *45*, W04401. [CrossRef]

64. Singh, U.K.; Ahmad, Z.; Kumar, A.; Pandey, M. Incipient motion for gravel particles in cohesionless sediment mixtures. *Iran. J. Sci. Technol. Trans. Civ. Eng.* **2019**, *43*, 253–262. [CrossRef]
65. Zarco-Tejada, P.J.; Diaz-Varela, R.; Angileri, V.; Loudjani, P. Tree height quantification using very high resolution imagery acquired from an unmanned aerial vehicle (UAV) and automatic 3D photo-reconstruction methods. *Eur. J. Agron.* **2014**, *55*, 89–99. [CrossRef]
66. Chow, V.T. *Open Channel Hydraulics*; McGraw-Hill: New York, NY, USA, 1959.
67. Maidment, D.R. *Handbook of Hydrology*; McGraw-Hill: New York, NY, USA, 1993.
68. Simons, D.B.; Li, R.M.; Al-Shaikh-Ali, K.S. Flow resistance in cobble and boulder riverbeds. *J. Hydraul. Div.* **1979**, *105*, 477–488.

 *remote sensing*

Article

Rainfall Monitoring Based on Next-Generation Millimeter-Wave Backhaul Technologies in a Dense Urban Environment

Congzheng Han [1,2,*], Juan Huo [1,2], Qingquan Gao [1], Guiyang Su [3] and Hao Wang [1,2]

[1] Key Laboratory of Middle Atmosphere and Global Environment Observation,
 Institute of Atmospheric Physics, Chinese Academy of Sciences, Beijing 100029, China;
 huojuan@mail.iap.ac.cn (J.H.); gqq54515@126.com (Q.G.); wanghao114@mails.ucas.ac.cn (H.W.)
[2] University of Chinese Academy of Sciences, Beijing 100049, China
[3] Zhejiang Provincial Meteorological Observatory, Hangzhou 310002, China; sgy52401314@126.com
* Correspondence: c.han@mail.iap.ac.cn

Received: 31 January 2020; Accepted: 17 March 2020; Published: 24 March 2020

Abstract: High-resolution and accurate rainfall monitoring is of great importance to many applications, including meteorology, hydrology, and flood monitoring. In recent years, microwave backhaul links from wireless communication networks have been suggested for rainfall monitoring purposes, complementing the existing monitoring systems. With the advances in microwave technology, new microwave backhaul solutions have been proposed and applied for 5G networks. Examples of the latest microwave technology include E-band (71–76 and 81–86 GHz) links, multi-band boosters, and line-of-sight multiple-input multiple-output (LOS-MIMO) backhaul links. They all rely on millimeter-wave (mmWave) technology, which is the fastest small-cell backhaul solution. In this paper, we will study the rain attenuation characteristics of these new microwave backhaul techniques at different mmWave frequencies and link lengths. We will also study the potential of using these new microwave solutions for rainfall monitoring. Preliminary results indicate that all the test mmWave links can be very effective for estimating the path-averaged rain rates. The correlation between the mmWave link measurement-derived rain rate and the local rain gauge is in the range of 0.8 to 0.9, showing a great potential to use these links for precipitation and flood monitoring in urban areas.

Keywords: rainfall monitoring; remote sensing; rain rate estimation; 5G; millimeter-wave; E-band; LOS-MIMO

---

1. Introduction

Accurate and continuous monitoring of rainfall is very important in many applications. While measurements equipment such as satellites, radar, and weather stations are commonly used for rainfall monitoring, other opportunistic sources for relevant data are being exploited as we are living in the era of big data [1,2]. Big data research is pushing the boundaries of these new technologies and analytic tools, and one such important technology for providing weather data is the use of existing physical measurements in wireless microwave signals, such as the signal level in commercial cellular communication networks for near-ground rainfall monitoring [3]. Microwave backhaul links are used for communications between cellular base stations, and can also be used for measuring the path-averaged rain rate. Utilizing microwave backhaul links for environmental monitoring has also been recently mentioned as one of the Internet of Things (IoT) applications [4]. The densely deployed microwave links all of the world have great potential to be used to complement existing monitoring systems.

In the telecommunications industry, to meet the ever increasing demand in consumer data traffic, many countries have already started the deployment of 5G networks. Microwave backhauling is widely used in many frequency bands above 6 GHz and will also remain an essential medium for transport of 5G, in addition to fiber for macro radio deployments. Forty percent of backhaul connections are expected to be based on microwave by 2023, as reported in [5].

The radio spectrum is a scarce resource that is governed by national and international regulations. Operation of 5G networks requires enormous transmission capacity and ultra-low transmission latency [6], which bring great challenges to microwave transmission links [5]. Lower frequencies allow signals to transmit over longer distances and penetrate buildings better. At higher frequencies, signals have limited coverage, however because of much wider bandwidths they can achieve high capacity. The millimeter-wave (mmWave) technology ranging from 30 to 300 GHz is the key to enabling fast speed and high capacity backhauling in future wireless networks [7].

Governments around the world have allowed operations in the millimeter bands for backhauling, often for little or no licensing fee. Traditional backhaul communications are typically used throughout the 6–42 GHz frequency bands, but they are becoming increasingly popular at various mmWave frequency bands in the 50 GHz, 60 GHz, E-band (71–76 and 81–86 GHz), and 92–95 GHz bands throughout the world [8,9]. Even higher frequencies may be of interest to support the evolution of mobile broadband backhauling beyond 2020, such as 92–114.5 (W-band) and 141–174.8 GHz (D-band) frequency ranges [10,11].

A typical cellular network covers a large area and includes conditions ranging from urban canyons to open rural land. Depend on population density and propagation characteristics, geographical land is classified into one of four categories: dense urban, urban, suburban, and rural. For 5G networks, backhaul links in different frequency bands are adopted for different environments to achieve high capacity [5]. For densely populated areas (categorized as "dense urban"), the E-band (70 or 80 GHz) is favorable for links over a few kilometers and offering high capacity in the 10 Gbps band. The W-band (92–114.5 GHz) and D-band (130–174.7 GHz) are currently under investigation and high millimeter frequency bands will be able to support 40 Gbps capacity over about a kilometer range. Microwave links for urban environments typically have short distances and high capacity demand. An E-band link is suitable in these scenarios. In suburban areas, the link length increases and capacity is lower compared to dense urban and urban areas. Traditional bands (e.g., 6 to 42 GHz), multi-carrier, and multi-band, (mid-band, 15–23 GHz) with E-band solutions can be deployed. The range is typically 8 km. For rural environments, the link length increases further, while end site capacity decreases. For these environments, the traditional microwave band is preferred. The range is typically around 15 km. In this article, examples of several latest microwave technologies, including E-band (71–76 and 81–86 GHz) links, line-of-sight multiple-input multiple-output (LOS-MIMO) backhaul links, and multi-band solutions, are investigated.

E-band can provide double the 5 GHz bandwidth, offering a 10 GHz aggregate spectrum (71–76 and 81–86 GHz), enabling Gbps data rates. An 1.4-km long E-band link, which was tested by Ericsson and Deutsche Telekom, has demonstrated a data transmission rate of 40 Gbps, with a round-trip latency performance of less than 100 ms [12]. This is about four times greater data throughput compared to current mmWave backhaul links. The outdoor small cell E-band backhaul links can be rapidly deployed everywhere, including street lamps, rooftops, and the sides of buildings. E-band is becoming an essential backhauling band with high global alignment, which is also expected to facilitate dense mmWave 5G deployments on street-level sites. However, signals in mmWave frequencies are known to suffer from large propagation loss and rain attenuation is one of the main limiting factors [13,14]. As a result, the E-band links are used for high capacity transmission but at shorter distances compared to traditional bands, and can generally be applied to lengths of up to 3 km.

The emerging concept of carrier aggregation enables a much more efficient use of diverse backhaul spectrum assets. As it is easier to obtain wider channels at higher frequencies, we can aggregate a low frequency carrier for availability and a high frequency carrier for capacity. A multi-band solution combining E-band with traditional bands can increase the transmission distance. The traditional band links are used to guarantee the availability of high-priority services and support transmission distances of 3 to 10 km. This will allow the use of E-band to provide transmission for 5G in much wider geographical areas. A commonly used combination is 18–42 GHz bands and E-band (70 or 80 GHz) for distances up to 5 km (dense urban and urban environments), and 6-15 and 18–42 GHz bands for longer transmission ranges (suburban and rural environments). In this article, we will study the impact of atmospheric conditions on E-band backhaul links in city environments, especially rain attenuation. A 38 GHz mmWave backhaul link deployed in the same region will also be studied for comparison analysis.

The different forms of antenna technology refer to single or multiple inputs and outputs. When there are more than one antenna at the transmit side and receive side of the radio link, this is referred to as a multiple-input multiple-output (MIMO) system. MIMO can be used to provide improvements in both channel robustness and channel throughput compared to a single-input single-output (SISO) system, where there is a single antenna at the transmit side and receive side of the radio link [15,16]. MIMO has been widely used in wireless local area networks (WLANs), long-term evolution (LTE) mobile networks, and fifth generation cellular systems. MIMO with a spatial multiplexing scheme allows capacity to increase almost linearly with the number of antennas. Recently, MIMO technology has been applied to increase transmission rates in point-to-point backhaul links in mmWave bands for next-generation wireless backhaul networks [17–20]. This is referred to as a line-of-sight multi-input multi-output (LOS-MIMO) communication system. Most existing studies of the impact of rain on signal attenuation are for SISO microwave links. Signal attenuation in a MIMO backhaul link due to rain and other meteorological conditions are yet to be studied.

Large signal attenuation can occur due to heavy rain and can severely affect the mmWave link quality. Modeling and measurements of mmWave attenuation due to rainfall for near-ground communication links have been addressed in recent studies and are considered very important topics [21–26]. A power law empirical mathematical model relating the rain rate and rain-induced signal attenuation is given by International Telecommunication Union (ITU) Recommendation P. 838-3 and other relevant papers [27–29]. This model is used in the design of reliable communication systems. Recently, it has been suggested that microwave links in cellular networks can be considered as passive weather monitoring sensors, and a power law model relating the rain and rain attenuation can be adopted for rainfall estimation [30–32]. This approach exploits the fact that the strength of electromagnetic signals is weakened by certain weather conditions, especially rain. It makes microwave linking a potential tool for monitoring rainfall conditions with high temporal and spatial resolution. There is significant potential to increase the number of observation points and improve the quality of weather services, including forecasting, now-casting, flood warnings, and hydrological measurements. The most powerful impact is expected in developing countries and regions where no other measurements currently exist. Making use of the existing commercial wireless networks is equivalent to deploying a very high density of weather monitoring sensors and forming wireless environmental sensor networks (WESN) [30,33] all over the world.

The major contributions in this paper are (1) studying and comparing the rain attenuation characteristics of latest mmWave backhaul links and (2) studying the performance of rain rate estimation based on SISO and MIMO links at different frequencies, using existing measurements of the received signal level of the mmWave backhaul links.

This paper is organized as follows. Section 2 presents a brief summary of characteristics of mmWave propagation, the method of rain rate retrieval from the received signal level of mmWave links, and the setup of outdoor test links. Section 3 presents the experimental results on signal variation in sunny and rainy weather, wet antenna effects and the performance of rain rate retrieval studies using the latest mmWave backhaul test links. Then, the uncertainties in the experiment and the potentials of the proposed technology in supporting rainfall and flood monitoring in urban areas are discussed in Section 4. Finally, we summarize the work in Section 5.

2. Materials and Methods

2.1. Millimeter-Wave Propagation

For a point-to-point LOS mmWave link, the received power P_B (dBm) may be related to the transmitted power P_T (dBm); the antenna gains G_T (dBi) and G_R (dBi); and the propagation path loss (*PL*), atmospheric loss (*AL*) and other losses (*OL*). The link can be written as:

$$P_B = P_T + G_T + G_R - PL - AL - OL \tag{1}$$

The propagation path loss can be expressed as [34]:

$$PL(f_c, d) = 32.4 + 20log10(f_c) + 10nlog10(d/d_0) + \chi_\sigma, d \geq 1m \tag{2}$$

where f_c denotes the carrier frequency in GHz, d is the transmitter and receiver separation distance, the reference distance d_0 is 1 m, and n represents the path loss exponent. Here, χ_σ is a zero-mean Gaussian random variable with a standard deviation σ in dB.

2.2. LOS-MIMO-Based mmWave Backhaul System

The principle of a 2 x 2 LOS-MIMO microwave backhaul link is to design a MIMO channel with a phase difference of 90 degrees between short and long paths to make the signal streams orthogonal to each other. The channel is denoted by a $N \times M$ matrix **H**, and each element represents the channel from the mth transmit (Tx) antenna to the receive (Rx) antenna nth. Note that $N = M = 2$ in our measurement setup. Each element of the channel matrix can be written as $H_{mn} = e^{j\theta_{mn}}$, where θ_{mn} is the phase of the sub-channel. For the phase of the sub-channel from the nth transmit antenna to the mth receive antenna, $\theta_{mn} = 2\pi r_{mn}/\lambda$, where λ is the wavelength and r_{mn} is the propagation distance between the transmit antenna n and receive antenna m [19,20]. This can be achieved by designing the antenna separation distance at the transmitter d_1 and receiver d_2 to fulfill, $d_1 \times d_2 = \lambda L/N$, where L is the path length between the transmit site and receive site. Let **X** and **Y** denote the transmit and receive signal vector, respectively. The $N \times 1$ received signal vector can be written as:

$$\mathbf{Y} = \sqrt{P_R}\mathbf{HX} + \mathbf{W}$$

$$\begin{bmatrix} Y_1 \\ Y_2 \end{bmatrix} = \sqrt{P_R} \begin{bmatrix} H_{11} & H_{12} \\ H_{21} & H_{22} \end{bmatrix} \begin{bmatrix} X_{A1} \\ X_{A2} \end{bmatrix} + \begin{bmatrix} W_1 \\ W_2 \end{bmatrix} \tag{3}$$

where **W** is the $N \times 1$ complex additive white Gaussian noise vector and its variance equals No; P_R is the average received power, expressed in watts, where $P_R = 10^{P_B/10}$.

In practice, a sub-optimal linear zero-forcing algorithm can be applied to simply invert the channel and independently decode the data streams at the receiver to recover the spatially multiplexed signals:

$$\mathbf{G} = P_R^{-1/2}\mathbf{H}^+ = P_R^{-1/2}\left(\mathbf{H}^H\mathbf{H}\right)^{-1}\mathbf{H}^H \tag{4}$$

The character $^+$ denotes the pseudo-inverse operation. By applying the pseudo-inverse of the channel matrix to the received signal we get:

$$\begin{aligned} \widetilde{\mathbf{X}} = \mathbf{G}\left(P_R^{1/2}\mathbf{HX} + \mathbf{W}\right) &= \mathbf{X} + \mathbf{GW} \\ &= \mathbf{X} + P_R^{-1/2}\mathbf{H}^+\mathbf{W} \end{aligned} \tag{5}$$

The SNR after interference cancellation for the ith sub-channel is given as:

$$SNR_i = \frac{P_R}{N_0\left[\left(\mathbf{H}^H\mathbf{H}\right)^{-1}\right]_{i,i}} \tag{6}$$

Here, $\left[\left(\mathbf{H}^H\mathbf{H}\right)^{-1}\right]_{i,i}$ refers to the (i, i)th elements of $\left(\mathbf{H}^H\mathbf{H}\right)^{-1}$. The independent and identical (i.i.d) MIMO channel capacity, assuming equal transmit power, is given as:

$$C = \sum_{i=1}^{N} \log_2(1 + SNR_i) \tag{7}$$

where $(\cdot)^H$ denotes the Hermitian transpose.

2.3. Atmospheric Attenuation

Atmospheric attenuation and weather effects are important for mmWave propagation. The atmospheric loss is generally defined in terms of decibel (dB) loss per kilometer of propagation. Since the fraction of the signal loss is a strong function of the distance travelled, the actual signal loss experienced by a specific mmWave link due to atmospheric effects depends directly on the length of the link. A simple model describing the attenuation of mmWave for the range of 1 to 100 GHz through atmosphere AL can be described as follows:

$$AL = A_r + A_v + A_o + A_p \text{ (dB)} \tag{8}$$

which primarily includes the attenuation effects of dry air (including oxygen), humidity, fog, and rain. Here, A_r refers to the attenuation caused by rain, A_v represents the water vapor attenuation, A_o represents the attenuation due to dry air, and A_p is the attenuation as a result of other-than-rain precipitation (i.e., fog, sleet, snow).

There are other possible causes of losses (OL), such as the coaxial cable loss at the transmitter and receiver; temperature and water vapor affecting the stability of the transmit and receive signal terminals (equipment, circuits, etc.); wetness of the transmit and receive antenna surface causing considerable attenuation; and anything that obstructs the LOS channel introducing additional loss.

2.3.1. Water Vapor Attenuation

Attenuation due to absorption by oxygen and water vapor is always present and should be included in the calculation of total propagation loss at frequencies above approximately 10 GHz. For the millimeter frequency range, the resonance lines for water vapor and oxygen are at 22.3, 183.3, 323.8 GHz; and 57–63 and 118.74 GHz, respectively.

To illustrate the electromagnetic signal attenuation due to dry air and water vapor, Figure 1 was plotted based on the equations given in [35], for a given barometric pressure and temperature. The first excess attenuation occurs at around 22 GHz due to water vapor, and the second at 60 GHz due to oxygen. Oxygen absorption has a maximum attenuation at 60 GHz and contributes to 7–15 dB/km in the received signal strength at the frequency range of 57–63 GHz. For 32 and 38 GHz, the signal is mainly affected by water vapor, and the attenuation is less than 0.15 dB for a 1 km link length. For signals in E-band, the attenuation due to humidity can reach approximately 0.5 dB/km [24,35].

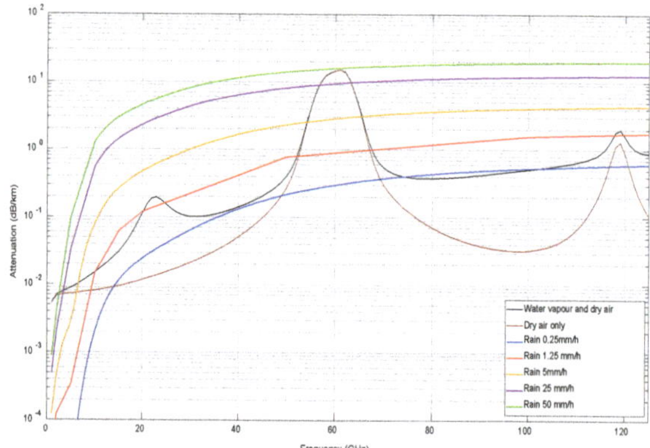

Figure 1. Frequency-dependent attenuation of electromagnetic radiation in standard atmosphere (barometric pressure 1013 mbar, temperature 15 °C, water vapor density of 7.5 g/m^3) and rain attenuation in dB/km at various rainfall rates.

2.3.2. Rainfall Effects on Radio Signals

A power law empirical model is often used in the calculation of rain-induced attenuation A_r and the average rain rate R [28,29] along the path:

$$A_r = aR^b L \text{ (dB)} \tag{9}$$

where the constants a and b are related to frequency, rain temperature, the rain drop size distribution, and polarization, depending on the rain attenuation model. In our study, L (km) is the length of the microwave link and A_r is the overall signal attenuation induced by rain between the transmitter and receiver. A set of commonly used power law coefficients can be found in International Telecommunication Union (ITU) Recommendation P. 838-3 [28]. The power law coefficients for vertical polarization (a_V, b_V) and horizontal polarization (a_H, b_H) at different frequencies are summarized in Table 1. Assuming vertical polarization, the power law coefficients a and b used in our measurement are given in Table 1.

Table 1. Power law coefficients for different frequencies.

Frequency	a_H	a_V	b_H	b_V	a	b
32 GHz	0.2778	0.2646	0.9302	0.8981	0.2646	0.8981
38 GHz	0.4001	0.3844	0.8816	0.8552	0.3844	0.8552
72 GHz	1.0618	1.0561	0.7293	0.7171	1.0561	0.7171
82 GHz	1.1946	1.1915	0.7077	0.6988	1.1915	0.6988

The rain attenuation values for the considered millimeter frequency band in this study at various rainfall rates (R) are given in Table 2. We categorized the rainfall intensity into six groups, including very light rain (rain rate < 1mm/h), light rain (1 mm/h ≤ rain rate < 2 mm/h), moderate rain (2 mm/h ≤ rain rate < 5 mm/h), heavy rain (5 mm/h ≤ rain rate < 10 mm/h), very heavy rain (10 mm/h ≤ rain rate < 20mm/h), extreme heavy rain (rain rate ≥ 20 mm/h). The theoretical rain-induced signal attenuation per kilometer based on Equation (9) for our considered millimeter frequencies is presented in Figure 2a. Compared to other atmospheric factors, atmospheric attenuation due to rain is one of the most noticeable components of excess losses at our considered frequencies. It is not important for low frequency bands, but rain affects links in millimeter frequency ranges, especially for higher frequencies. For increasing rain rate, the rain attenuation experienced by E-band links becomes more severe compared to the 32 and 38 GHz links. For a very heavy rain event, the rain-induced signal attenuation can be up to 9.7 dB for the 82 GHz link at a rain intensity of 20 mm/h. Figure 2b gives the theoretical rain attenuation for our measurement setup. Both 32 and 38 GHz links are 7-km long, and both 72 and 82 GHz are 3 km long. Based on the power law coefficients given by ITU-R P. 838-3, a 7-km long 38 GHz link experiences similar or less signal attenuation compared to a 3 km E-band link for rain intensity lower than 7 mm/h. At 32 GHz, the rain attenuation is lower compared to an E-band signal, even if the link has more than double the deployment length.

Table 2. Signal loss due to rain (dB/km).

Description	Rain Rate (mm/h)	Signal Loss (dB) Per Kilometer			
		32 GHz	38 GHz	72 GHz	82 GHz
Very light rain	$R < 1$	< 0.3	< 0.4	< 1.1	<1.2
Light rain	$1 \leq R < 2$	< 0.5	< 0.7	< 1.7	<1.9
Moderate rain	$2 \leq R < 5$	< 1.1	< 1.5	< 3.3	<3.7
Heavy rain	$5 \leq R < 10$	< 2.1	< 2.75	< 5.5	<6.0
Very heavy rain	$10 \leq R < 20$	< 3.9	< 5.0	< 9.1	<9.7
Extreme rain	$R \geq 20$ (e.g., 50)	≥ 3.8 (8.9)	≥ 5.0 (10.9)	≥ 9.1 (17.5)	≥ 9.7 (18.3)

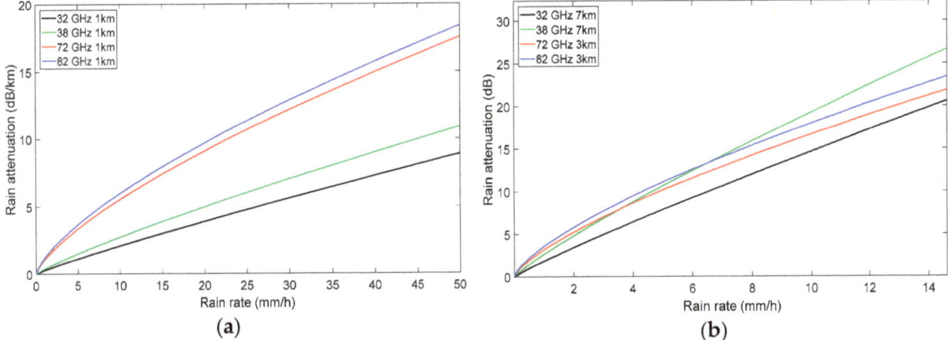

Figure 2. The theoretical rain-induced signal attenuation for various rain rates at different frequencies: (**a**) per kilometer; (**b**) over a 7-km long 32 GHz 2 × 2 line-of-sight multiple-input multiple-output (LOS-MIMO) link, a 7-km long single-input single-output (SISO) 38 GHz link, a 3-km long 72 GHz link, and a 3-km long 82 GHz test link using our measurement scenario.

2.3.3. Rain Rate Estimation Using the Receive Signal Levels from Millimeter-Wave Links

The use of microwave links for near-ground environmental monitoring is a new technology, and it has shown to be an effective tool for rainfall monitoring in over 20 countries. The method of retrieving the rain rate from the rain-induced attenuation in the received signal level is based on the power law model in Equation (9) from the recommendations of the ITU-R P. 838-3:

$$R = \sqrt[b]{\frac{A_r}{aL}} \; (\text{mm/h}) \tag{10}$$

For a vertical polarized setup, the power law coefficients a and b for the frequencies considered in this study are presented in Table 1, derived from [28]. Therefore, the average rain rate along a link can be derived from the microwave link rain-induced attenuation. In order to determine the signal attenuation caused by rain, we will need to choose a reference level P_{ref}, which can be calculated using the average of the received power in the previous 3 hours in dry weather before rain [36,37]. If there are I observations, then the rain attenuation for the ith observation becomes:

$$A_{r,i} = P_{ref} - P_{R,i} \tag{11}$$

For the case of the LOS-MIMO system, we assume that link length L is the same for all MIMO data streams.

2.4. Outdoor Measurement

We present here a summary of the measurements from three outdoor test mmWave links and local rainfall measurements using rain gauges. These measurements are also used to validate the accuracy of rain rate estimation.

The locations of the three test links are illustrated in Figure 3. The measurement setup parameters are given in Table 3. The transmitter and the receiver of the 32 GHz LOS-MIMO link and the 38 GHz link were closely installed between site A and B. Both links had a length of approximately 6.87 km. One side of the two links was on the roof of the Ericsson building (site A), close to water. Both the 32 GHz LOS-MIMO link and 38 GHz SISO link were operated in a line-of-sight environment. The 32 GHz LOS-MIMO link was horizontally deployed, with antenna separation at both sites. The antenna separation at one end (site B) was fixed and installed at 5 m, whereas antennas at the other end (site A) were installed on tripods, with an antenna separation distance of 7.93 m. The details of the measurement setup can be found in [38]. Here, the radio link operating in the forward transmission direction is referred to as link 1, while the radio link operating in the opposite transmission direction is referred to as link 2. The 32 GHz LOS-MIMO link employed 2 antennas at both the transmit and receive sides, while 2 data streams were transmitted in the forward direction and in the opposite direction over the radio link. Altogether, 4 data streams were in transmission simultaneously. Note that the data used in this study from the LOS-MIMO test link is recorded at the receiver without post-processing, but for practical deployment, the received data streams should be decoupled using carrier-to-interference (C/I) measurement. Since the rain attenuation is mainly distance dependent as shown in Equation (9), the actual wanted and interference signal power will not impact on the accuracy of rain retrieval analysis.

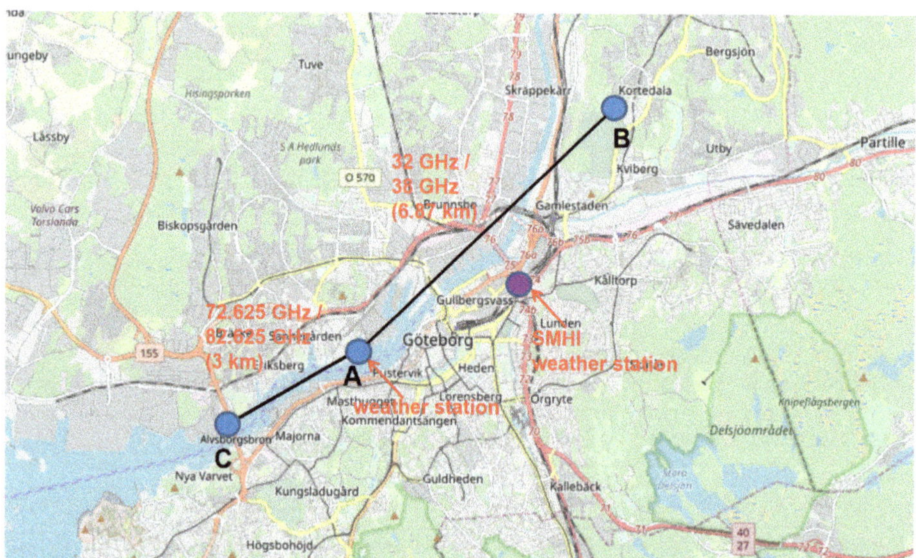

Figure 3. The measurement setup and locations in Gothenburgh, Sweden. SMHI, Swedish Meteorological and Hydrological Institute.

Table 3. Outdoor measurement parameters.

Parameter	32 GHz	38 GHz	72.625 GHz	82.625 GHz
Sampling interval	30 s			
Antenna type	Cassegrain antenna			
Location	57°42′18.97″ N, 11°56′29.67″ E; 57°44′52.8″ N, 12°1′26.4″ E		57°42′18.97″ N, 11°56′29.67″ E; 57°41′20.04″ N, 11°54′10.76″ E	
Link length	6.87 km		3 km	
Setup	MIMO	SISO	SISO	SISO
Antenna no.	2×2	1×1	1×1	1×1
Antenna Separation	5 m; 7.93 m	N/A	N/A	N/A
Tx power	5 dBm	15 dBm	7 dBm	7 dBm
Tx antenna gain	43.6 dBi	40.3 dBi	50.5 dBi	50.5 dBi
Tx half power beam width	0.5°	0.5°	0.5°	0.5°
Tx polarization	V	V	V	V
Rx antenna gain	43.6 dBi	40.3 dBi	50.5 dBi	50.5 dBi
Rx half power beam width	0.5°	0.5°	0.5°	0.5°
Rx polarization	V	V	V	V

For the 38 GHz SISO link, one data stream was transmitted in the forward and reverse directions, and there were 2 data streams transmitting instantaneously over the radio link.

For the 3 km SISO E-band link, one end was also installed on the roof of the Ericsson building at site A and was deployed between sites A and C. The geographic locations of the E-band links are listed in Table 3. In the forward transmission the radio link operates at 72.625 GHz, while in the reverse transmission the radio link operates at 82.625 GHz.

All the mmWave links have a sampling interval of 30 seconds. Rain, humidity, temperature, air pressure, and wind information were provided by a weather station equipped with a rain gauge located at the rooftop of the Ericsson building. The accuracy of the rain gauge is of the order +/− 3% [39]. We used the measurement from this rain gauge for the analysis in the results section. As this weather station was located on the rooftop, we also selected the closest rain gauges operated by the Swedish Meteorological and Hydrological Institute (SMHI). The SMHI rain gauge reported the cumulative rain amount in mm over 1 day and its location is indicated in Figure 3.

3. Measurement Results

During the outdoor trial measurements, the received signal level and path attenuation were recorded in changing weather conditions for the three mmWave links at 32 GHz, 38 GHz, and E-band ranges. All the links were in operation at the same time.

3.1. Rainfall Effects

The received signal variation and the rain intensity on 7 and 11 June 2017 are presented in Figures 4 and 5, respectively. Assuming other losses (wet antenna attenuation, water vapor attenuation, etc.) are the same, the average signal attenuation values over 1 km distance are also compared in Figures 4b and 5b, which show more clearly the impact of rain on links at different frequencies. The measured rain attenuation result is consistent with the theoretical predictions in Table 2. Although the 32 and 38 GHz links were built over a much longer distance compared to the E-band links, the rain-induced attenuation in the 32 and 38 GHz links was lower. This difference in rain-induced attenuation between the lower and higher frequency mmWave link becomes more significant as the rain intensity increases. The 32 GHz and 38 GHz links are more robust to poor weather conditions compared to the E-band link, although they are deployed over longer distances. The attenuation of the links, especially the E-band link, was a lot more severe than expected in late evening on 7 and 11 June. One possible reason is that the rain gauge provides a point measurement, while the measured signal attenuation is caused by the rain along the link. Although one side of all links is located on top of the same building, the links are built across different and wide areas, with a separation distance of up to 10 km. The rain intensity that each link experienced along the path could be very different from the rain gauge measurement, and it could contribute to the difference in attenuation value. In addition, the rain gauge that was used for the analysis was located on the rooftop of a building. There could be a significant under catch of rainfall in the gauge especially during windy conditions, and more rainfall could have been detected by the mmWave link.

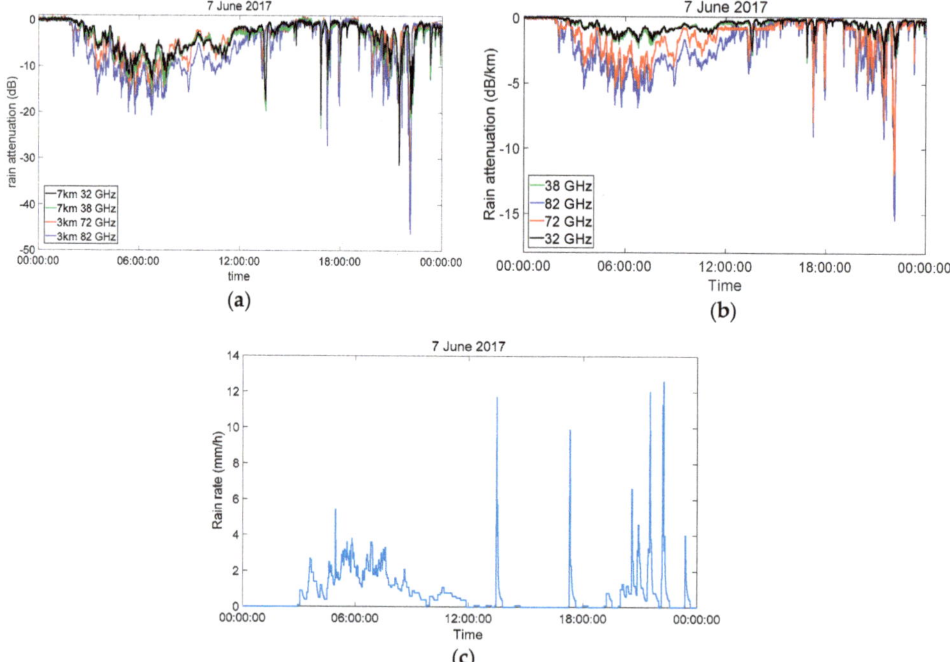

Figure 4. (**a**) Received signal variation of the test links on June 7, 2017; (**b**) received signal variation of the test links averaged over 1 km distance; (**c**) rain rate monitored by a rain gauge.

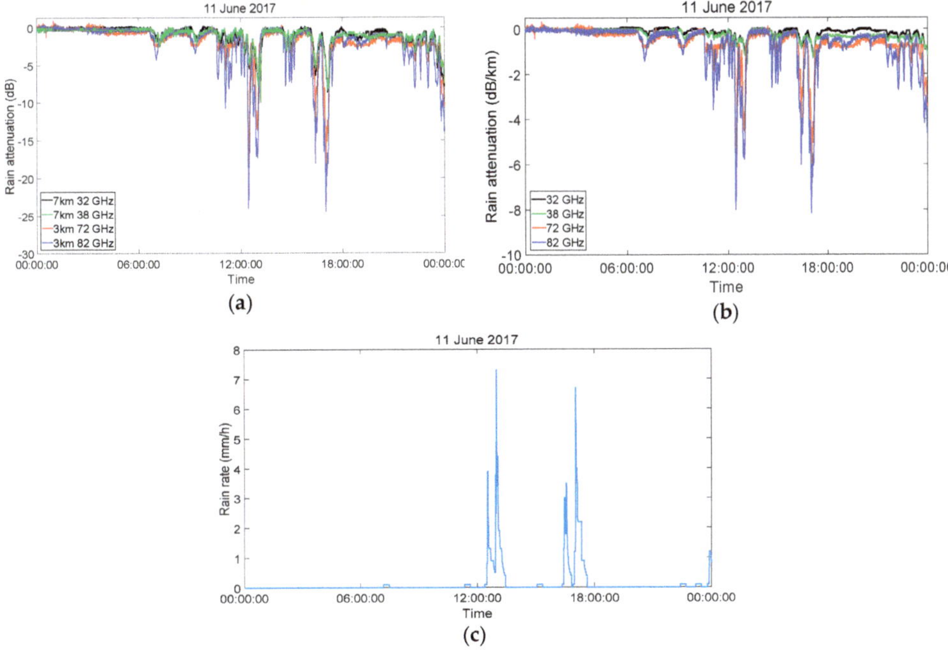

Figure 5. (**a**) Received signal variation of the test links on June 11, 2017. (**b**) received signal variation of the test links averaged over 1 km distance. (**c**) rain rate monitored by a rain gauge.

3.2. Water Vapor Attenuation

As discussed in the previous section, change in water vapor level may also cause additional attenuation. Atmospheric attenuation of signal level due to dry air and water vapor is related to the air pressure, temperature, and the water vapor density. For the dry period from 13 to 15 June 2017, the changes in temperature, air pressure, and humidity level are presented in Figure 6a, and the variations of the received signal level from the test links are also given in Figure 6. During these sunny days, it can be seen that the received signal level also varies over time as a result of atmospheric effects. The variation of temperature and humidity is inversely related with a correlation coefficient of −0.9. Attenuation from water vapor is a function of the pressure p (hPa), temperature T (°C), and the water vapor density ρ (g/m^3) [35]. For the frequencies considered in this study, the signal attenuation due to oxygen absorption is negligible. The attenuation due to changes in water vapor density at 32 and 38 GHz is very low, approximately up to 1 dB for a 7-km long link, as shown in Figure 6b,c. As the frequency increases to the E-band, variation in water vapor density can contribute over 0.45–0.55 dB/km for E-band signals, and therefore a total of 1.35–1.65 dB for the 3 km link, which is illustrated in Figure 6d.

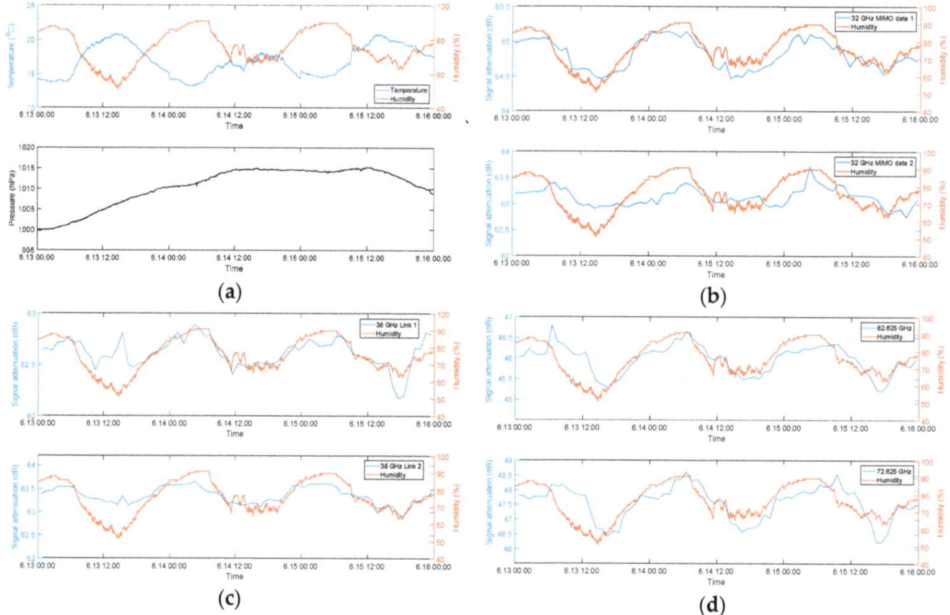

Figure 6. (**a**) The variations in temperature, humidity, and pressure during 13-15 June, 2017. (**b**) The variations in the signal attenuation and humidity of the 32 GHz link (link 1, data streams 1 and 2). (**c**) Links 1 and 2 at 38 GHz; (**d**) The 72.625 and 82.625 GHz links.

3.3. Data Post-Processing and Uncertainties

The determination of the baseline level and wet antenna attenuation are very important for accurate estimation of rain rates from the mmWave links [40–42]. Here, we consider the reference level to be the average received signal strength over 3 hours in dry weather before rain. Subtracting the baseline level, also called the reference level, from the actual received signal levels gives the rain-induced, path-integrated attenuation, which can be transformed into the path averaged rain rate.

During rainy periods, the dampening of the radomes of the antenna causes attenuation, and this additional attenuation factor is known as the wet antenna effect [40]. The wet antenna effect has been shown to be consistent for a specific microwave link, but varies from link to link. Therefore, it has been suggested that the wet antenna attenuation depends on the specific link properties, such as the signal polarization, frequency, and the radome material, meaning each link needs to be examined individually. For a rainfall event lasting for a long period of time, the wet antenna attenuation is expected to increase with increasing thickness of water film on the antenna. In addition to wet antenna attenuation, water vapor may also cause additional variation at high frequencies, as shown in Section 3.2. As discussed in [41,42], bias due to the instability of the transmit power of commercial microwave backhaul equipment could be up to 1.6 dB, therefore causing more attenuation to the received signal level. Therefore, hardware (radio, antenna) and alignment possibly also contribute to this difference. This could be studied as future work if long term measurement is carried out and analyzed.

For the experiment period, the measured statistics of rain attenuation versus calculated rain attenuation, based on Equation (9), are plotted in Figure 7. Each point represents the rain rate and measured attenuation value from the link measurement over a 15 min interval. The correction factor, which mainly accounts for wet antennas, is applied to the measurement data. As the water vapor attenuation is insignificant, it is combined in one correction factor. The distribution of rain attenuation values for increasing rain rate before and after applying the correction factor is presented.

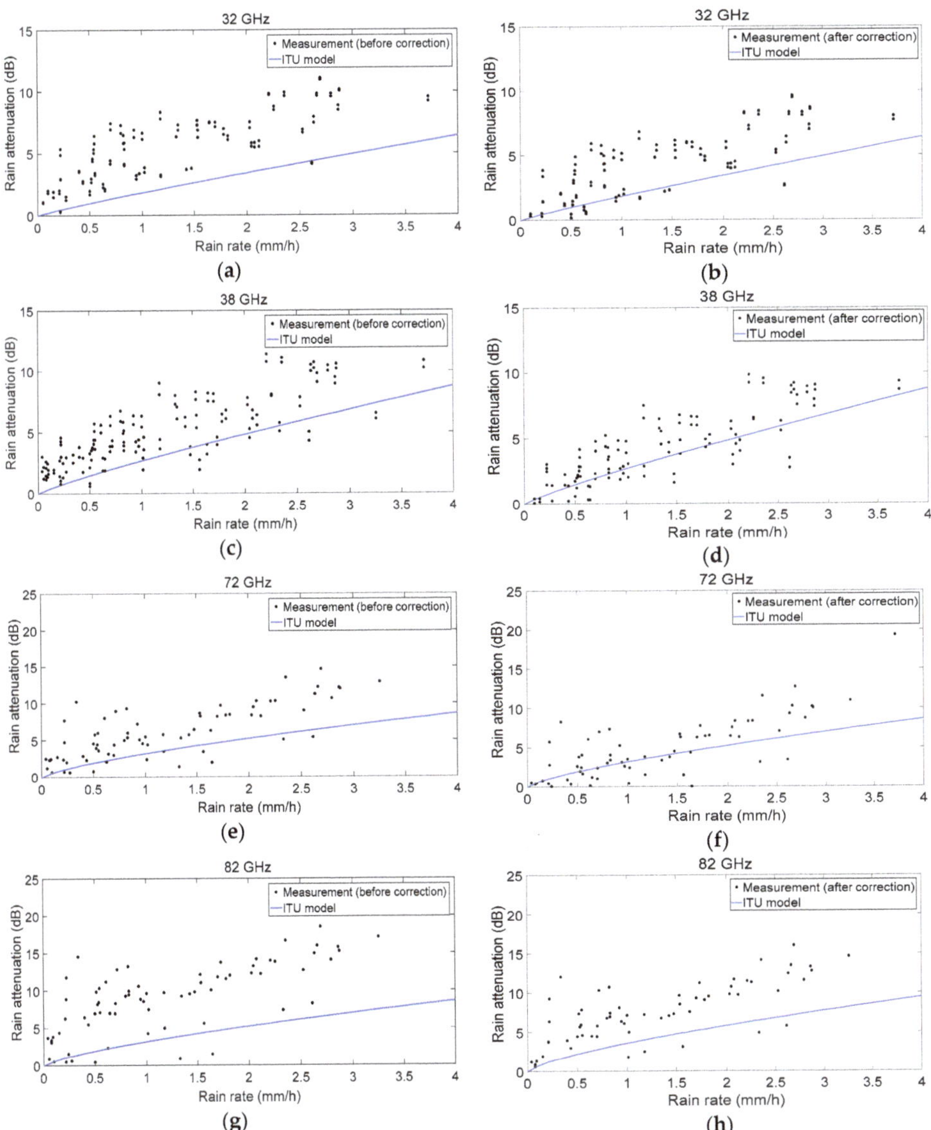

Figure 7. Rain attenuation statistics from the measurement before correction and after correction in comparison with the calculated rain attenuation, using the ITU model from Equation (9) for different frequencies: (**a**,**b**) 32 GHz; (**c**,**d**) 38 GHz; (**e**,**f**) 72 GHz; (**g**,**h**) 82 GHz.

3.4. Rain Rate Estimation

We evaluate the rainfall estimates from mmWave test links through three metrics—the Pearson correlation coefficient, the root mean square difference, and the bias.

The linear dependence of the time series data of average rain attenuation obtained from the link measurement $X = A_r$ and rain rate measurement $Y = R$ from the rain gauge is estimated by calculating the correlation coefficient of the variables. If each variable has I averaged observations, the Pearson correlation coefficient is calculated as:

$$r(A_{r,i}, R_i) = r(X_i, Y_i) = \frac{1}{I-1} \sum_{i=1}^{I} \left(\frac{X_i - \mu_X}{\sigma_X} \right) \left(\frac{Y_i - \mu_Y}{\sigma_Y} \right) \tag{12}$$

where μ_X and σ_X are the mean and standard deviation of X, respectively, and μ_Y and σ_Y are the mean and standard deviation of Y. Here, r ranges from -1 to $+1$. A high correlation coefficient value shows stronger relation between two data sets. On 7 and 11 June 2017, the signal power attenuation is mainly caused by rainfall and the values are highly correlated, resulting in an average correlation coefficient greater than 0.8. The strong correlation between the receive signal attenuation and rain rate during the measurement period indicates that it is possible to retrieve the rain rate from the receive signals of the mmWave links.

After applying the correction to the signal attenuation, the rain rate is estimated on a 15 min basis using Equation (10). The time series data of average rain rate derived from the mmWave link measurement $X = R_{link}$ is then compared with the rain rate measurement $Y = R$ from rain gauge, based on Equation (12). Figures 8 and 9 show the comparison between the link-derived rain rate estimation and rain gauge measurement. The accuracy of the rain estimation using different links is presented in Table 4. The root mean square difference (RMSD) was also computed for accuracy analysis according to the following formula:

$$RMSD = \sqrt{\frac{1}{I} \sum_{i=1}^{I} (X_i - Y_i)^2} \ (mm/h) \tag{13}$$

The bias is a measure of the average error between the link estimate rain rate and the rain gauge measurements, and it can be calculated using the following formula:

$$Bias = \frac{1}{I} \sum_{i=1}^{I} (X_i - Y_i) \tag{14}$$

The rain rates derived from the three mmWave links are closely related to the observed rain rate recorded by the weather stations, with a very good accuracy. The RMSD is found to be in the range of 0.36–1.00 mm/h for the test links.

Using the latest introduction of MIMO mmWave links, the number of rain estimation values grows linearly with the minimum number of transmitter and receiver antennas of the MIMO link compared to the case of a single rain rate estimation value from a SISO backhaul link. Both the forward and reverse links can be used for rainfall estimation, and the 4 data streams in 2 × 2 MIMO links can effectively provide up to 4 rain rate estimation values over the path. All the data streams can contribute to understanding of the statistics of the local rain rates.

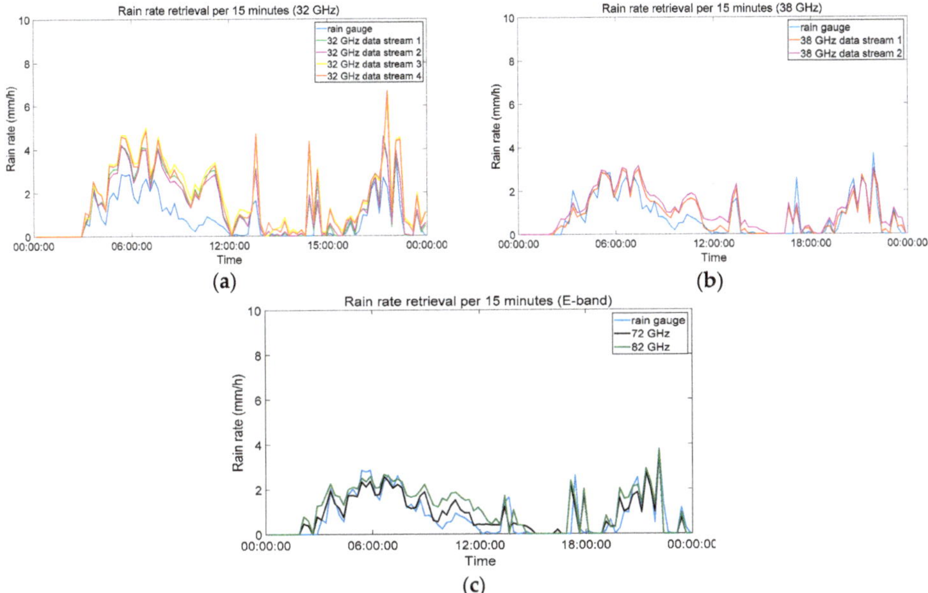

Figure 8. Average rain rate per 15 min derived from the signal link compared with the rain gauge measurement on 7 June 2017: (**a**) 32 GHz link; (**b**) 38 GHz link; (**c**) E-band link.

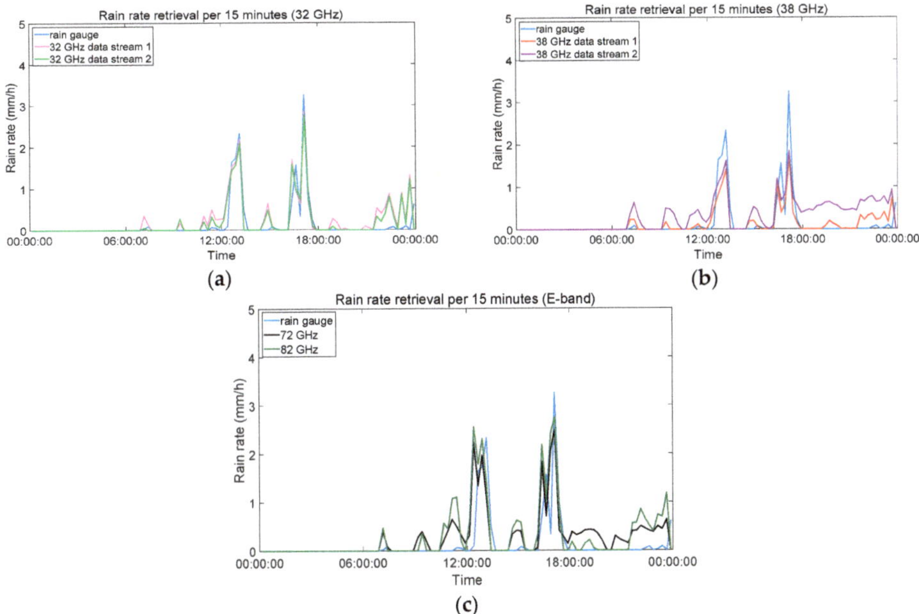

Figure 9. Average rain rate per 15 min derived from the signal link compared with the rain gauge measurement on 11 June 2017: (**a**) 32 GHz link; (**b**) 38 GHz link; (**c**) E-band link.

Table 4. Accuracy analysis of rain retrieval from the three test links.

Frequency	Data Stream	Correction	Correlation	RMSD	BIAS
32 GHz	1 (link 1)	1.5 dB	0.87	0.95	0.60
	2 (link 1)	1.5 dB	0.88	0.84	0.56
	1 (link 2)	1.5 dB	0.84	1.00	0.55
	2 (link 2)	1.5 dB	0.85	0.94	0.49
38 GHz	1 (link 1)	1.0 dB	0.90	0.36	0.06
	1 (link 2)	1.0 dB	0.88	0.48	0.25
73 GHz	1 (link 1)	1.5 dB	0.83	0.48	0.12
83 GHz	1 (link 2)	1.75 dB	0.80	0.61	0.26

4. Discussion

It can be seen from the results that the link-experienced rain attenuation could be much higher than the ITU model [28]-predicted attenuation value calculated using the rain gauge records. One possible reason was the location of the rain gauge. The weather station used in this study was located on the rooftop of a business building. A non-rooftop weather station operated by SMHI near the measurement area is available, but the data was not collected very frequently, making comparison with the microwave link derived rain rate at 15 minutes interval difficult. The daily rainfall amounts recorded by the SMHI rain gauge were 39 and 9 mm on 7 and 11 June, respectively, while the recorded daily rainfall values by the rain gauge used for this study was 17.4 and 7 mm, respectively, during these two days. Compression of atmospheric streamlines will produce significant Bernoulli lifting and cause under catch of rainfall in the rain gauge located on the rooftop. The actual rainfall along the links was greater than the rain gauge measurements, which was also indicated by the bias calculation results. The positive bias values show that the rainfall estimated based on the attenuation of the mmWave link signals is higher than the rain rate recorded by the rain gauge. As expected, the mmWave links detected more rainfall than the rooftop rain gauge, which underestimated the rainfall amount.

The widely deployed microwave links from existing cellular networks have become installation-free facilities, and they can be particularly useful for applications in urban hydrology and in supporting monitoring in flood-prone urban areas. Rain gauges have high accuracy but the measurement data is collected at the point scale. Because of the temporal and spatial variability of rainfall, especially in the heavy rain and flooding events, the changes of rain rate occur at very short time intervals. Even in dense urban areas, the density of rain gauges is often not sufficient to capture significant variation in observed rainfall. Radars and satellites can monitor over much wider areas, but the estimates from those sources are less accurate at near-ground levels, and the temporal and spatial resolution is not sufficient for flood monitoring purposes. The measurements presented here records a received signal level measurement every 30 seconds, producing 10 rain rate estimates within 5 minutes, which is usually the time resolution of weather radars. Microwave backhaul links can measure near-ground rain rates more accurately with high time resolution, which could assist in flash flood warnings [43–45], as the wireless networks exist over large regions of land, including complex topographies, where traditional monitoring equipment cannot be easily installed. Therefore, this technology could also be very useful in cities that lack monitoring of rainfall by radar. As more mmWave links are expected to be widely deployed in cities, those densely deployed links can all be used for rainfall monitoring, and the amount of rainfall estimation data will be impressively large. These links can be further processed using standard interpolation methods to create rainfall maps, or combined with other existing monitoring networks based on radars and rain gauges to produce high temporal and spatial resolution rainfall maps [46].

5. Conclusions

New mmWave technologies, including E-band, multi-band boosters, and LOS-MIMO, have been recently introduced to meet the global demand on microwave backhaul capacity and 5G network build-out. The large bandwidth available at mmWave frequencies in the 30-300 GHz range will enable very high connection speed and capacity. With large available bandwidth, E-bands (71–76 and 81–86 GHz) are considered a strategic solution for 5G heterogeneous networks. While E-band links are generally used up to 3 km, a multi-band technology combining E-band with traditional lower frequency bands is suggested for longer distance deployment. Supporting transmission of multiple data streams simultaneously, LOS-MIMO is the latest wireless backhaul technology to significantly increase the capacity of short-range, point-to-point, mmWave line-of-sight connections. In this paper, these new wireless backhaul technologies are studied using outdoor test links deployed in the same region. The trial was performed in Gothenburgh, Sweden, over the Ericsson premises in 2017.

Real time path attenuation caused by changing weather conditions were monitored and recorded for a 32 GHz 2 × 2 LOS-MIMO link and a 38 GHz SISO link over 6.87 km, as well as an E-band microwave backhaul link over 3 km. The measurement records showed that the rain-induced attenuation of all the three test links are closely related to the variation of the rain rate, with an average correlation value greater than 0.8. The 32 GHz LOS-MIMO link and 38 GHz link were deployed in the same location; the rain attenuation was similar for the two links, but the LOS-MIMO has greater capacity compared to a SISO link. While the 32 and 38 GHz links were built over a much longer distance than the E-band link, but they were less affected by atmospheric attenuation. For light rainfall, the difference in the signal attenuation observed in three test links is less significant. As the rain intensity increases, over a relatively short deployment of 3 km, for the 82 GHz link, the rain attenuation could be greater than 40 dB for a heavy rain event. The high rain attenuation restricts the use of E-band links over longer distances. While E-band backhaul links can achieve high throughput, the coverage is limited; lower frequency links are more robust and can be deployed over much longer distances. Multi-band booster technology pairs a higher frequency link with a lower frequency link, meaning the capacity and coverage requirements for the next-generation microwave backhaul links can be met compared to a single frequency microwave link.

Accurate rain monitoring of precipitation is of great importance to many applications, including meteorology, hydrology, agriculture, and flood monitoring. Microwave backhaul link is considered as a new tool for near-ground rainfall monitoring. We examined the accuracy of using these new mmWave backhaul technologies for rain rate estimation. Additional attenuation due to effects of water film on the antenna surface and other atmospheric conditions, such as humidity, needs to be considered for rain rate estimation for improved accuracy. In real deployments, the measured rain attenuation of the test links is found to be 1–1.75 dB higher than calculated rain attenuation based on the ITU model. We have applied data post-processing and attenuation correction to the received signal level measurement. The derived rain rate from all the links have been shown to be very good compared to the rain rate recorded by the weather stations located in the measurement site. These additional weather data obtained from commercial cellular networks will be particularly useful for big data analysis. Furthermore, mmWave backhauls are expected to be widely used for 5G and smart city networks in cities and densely populated areas, and there is a great potential to use these links for precipitation and flood monitoring in urban areas.

Author Contributions: Conceptualization, C.H.; methodology, C.H.; software, C.H.; validation, G.S., H.W.; formal analysis, C.H.; investigation, J.H.; resources, C.H.; data curation, Q.G.; writing—original draft preparation, C.H.; writing—review and editing, C.H.; visualization, C.H.; supervision, C.H.; project administration, J.H.; funding acquisition, C.H. All authors have read and agreed to the published version of the manuscript.

Funding: This research was funded by National Natural Science Foundation of China, grant number 41605122, 41775032; Chinese Academy of Sciences President's International Fellowship Initiative for Visiting Scientists, grant number 2018VTA0013.

Acknowledgments: The authors would like to thank Ericsson gratefully for giving them access to data. The authors would also like to thank Hagit Messer and Pinhas Alpert's research groups for helpful advice.

Conflicts of Interest: The authors declare no conflict of interest.

References

1. McCabe, M.F.; Rodell, M.; Alsdorf, D.E.; Miralles, D.G.; Uijlenhoet, R.; Wagner, W.; Lucieer, A.; Houborg, R.; Verhoest, N.E.; Franz, T.E.; et al. The future of earth observation in hydrology. *Hydrol. Earth Syst. Sci.* **2017**, *21*, 3879. [CrossRef] [PubMed]
2. Tauro, F.; Selker, J.; Van De Giesen, N.; Abrate, T.; Uijlenhoet, R.; Porfiri, M.; Manfreda, S.; Caylor, K.; Moramarco, T.; Benveniste, J.; et al. Measurements and observations in the XXI century (MOXXI): Innovation and multi-disciplinarity to sense the hydrological cycle. *Hydrol. Sci. J.* **2018**, *63*, 169–196. [CrossRef]
3. Gazit, L.; Messer, H. Advancements in the statistical study, modeling, and simulation of microwave-links in cellular backhaul networks. *Environments* **2018**, *5*, 75. [CrossRef]
4. Muller, C.L.; Chapman, L.; Johnston, S.; Kidd, C.; Illingworth, S.; Foody, G.; Overeem, A.; Leigh, R.R. Crowdsourcing for climate and atmospheric sciences: Current status and future potential. *Int. J. Climatol.* **2015**, *35*, 3185–3203. [CrossRef]
5. Ericsson Mobility Report. December 2018. Available online: https://www.ericsson.com/assets/local/microwave-outlook/documents/ericsson-microwave-outlook-report-2018.pdf (accessed on 31 January 2020).
6. Ji, B.; Li, Y.; Chen, S.; Han, C.; Li, C.; Hong, W. Secrecy Outage Analysis of UAV Assisted Relay and Antenna Selection for Cognitive Network under Nakagami-m Channel. *IEEE Trans. Cogn. Commun. Netw. Spec. Issue AI-based Licens./Unlicensed Spectr. Interoperability Future Mob. Wirel. Syst.* **2020**, 1–11. [CrossRef]
7. Dehos, C.; González, J.; Domenico, A.; Kténas, D.; Dussopt, L. Millimeter-wave access and backhauling: The solution to the exponential data traffic increase in 5G mobile communications systems. *IEEE Commun. Mag.* **2014**, *52*, 88–95. [CrossRef]
8. Studies on Frequency-Related Matters for International Mobile Telecommunications Identification Including Possible Additional Allocations to the Mobile Services on a Primary Basis in Portion(s) of the Frequency Range Between 24.25 and 86 GHz for the Future Development of International Mobile Telecommunications for 2020 and Beyond. Resolution 238 (WRC-15). Available online: https://www.itu.int/dms_pub/itu-r/oth/0c/0a/R0C0A00000C0014PDFE.pdf (accessed on 31 January 2020).
9. Agenda and Reference (Resolutions and Recommendations), ITU. August 2017. Available online: https://www.itu.int/dms_pub/itu-r/oth/14/02/R14020000010001PDFE.pdf (accessed on 31 January 2020).
10. Frequency Band Review for Fixed Wireless Services. Final Report Prepared for Ofcom, Document 2315/FLBR/FRP/3. 2011. Available online: http://stakeholders.ofcom.org.uk/binaries/consultations/spectrum-review/annexes/report.pdf (accessed on 31 January 2020).
11. Chen, X. OFDM based multi-node transmission in the presence of phase noises for small cell backhaul. *IEEE Commun. Lett.* **2017**, *21*, 1207–1210. [CrossRef]
12. Daniels, G. Deutsche Telekom Achieves Fiber-like Results with MmWave Wireless Backhaul for 5G. 11 January 2019. Available online: https://www.telecomtv.com/content/news/deutsche-telekom-achieves-fiber-like-results-with-mmwave-wireless-backhaul-for-5g-33773/ (accessed on 31 January 2020).
13. Sen, R.; Singh, M.P. Effect of rain on millimeter-wave propagation—A review. *AIP Conf. Proc.* **2007**, *923*, 45–76.
14. Han, C.; Duan, S.; Bao, L.; Zhang, G.; Ji, B.; Ran, L. E-Band Link for Next Generation Small-Cell Backhaul in Dense Urban Environment. In Proceedings of the IEEE VTC 2019 Fall, Honolulu, HI, USA, 22–25 September 2019; pp. 1–2.
15. Han, C.; Harrold, T.; Armour, S.; Krikidis, I.; Videv, S.; Grant, P.M.; Haas, H.; Thompson, J.S.; Ku, I.; Wang, C.-X.; et al. Green radio: Radio techniques to enable energy efficient wireless networks. *IEEE Commun. Mag. Spec. Issue Green Commun.* **2011**, *49*, 46–52. [CrossRef]
16. Nicolaou, M.; Han, C.; Beh, K.C.; Armour, S.; Doufexi, A. MIMO techniques for green radio guranteeing QoS. *IEEE J. Commun. Netw. Spec. Issue Green Radio Energy Effic. Wirel. Netw.* **2010**, *12*, 130–139.
17. Sacchi, C.; Rahman, T.F.; Hemadeh, I.A.; El-Hajjar, M. Millimeter-wave transmission for small-cell backhaul in dense urban environment: A solution based on MIMO-OFDM and space-time shift keying (STSK). *IEEE Access* **2017**, *5*, 4000–4017. [CrossRef]
18. Hansryd, J.; Edstam, J. Microwave capacity evolution. *Ericsson Rev.* **2011**, *88*, 22–27.
19. Rundstedt, K.; Bao, L.; Krishnan, R.; Olsson, B.-E.; Eriksson, T. On Field Measurements and Modelling of 2×2 Microwave LOS-MIMO Systems. In Proceedings of the IEEE Global Communications Conference (GLOBECOM), San Diego, CA, USA, 6–10 December 2015; pp. 1–6.

20. Bøhagen, F.; Orten, P.; Øien, G. optimal design of uniform rectangular antenna arrays for strong line-of-sight MIMO channels. *J. Wirel. Commun. Netw.* **2007**, *2007*, 045084. [CrossRef]

21. Zhao, Q.; Li, J. Rain attenuation in millimeter wave ranges. In Proceedings of the International Symposium on Antennas, Propagation, & EM Theory, Guilin, China, 26–29 October 2006.

22. Coldrey, M.; Allasia, A.; Bao, L.; Boch, E.; Ferrari, G.; Gentina, D.; Putkonen, J.; Sutton, A.; Yigal, L.; Zein, N. *Maturity and Field Proven Experience of Millimeter Wave Transmission*; ETSI White Paper: Sophia Antipolis, France, 2015.

23. Forknall, N.; Cole, R.; Webb, D. Cumulative Fading and Rainfall Distributions for a 2.1 km, 38 GHz, Vertically Polarized, Line-of-Sight Link. *IEEE Trans. Antennas Propag.* **2008**, *56*, 1085–1093. [CrossRef]

24. Han, C.; Duan, S. Impact of atmospheric parameters on the propagated signal power of millimeter-wave bands based on real measurement data. *IEEE Access* **2019**, *7*, 113626–113641. [CrossRef]

25. Niu, Y.; Li, Y.; Jin, D.; Su, L.; Vasilakos, A.V. A survey of millimeter wave (mmWave) communications for 5G: Opportunities and challenges. *Wirel. Netw.* **2015**, *21*, 2657–2676. [CrossRef]

26. Han, C.; Duan, S. The study on characteristics of rain attenuation along 28 ghz and 38 ghz line-of-sight millimeter-wave links. In Proceedings of the 2019 URSI Asia-Pacific Radio Science Conference (AP-RASC), New Delhi, India, 9–15 March 2019; pp. 1–3.

27. Liebe, H.J.; Hufford, G.A.; Cotton, M.G. Propagation modeling of moist air and suspended water/ice particles at frequencies below 1000GHz. In Proceedings of the Electromagnetic Wave Propagation Panel Symposium, AGARD Conference Proceedings, Palma, Spain, 17–20 May 1993.

28. ITU-R P. 838–3 (International Telecommunication Union Radio Communication Bureau Propagation Recommendation). *Specific Attenuation Model for Rain for Use in Prediction Methods*; ITU-R: Geneva, Switzerland, 2005.

29. Medhurst, R.G. Rainfall attenuation of centimeter waves: Comparison of theory and measurement. *IEEE Trans. Antennas Propag.* **1965**, *13*, 550–564. [CrossRef]

30. Messer, H. Capitalizing on cellular technology—Opportunities and challenges for near ground weather monitoring. *Environ. Sci. Technol.* **2018**, *5*, 73. [CrossRef]

31. Uijlenhoet, R.; Overeem, A.; Leijnse, H. Opportunistic remote sensing of rainfall using microwave links from cellular communication networks. *Wiley Interdiscip. Rev. Water* **2018**, *5*, e1289. [CrossRef]

32. Chwala, C.; Kunstmann, H. Commercial microwave link networks for rainfall observation: Assessment of the current status and future challenges. *Wiley Interdiscip. Rev. Water* **2018**, *6*, e1337. [CrossRef]

33. Zhang, G.; Han, C.; Ji, B.; Shi, C.; Xie, P.; Yang, L. A new multiple-symbol differential detection strategy for error-floor elimination of IEEE 802.15.4 BPSK receivers impaired by carrier frequency offset. *Wirel. Commun. Mob. Comput.* **2019**, *2019*, 5409612. [CrossRef]

34. Rappaport, T.S.; Xing, Y.; MacCarney, G.R.; Molisch, A.F.; Mellios, E.; Zhang, J. Overview of millimeter wave communications for fifth-generation (5G) wireless networks-with a focus on propagation models. *IEEE Trans. Antennas Propag.* **2017**, *65*, 6213–6230. [CrossRef]

35. ITU-R P. 676–10 (International Telecommunication Union Radio communication Bureau Propagation Recommendation). *Attenuation by Atmospheric Gases*; ITU-R: Geneva, Switzerland, 2013.

36. Overeem, A.; Leijnse, H.; Uijlenhoet, R. Retrieval algorithm for rainfall mapping from microwave links in a cellular communication network. *Atmos. Meas. Tech.* **2016**, *9*, 2425. [CrossRef]

37. Han, C.; Bi, Y.; Duan, S.; Lu, G. Rain rate retrieval test from 25 GHz, 28 GHz, and 38 GHz millimeter-wave link measurement in Beijing. *IEEE J. Sel. Top. Appl. Earth Obs. Remote Sens.* **2019**, *12*, 2835–2847. [CrossRef]

38. Bao, L.; Olsson, B.-E.; Hansryd, J. On measurements of availability penalty due to antenna separation in a 2 × 2 LOS-MIMO Link. In Proceedings of the EuCAP, London, UK, 9–13 April 2018.

39. Davis Wireless Vantage Pro2 Weather Station 6322. Available online: https://www.davisnet.com/weather-monitoring/ (accessed on 31 January 2020).

40. Ostrmetzky, J.; Raich, R.; Bao, L.; Hansryd, J.; Messer, H. The wet-antenna effect—A factor to be considered in future communication networks. *IEEE Trans. Antennas Propag.* **2018**, *66*, 315–322. [CrossRef]

41. Leijnse, H.; Uijlenhoet, R.; Berne, A. Errors and uncertainties in microwave link rainfall estimation explored using drop size measurements and high resolution radar data. *J. Hydrometeorol.* **2010**, *11*, 1330–1344. [CrossRef]

42. Zinevich, A.; Messer, H.; Alpert, P. Prediction of rainfall intensity measurement errors using commercial microwave communication links. *Atmos. Meas. Tech.* **2010**, *3*, 1385–1402. [CrossRef]

43. Raich, R.; Alpert, P.; Messer, H. Vertical prediction estimation using microwave links in conjunction with weather radar. *Environments* **2018**, *5*, 74. [CrossRef]
44. Eshel, A.; Messer, H.; Osttrometzky, J.; Raich, R.; Alpert, P.; Laronne, J.B. On the use of measurements from a commercial microwave link for evaluation of flash floods in arid regions. *Atmos. Chem. Phys. Discuss.* **2017**. [CrossRef]
45. Hoedjes, J.C.B.; Kooiman, A.; Maathuis, B.H.P.; Said, M.Y.; Becht, R.; Limo, A.; Mumo, M.; Nduhiu-Mathenge, J.; Shaka, A.; Su, B. A conceptual flash flood early warning system for Africa, based on terrestrial microwave links and flash flood guidance. *ISPRT Int. J. Geo-Inf.* **2014**, *3*, 584–598. [CrossRef]
46. Liberman, Y.; Samuels, R.; Alpert, P.; Messer, H. New algorithm for integration between wireless microwave sensor network and radar for improved rainfall measurement and mapping. *Atmos. Meas. Tech.* **2014**, *7*, 3549–3563. [CrossRef]

MDPI

St. Alban-Anlage 66

4052 Basel

Switzerland

Tel. +41 61 683 77 34

Fax +41 61 302 89 18

www.mdpi.com

Remote Sensing Editorial Office

E-mail: remotesensing@mdpi.com

www.mdpi.com/journal/remotesensing